Microbial Ecology of the Soil and Plant Growth

Microbial Ecology of the Soil and Plant Growth

Pierre Davet

Directeur de Recherches INRA
Ecole Nationale Supérieure Agronomique de Montpellier
France

CRC Press
Taylor & Francis Group
Boca Raton   London   New York

CRC Press is an imprint of the
Taylor & Francis Group, an informa business

CRC Press
6000 Broken Sound Parkway, NW
Suite 300, Boca Raton, FL 33487
270 Madison Avenue
New York, NY 10016
2 Park Square, Milton Park
Abingdon, Oxon OX14 4RN, UK

ISBN 1-57808-303-6 (Paperback)
ISBN 1-57808-180-7 (Hardcover)

Published by arrangement with INRA, Paris

Ourage publié avec le concours du Ministère Français chargé de la culture-Centre national du livre
This work has been published with the help of the French Ministère de la Culture-Centre national du livre.

© 2004 Copyright reserved

Translation of: *vie microbienne du sol et production végétale*, INRA, Paris, 1996.
         Updated by the Author for the English edition in 2001.
French edition: © INRA, Paris, 1996
         ISBN 2-7380-0648-5

ISBN 978-1-57808-303-9

Preface

The first living beings to emerge permanently from the primordial ocean about 400 million years ago to colonise and conquer the emergent land were chlorophyllous organisms. They began assembling terrestrial organic molecules on still arid rock and in deserts, using water, mineral salts and atmospheric carbon. Their descendants, always frugal, are today still the source of all life on the surface of the Earth. Plants are indeed the **primary producers** of biomass because of their ability to transform solar energy into chemical energy, synthesising all the organic molecules they need. These molecules and the energy they contain, are indispensable elements in the development of a succession of non-photosynthetic organisms. Herbivorous animals as well as parasitic organisms use this material to sustain themselves. Many of these **primary consumers** later serve as food for **secondary consumers,** which eventually become victims of predators and other parasites.

Only a small part of the carbon incorporated in organic matter is mineralised by respiration of plants and animals. The remainder, made up of unconsumed plant debris and the excreta and carcasses of animals, accumulates on the soil surface. For instance, 1 ha of forest soil receives 1 to 4 tons of organic matter (dry mass) each year in temperate zones and 12 to 20 tons in tropical humid zones. However, plants are incapable of utilising the elements immobilised in these wastes; they can only assimilate mineral salts. It is here that **decomposer** microorganisms intervene. Capable of breaking down the most complex molecules, they ensure that the residual organic matter is returned to the mineral state. Without them, organic remains would accumulate until exhaustion of atmospheric carbon dioxide, immobilise the fundamental mineral elements and thereby preclude plant development and consequently life. The decomposers therefore act as recyclers, a role which is essential in the functioning of ecosystems (Figure).

The essential concern of an agronomist is the primary productivity of a given area, which depends on climate and richness of the soil with respect to elements assimilable by plants. This is partly governed by proper recycling of organic matter: an ecosystem, natural or cultured, can only function normally if its microbial populations are active. Soil microorganisms mostly discharge their functions so well that one tends to forget how important

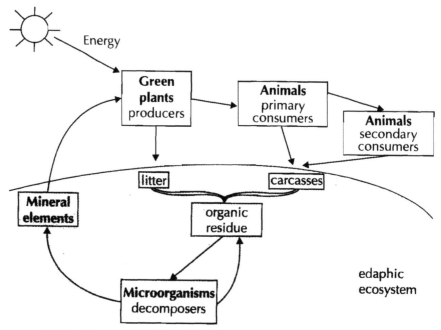

Fig. Trophic groups and transfer of energy in the terrestrial ecosystem.

their role is. It so happens sometimes, however, that due to natural factors or untimely human intervention, the mechanism slows down or speeds up. Then a rapid drop in fertility and degradation of the soil structure is observed.

Unfortunately, not all soil microorganisms are decomposers of dead organic matter. Some are also consumers and depend directly on living matter. The roots of plants are thus exposed to a large number of parasites against which defence is difficult. On the other hand, other microorganisms have gone beyond the state of parasitism to establish mutually beneficial relationships with their host plants. Microfauna and microflora can themselves be victims of microbial predators which play an important role in the equilibrium of the communities.

Good plant development, healthy cultures and, consequently, production of food resources, depend for numerous reasons on the quality of soil microbial life. The purpose of this book is to highlight some of these interactions. Most of the phenomena are the result of complex equilibria between the partners. Their interpretation requires knowledge of the framework in which they are produced. Therefore, the edaphic ecosystem (from Greek *edaphos* = soil) is considered first. An ecosystem is constituted of all communities of living organisms (as well as their interrelations) occurring in a determined space, defined by its physicochemical characteristics and its environmental conditions. After describing the main groups of soil inhabitants and the framework within which they develop, we shall consider their relations with their environment, then their relations between themselves and with plants, especially with cultivated plants. Lastly, measures that should be undertaken to derive the best from auxiliary soil microorganisms, or to eliminate harmful organisms, are examined. The general objective is thus to ascertain what can be done to manipulate microbial equilibria in the soil to improve plant productivity.

An ecosystem requires a permanent external source of energy irrespective of the level at which it is considered. On the Earth, the unique and irreplaceable energy source is the sun (Figure). At least this was the notion held until animal communities living in absolute obscurity at ocean depths, close to hydrothermal sites along oceanic dorsals, were discovered in 1977. It was realised then with great amazement that the organic matter used by these populations is not of photosynthetic origin; rather the oxidation of hydrogen sulphide of the thermal waters provides all the energy required for the functioning of the system. This oxidation reaction is assured by microorganisms living in close association with invertebrates, which constitute the lower end of a long food chain.

An even greater surprise for microbiologists was the discovery in 1986 in Movilé, Rumania, of a subterranean cave completely isolated from the external world for several million years. Live microorganisms, molluscs and arthropods were found in this cave. Here again, the energy involved in the functioning of the ecosystem is not of photosynthetic origin but provided from oxidation by bacteria (*Thiobacillus* and *Beggiatoa*) of sulphides in underground waters.

SOME GENERAL WORKS ON SOIL MICROBIOLOGY

Atlas R.M. and Bartha R. 1998. Microbial Ecology: Fundamentals and Applications. Benjamin Cummings, Redwood City, CA.
Coleman D.C. and Crossley Jr. D.A. 1996. Fundamentals of Soil Ecology. Academic Press, NY.
Elsas J.D. van, Trevors J.T. and Wellington E.M.H. (eds.). 1997. Modern Soil Microbiology. Marcel Dekker Inc., NY.
Lynch J.M. 1983. Soil Biotechnology: Microbiological Factors in Crop Productivity. Blackwell Scientific Publ., Oxford.
Metting F.B. (ed.). 1993. Soil Microbial Ecology. Applications in Agricultural and Environmental Management. Marcel Dekker Inc., NY.
Paul E.A. and Clark F.E. 1996. Soil Microbiology and Biochemistry. Academic Press, NY (2nd ed.).
Sylvia D.M., Fuhrmann J.J., Hartel P.G. and Zuberer D.A. 1998. Principles and Applications of Soil Microbiology. Prentice Hall, Upper Saddle River, N.J.
Tate R.L., III. 2000. Soil Microbiology. John Wiley and Sons, NY (2nd ed.).

Acknowledgements

Many friends and colleagues helped me during the preparation of the first drafting, particularly Claude Alabouvette, Marie-Madeleine Coûteaux, Jean Dalmasso, Eric Ducelier, Michèle Frésia, Silvio Gianinazzi, Jean Guttierez, Thierry Heulin, Christian Martin, Daniel Mousain, Jacques Schimt and Pierre Signoret. Part of the photographic illustration was also kindly furnished by Bernard Digat, Jean-Jacques Drevon, Serge Kreiter, Philippe Normand and Christine Poncet.

When I revised the text in view of the English edition, Claude Alabouvette, Julien Davet, Jean-Jacques Drevon, Jean-Loup Notteghem and Pierre Signoret also gave me an appreciated assistance.

I am pleased to express here my gratitude to all of them.

Contents

PART III
POSSIBILITIES OF INTERVENTION

Part I
The Soil Medium

Soil serves as the medium for anchorage of roots and the source of water and nutrients necessary for growth and development of plants. Soil is a thin pellicle on the surface of the Earth's crust, formed over geological time by the slow transformation of the first parent rocks due to physical, chemical and biological action, a continuous process that can be observed even today. The genesis and evolution of soil have been studied by pedologists and its fertility by agronomists. In Part I, we consider primarily those aspects of this particular medium which have a direct relation with microbial life.

Soil is a porous mineral medium with gases and liquids circulating in it. Hence, it can be divided into three physical 'compartments': solid, liquid and gaseous. However, soil is not just a physicochemical substrate, it also supports life, the creator of organic matter. To describe soil properly, therefore, another 'compartment', namely the organic, must be added to the three aforesaid fractions. This organic compartment can be separated into living, metabolically active, and dead organic matter.

In the final analysis, therefore, soil is an ensemble of five different fractions (Figure).

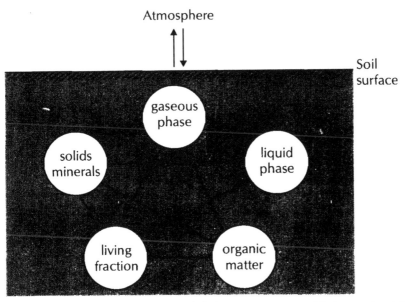

Figure   The five soil compartments. Each link represents a possible interaction (after Morel, 1989).

The transfer of energy and substances occurs continuously not only among these compartments, but also between each and the external environment: gaseous exchanges, temperature exchanges, rainfall-evaporation-drainage, addition of animal and plant residue, etc.

So in successively examining each component of this ensemble, it should not be forgotten that none can in fact be considered independent of the others.

1

Inert Components

SOLID MINERAL FRACTION

The solid mineral fraction alone constitutes 93 to 95% of the total soil weight. It is composed of particles varying in size, depending on the degree of fragmentation of the original parent rock. The particles can be classified according to their volume as pebbles, stones, gravel, coarse sand, fine sand and silt (Table 1.1). Alteration and transformation of these basic mineral elements give rise to another category of particles, still finer than silt (less than 2 µm), namely clays. The granulometric composition (i.e., the relative proportions of sand, silt and clay, estimated after the breakdown of aggregates and dispersal of clay) characterises the soil **texture**. Some of the physical properties of soil depend upon its texture. For instance, sand or silty sand is more permeable to water and easier to work than a soil of clayey silt or clayey texture.

Among the elementary particles of the soil, clays play a role so important from the biological perspective that their detailed study is indispensable.

Table 1.1 Classification of mineral particles of soil according to size

Category	Utmost diameters
Gravel	from 2 to 20 mm
Coarse sand	from 0.2 to 2 mm
Fine sand	from 0.02 to 0.2 mm
Silt	from 0.002 to 0.02 mm
Clay	less than 0.002 mm

Clays

Very simply, clays are minerals composed of layers of microcrystalline structure stacked one upon the other. Some of these layers may be strongly bonded by hydrogen, as in the case of illite and kaolinite. In other types of clays, cohesion is effected by relatively weak forces (Van der Waals force, for example). The layers can thus be separated and molecules of water, containing dissolved substances, can permeate the interfoliate spaces, making the clay swell. This is true of smectites, the group to which, for example, montmorillonite and beidellite belong.

Each layer is itself made up of two or three superposed sublayers. In each sublayer, the centre of a geometric three-dimensional figure is occupied by a metallic ion and the vertices hold oxygen or hydroxyl ions (Fig. 1.1). Theoretically, the ions are so arranged that the positive charge of the central

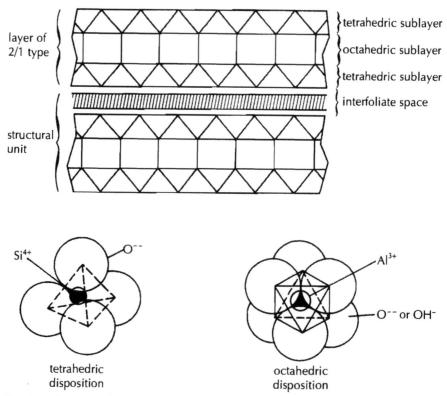

Fig. 1.1   Structural diagram of a clay of 2/1 type, such as illite or montmorillonite. Each layer of this clay comprises 2 tetrahedric sublayers and an octahedric sublayer. In the tetrahedric disposition, an Si^{4+} ion (in green) occupies the centre of a tetrahedron at the vertices of which are O^{--} or OH^- ions. In both cases, the excess negative charges are compensated by the positive charges of the neighbouring cations as these structures are placed side by side to form a sublayer in which some of the vertices are common. The layers are separated by gaps, the interfoliate spaces. The ensemble constituted by a layer and an interfoliate space constitutes the structural unit of clay (Morel, 1989).

one (Si^{4+}, Al^{3+} or Fe^{3+}) is completely neutralised by the negative charges of the anions. However, these microcrystals are not perfect and it often happens that the central metallic ions are replaced by cations of lower valency: for example, Al^{3+} can replace Si^{4+} and Fe^{2+} or Mg^{2+} can substitute for Al^{3+}. The equilibrium is then not so stable and negative charges appear at the surface of the layers. Some bonds at the margins of the layers can be ruptured. The presence of hydroxyl ions (in an alkaline medium) can then lead to the appearance of negative charges and the Al^{3+} ions (in an acidic medium) the appearance of positive charges on the ridges of particles. Furthermore, some clays formed on volcanic ash (allophanes) carry positive charges on the surface.

Because of their charge and their fineness, such particles form a stable colloidal suspension when suspended in water, and remain dispersed indefinitely. This stability is due to repulsion of negative charges directed towards each other. When an electrolytic solution is added to the water, the particles precipitate—a phenomenon known as flocculation. One can consider in the first approximation that, under the effect of the electrolyte the repulsive forces between particles have been attenuated and that attractive forces between atoms have prevailed (Van der Waals forces). The nature and concentration of the electrolyte are the main parameters to be considered while studying the phenomena of flocculation.

Another very important property of the electronegative charge of clays is the ability to form electrostatic bonds with protons, metallic cations and, in a general way, with all positively charged particles. These bonds are reversible and the fixed cations can be replaced by other cations in the aqueous soil phase; they are **exchangeable** cations. The **exchange capacity** of cations, expressed in milliequivalents (meq) per 100 g, is a property characteristic of each type of clay. For example, it is much higher for a smectite (more than 100 meq/100 g) than for a clay such as kaolinite (less than 10 meq/100 g).

In addition to the electrostatic bonds, other forces can manifest on the surface of the clay layers: hydrogen bonds, Van der Waals forces, coordination bonds etc. All these forces of attraction are sometimes designated by the general term **sorption**.

Consequences of Properties of Clays

Exchange Capacity and Buffering Effect

The constitution of clays and their ability to exchange H^+ ions for other cations confer upon them a buffering power which helps the organisms present in the liquid phase of the soil to withstand sudden changes in pH. The higher the exchange capacity, the higher the buffering effect. For instance, the buffering effect of montmorillonite is higher than that of kaolinite. More generally, the absorbent capacity of clays allows them to build a significant reserve of mineral elements and ensures their regular provisioning to the liquid soil phase. This phenomenon is primarily concerned with plant nutrition and has been intensively studied under this heading. However, it also concerns soil microorganisms.

Water Retention

Because of the multiple bonds (chemical or electrostatic) likely to be established between the layers or between the cations of interfoliate spaces and the water molecules, clays are capable of storing large quantities of water in a reversible fashion (Fig. 1.2). This hydration capacity is especially high in clays which have particularly large surfaces (swelling clays). This feature, which attentuates the irregularities in rainfall due to varying climate, is also favourable to microbial life.

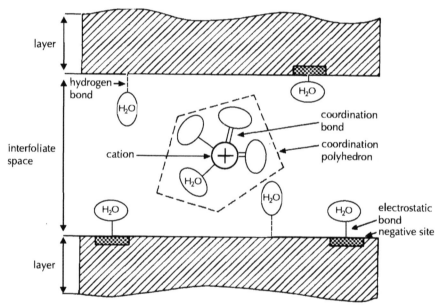

Fig. 1.2 Different types of bonds that can occur in interfoliate water (Morel, 1989).

Adsorption of Organic Compounds

Clays can retain not only mineral ions, but also organic molecules. This retention has two consequences: it prevents these compounds from being leached (most organic substances migrate slowly depth-wise) and thereby temporarily shelters them from degradation by microorganisms or by soil enzymes. Trapping of molecules between clay layers only partly explains this protective effect, as it can be demonstrated that a protein such as catalase is not biodegraded even when adsorbed on the external face of a clay particle. On the one hand, it seems that the terminal amino acids from which the exopeptidases could start cleavage of the protein would be masked by sorption; on the other hand, bonding with clay modifies the steric configuration of the molecule, making the endopeptidases incapable of recognising their specific sites of attack (Stotzky, 1980). This change in spatial disposition of the adsorbed catalase molecule seems to be confirmed by the observation that this enzyme is four times more active when bonded than when in a free state: in the new configuration resulting from adsorption, the active sites of the enzyme could be readily accessed by their substrates. The change in reactivity of the adsorbed molecules compared to their behaviour in solution, seems to be a rather general phenomenon of growing interest to chemists. Smaller molecules (amino acids, peptides) can also be adsorbed. Although more readily usable than proteins, for them to be transported into cells and metabolised, their affinity to permeases that assure their passage across the microbial membranes should be greater than their affinity for the clay substrate (Dashman and Stotzky, 1986). Utilisation of these molecules is therefore restricted only to species capable of desorbing them.

This adsorbed organic matter is not fixed for ever. Localised changes in the environment (modifications of pH, presence of certain cations), mechanical abrasion caused either by root growth, or due to microfauna, or tilling of the soil, can result in desorption of the molecules and their return to circulation in the soil solution. Alternation of dry-wet cycles or freeze-thaw cycles can produce the same effect (Stotzky, 1980).

One may thus consider that the clayey fraction of the soil behaves like a kind of warehouse in which part of the nutritive resources are kept in reserve, thus ensuring availability of nutrients to microorganisms as well as plant roots over a longer period of time.

Effect on Soil Structure

The liquid phase of the soil can be compared to a solution of electrolytes in which the most common cations are usually the Ca^{++} and H^+ ions. Clays are therefore in a flocculated state, which ensures good soil structure, water circulation and gas exchange. These conditions are conducive to the development of roots as well as microbial life.

INERT ORGANIC MATTER

Origin

Organic matter is produced by the activity of all the organisms present on the surface or inside the soil. Part of this organic matter is produced by living organisms, such as animal wastes, root exudates, plant litter and microbial polysaccharides. The remainder is made up of dead plant debris, animal carcasses, and lysed microbial cells. Most of the organic matter that reaches the soil is of plant origin. Annual leaf fall in a temperate forest constitues several tons of dry matter per hectare.

Soluble substances of low molecular weight are rapidly used and/or mineralised. The remainder is progressively transformed into very complex dark compounds of high molecular weight, globally termed **humus**. This term corresponds to no exact chemical substance but does designate a material whose properties are rather well defined.

Properties of Humus

Humus has properties very similar to those of clays but the mechanisms responsible for those properties sometimes differ.

Humic compounds can carry positive charges (created by the protonation of an amine group) but in general they are negatively charged (because of ionisation of the hydrogen of carboxyl and hydroxyphenol groups) and, like clays, can reversibly attract cations. The exchange capacity of humus is very high, on average 2 to 3 times higher than that of smectites. The phenomena which regulate the exchange processes in humus are more complicated than in clays, however, varying with the metallic ion and the pH. Because of this exchange capacity and the presence of R-COOH carboxyl groups, humic substances behave like weak acids and possess a great buffering capability. Humic acids have a capacity for adsorbing a large number of organic compounds and can store considerable quantities of water (up to 20 times their weight). Lastly, like clays, humic compounds are colloids. They occur in flocculated form in the presence of acids and mineral salts of the soil.

Humus-Clay Complex

Most soil organic matter (and sometimes all of it) is associated with clay with which it forms very stable complexes. Clay particles are usually enveloped in a pellicle of organic matter. Various types of chemical or physicochemical bonds ensure the cohesion of these complexes. The polymerised hydroxides of iron, aluminium or manganese, which are themselves positively charged, can further reinforce the stability of the complex and act as bridges between negatively charged particles.

SOIL STRUCTURE

Definition

The manner in which elementary mineral particles are disposed in relation to one another is a soil characteristic termed **structure**. The normal structure of a soil suitable for cultivation is fragmented and consists of small grains or **aggregates** agglomerated in clumps. The structure is good if in a dry soil exposed to rain, the aggregates do not shatter under the action of droplets. Disintegration of an aggregate into subunits which are subsequently dispersed can result in the formation of a crust and obturation of the pores in the top part of the profile. The stability of the structure depends in part on ionic and electrostatic phenomena; a soil rich in Ca^{++} ions is less easily dispersed than one rich in monovalent cations. Structure also depends on the constitution of the aggregates. Each small grain (macroaggregate) is formed by an assembly of basal elements of sizes smaller than or equal to 250 μm, the microaggregates.

Microaggregates

These represent the smallest coherent unit of the soil. They comprise grains of sand, silt and a few particles of clay and organic matter; they do not disperse in the presence of water. Their cohesion is due to the combined action of **mineral cements** and **organic binders**. The mineral cements are made of clays and metallic hydroxides. The organic binders consist of short-lived conpounds, mainly polysaccharides, produced directly in the presence of roots (mucigel) and microorganisms (bacterial capsules, mucus of algae and fungi), and more complex substances formed by microbial transformation of organic matter into humus, which contain numerous aromatic nuclei. These substances, associated with polyvalent metallic cations, form with clay very stable «ashlars». The hydrophobic nature of some colloids of the humus reinforce the resistance of microaggregates to the dispersive effect of water.

The major role played by the binders of microbial origin in the stability of aggregates has been confirmed by numerous studies. We cite here only the experiment of Hénin (1944) which involved the addition of glucose and ammoniacal sulphate to soil samples at the experimental station of Versailles. These simple sources of carbon and nitrogen had no direct effect on soil structure but stimulated microbial activity. Three weeks of incubation sufficed for a net improvement in soil structure as a result of the organic compounds synthesised by the soil microflora. This improvement resulted in an increase in percentage of stable aggregates proportional to the quantity of glucose introduced in the soil (Table 1.2). The contribution of microorganisms to the stability of soil structure is detailed in Chapter 5.

Table 1.2 Percentage of stable aggregates in soil samples of Versailles incubated for three weeks at 25% humidity in 10% $(NH_4)_2SO_4$ and gradually increased quantities of glucose (after Hénin, 1944)

Percentage of glucose added	0	0.05	0.1	0.5	1	3
Horizon A (superficial, organic)	7	7	6	12	22	45
Horizon B (developed, clayey)	2	2	4	10	20	40

Macroaggregates

The union of several microaggregates constitutes a macroaggregate. Quite likely, these aggregations as well as the formation of various-size macroaggregate lumps originate in part from mechanical phenomena. Macroaggregates measure between 250 μm and a few mm. They are combinations less durable and looser than microaggregates. The cohesion of these frameworks is ensured by mineral and organic cements and a network of very fine roots. Soil fungi also play an important role, as proven by the presence of high quantities of sterols (the chemical signatures of these microorganisms) in these structures.

Aggregates, a Heterogeneous Biotope

Microaggregates are not assembled in a jointed manner. Spaces exist between them as well as inside the mass that they constitute, through which air and soil solution can circulate. Fungi mostly localise on the exterior of this mass while bacteria adhere to both the internal and external faces. Bacteria are also found in the cavities which form inside the macroaggregates, either in a free state or bound to some organic debris (Figs. 1.3 and 1.4). If the pores that connect these cavities to the external medium are smaller than the bacteria, they remain captive until the mass disintegrates. This confinement can result in their being sheltered from predators such as amoebae or nematodes, thereby permitting development of colonies containing a large number of bacterial cells. If the pores are blocked by capillary water or

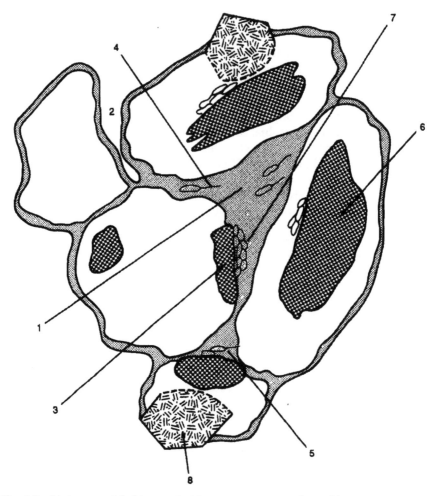

Fig. 1.3   Various possible biotopes inside a macroaggregate formed by 5 microaggregates made up of quartz grains, organic debris and an amorphous matrix. The mycelial filaments which consolidate the aforesaid are not depicted. 1 – liquid phase; 2 – gaseous phase; 3 – organic matter; 4 – aerobic site; 5 – microaerobic site; 6 – anaerobic site; 7 – bacterial colony (in green); 8 – quartz grain

Fig. 1.4   Bacteria developing on organomineral particles on the surface of a wound in a potato tuber (plate, B. Tivoli and E. Lemarchand, photo library, INRA).

by cements, the environment inside the aggregate rapidly becomes totally anaerobic.

Even in a small grain of soil therefore, very different biotopes coexist. The outer surface of organic matter, readily accessible, will be rapidly utilised by aerobic organisms without time for much evolution. Interiorly, where conditions are almost anaerobic, the organic matter will be slowly consumed by less active microflora and can undergo greater evolution. The microbial populations present in the **external compartment** are exposed to predators and subject to drastic variations of the environment. They can only survive if they have a high rate of growth and reproduction. Populations present in the **internal compartment**, submerged in a clay and polysaccharide bath, are protected from external environmental fluctuations, but to remain in that confined atmosphere they must be content with slowed development. However, the distribution of species between the two compartments corresponds only in part to ecological specialisation. Indeed, to a large extent, this distribution is solely passive: the bacteria, bonded to the substrate on which they develop, may or may not be enveloped or imprisoned in a microstructure. The difference between the internal and external compartments is thus more (at least in origin) a difference in metabolic activity than a qualitative difference in the microflora.

When the soil is cultivated, and especially when ploughed, a reduction in number of macroaggregates occurs accompanied by reduction in organic matter and biomass, principally mycelial biomass (Gupta and Germida, 1988). These results suggest that in reclaiming soil for cultivation, while aeration improves, more intense mineralisation of organic matter takes place, followed by reduction in microbial populations, especially of fungi, leading to instability of the macroaggregates.

LIQUID PHASE

Definition and Composition

Subsequent to rainfall or irrigation, not all the water received by the soil flows by gravity to the lower horizons. A more or less significant volume remains in the soil pores due to capillary action, and on the surface of the clay-humus complex due to binding forces discussed earlier. A given soil has a certain **capacity for water retention**. This varies according to soil texture; the higher the sand content of the soil, the lower its water retention capacity. Water retention also varies for a given soil depending on its degree of compaction. This retained water constitutes the liquid phase of the soil.

Soil water is not pure; it contains a goodly variety of substances in solution, such as mineral salts, organometallic complexes and organic compounds of various origins. The nature of these compounds can vary over time as a result of metabolic functioning of the vegetation and soil microorganisms themselves dependent on climatic conditions. The concentrations of solutes further varies depending on soil moisture. The osmotic pressure of the **soil solution** remains low, on average less than 0.1 MPa—equivalent to a very dilute solution. The absorbent clay-humus complex amortises in part too drastic variations.

Water Availability

The quantity of water in a soil sample can be expressed in grams of water per 100 g dry soil, which can be determined by measuring the mass of water lost by the sample after desiccation (at 105°C to obtain evaporation of bonded molecules of water). This is an easy and convenient method for comparing several states of wetness in the same soil. But it gives no idea of how much soil water is available for plants and microorganisms. Thus a water content of 15% roughly represents the retention capacity of a sandy-silt soil; water will be readily available in such a saturated soil. An identical water content in a clayey soil, with a much higher water retention capacity, contrarily results in commencement of plant wilting. So a measure which takes into account the effort required for extraction of water retained in the soil is needed.

To release a bonded molecule of water to the free state, energy is required; termed the total **hydric potential,** this energy has to compensate for the gravity potential (due to weight, and negligible), osmotic potential (insignificant, as just seen above) and the matrix potential.

In fact, that currently measured is the **matrix potential**. To visualise the effect of suction exercised by these forces in an unsaturated soil, the following apparatus can be devised (Fig. 1.5). A porous porcelain plug connected to a U tube is introduced into the soil. At the beginning of the experiment, the tube is filled with water to the level of the plug. The height of the water in

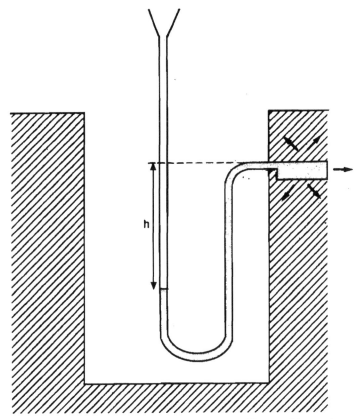

Fig. 1.5   Manometric tube showing the suction exercised by a soil beginning to dry up: pF = log h (cm).

the tube falls progressively until a level of equilibrium is attained; this displacement in water level is the result of action by the matrix forces.

The value of the difference in levels, h, expressed in cm, is an illustration of the matrix potential. However, to preclude manipulation of too many figures, the logarithm of the number, **pF**, is used.

The potential can also be expressed as a function of soil porosity by the following empirical formula:

$$pF = \log 0.15 - \log r$$

where r represents the range of the capillaries expressed in cm. Pores of radius 6 to 0.2 μm play the most important role in water retention.

Nevertheless, there is a tendency to replace the idea of pF by a normalised expression of the hydric potential in megapascals. These values are negative numbers since they refer to suction (Table 1.3).

Table 1.3 Relationship between different expressions of hydric potential of a soil

pF	Pressure* expressed		Reference values
	in bars	in MPa	
1	0.01	0.001	Saturated soil
2	0.1	0.01	
3	1	0.1	Retention capacity
4	10	1	
4.2	15	1.5	Wilting point
5	100	10	
5.8	650	65	Limit of microbial activity

*Pressure values are actually negative numbers since they correspond to suction.

HOW THE MATRIX POTENTIAL OF A SOIL IS MEASURED

A plate extractor is used in the laboratory. The sample, a thin layer of soil previously saturated with water, is placed on a porous porcelain sheet in a closed container and subjected to air pressure to squeeze out part of the water. The water content of the sample is then determined. The same operation is carried out several times on samples subjected to various pressures. In this manner one can construct a characteristic curve for the soil under study, which gives the relation between pressure exerted and quantity of water retained by the soil. Knowing the water content of the soil (easy to measure), the matrix potential can be deduced from the curve. Thus in the example given below, a water content of 20% corresponds to a matrix potential of 1 bar or 0.1 MPa.

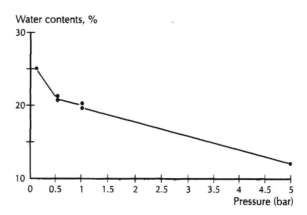

In small samples, very precise measurements can be obtained using microprobes with the Peltier effect. These are thermocouple psychrometers, which determine the tension of the water vapour under controlled temperature, from which the hydric potential can be deduced.

For in-situ soils, a tensiometer can be used. This apparatus consists of a porous plug fitted with water and introduced into the soil. The matrix forces exert suction on the water, leaving the plug to penetrate the soil. This results in a depression that can be measured with a manometer. Nevertheless, if the soil dries too much, the water column between the plug and the manometer may be interspersed with bubbles. For such dry soil, it is preferable to use a technique in which an electric current is passed between two electrodes placed side by side and enclosed in a small plaster cube in which the water content is in equilibrium with the water content in the soil. Resistance to the passage of current increases when moisture decreases. This apparatus gives satisfactory measures until the wilting point is reached.

Neutron probes, which can also be used in the field, measure the mass of water but not its tension. Practically, the quantity of hydrogen atoms present in a volume of soil is measured, assuming that the hydrogen content in the organic matter is negligible compared to that in the water.

Furthermore, osmotic potential can be estimated by measuring the electric conductivity of the soil solution since the relation between the logarithms of these two parameters is approximately linear.

Soil pH

Actually speaking, the pH of a soil refers to the pH of its aqueous phase wherein the presence of H_3O^+ ions is due to multiple reactions and permanent exchanges among this phase, the solid phase (absorbent complex) and the gaseous phase (carbonic gas). These reactions can be very different at any given moment between two very close sites in the same soil. Thus the mineralisation of organic nitrogen under the action of ammonifying bacteria results in the appearance of NH_4^+ ions and the pH increases in the immediate vicinity of these bacterial colonies. On the other hand, anaerobic fermentation of plant debris rich in glucides leads to production of organic acids, reducing local pH. Moreover, as already described, clay layers are capable of capturing protons and the pH in the immediate vicinity may be appreciably lower (by a unit or more) than that of the solution which bathes the clay.

The pH measured on a soil sample is therefore only a very approximate mean of the pH of all the microsites of that sample. Methods of measuring soil pH, though generally satisfactory for the agronomist, are not always sufficiently refined to enable the microbiologist to interpret all the phenomena observed, especially in the vicinity of the roots (see Chapter 7).

HOW THE pH OF A SOIL IS MEASURED

In practice, the pH of a dilution of the soil solution is measured.

The simplest method involves the addition of water to the soil and measurement of the pH of the supernatant using the electrode of a pH meter after agitation and decantation. The soil plus water mixture must be prepared in well-defined proportions: 1 part soil for 2.5 parts water by weight. The water must be deionised.

However, a significant quantity of protons is retained by the electronegative charges of the clay. Therefore, the water is sometimes replaced by a solution of electrolytes. Cations of the electrolyte interchange with the H^+ ions fixed on the adsorbent complex, which pass into the solution. The pH measured in such conditions is always less than the pH measured in water; the difference can be one unit and sometimes even more.

Currently, a normal solution of KCl is used in laboratory soil analysis. However, most Anglo-Saxon researchers prefer a solution of 0.01 M of $CaCl_2$, with a ratio of 1:2 soil:water.

The pH measured also depends on the quantity of CO_2 dissolved in the soil solution: a dry soil sample, in which air circulates easily, has a higher pH compared to a sample of the same soil collected after rain and tested immediately.

Hence it is imperative that the operating conditions be clearly stated. The Table below illustrates the extent of variations which can occur if the techniques of measurement are not standardised: it can be seen that depending on the method selected, the soil may be considered acidic, neutral or alkaline!

Effects of conditions of measurement on pH values of a clayey silt soil (after Smiley and Cook, 1972)

Conditions				Soil/liquid proportions			
Dilution liquid	Prior agitation	Location of electrode	Degree of aeration	Saturation	1/1	1/2	1/5
Water	Yes	Supernatant	Normal	7.7	7.8	8	8.2
Water	No	Supernatant	Normal		8.1	8.2	8.5
Water	No	Suspension	Normal		8.2	8.4	8.4
Water	Yes	Supernatant	10% CO_2		6.1		
Water	Yes	Supernatant	Dry soil		7.3		
KCl N	Yes	Supernatant	Normal	6.4	6.4	6.4	6.4
$CaCl_2$ 0.01 M	Yes	Supernatant	Normal	7.3	7.3	7.3	7.3

GASEOUS PHASE

Definition

The gaseous phase occupies the soil pores not already occupied by the liquid phase. In general, pores more than 75 μm in diameter are occupied by the gaseous phase whereas pores between 0.2 and 30 μm are filled with capillary water, while intermediate canals constitute a mixed domain. The finest pores are mostly formed by microspaces existing between mineral particles and between irregularly stacked aggregates. Pores of larger diameter mainly consist of canals that remain in place after the roots die and the galleries of burrowing animals. Earthworms and ants play a significant role in the constitution and maintenance of a network for distribution of air throughout the arable soil layer. Pedofauna respire directly in the soil atmosphere whereas gaseous exchanges of roots and microorganisms, usually covered with a liquid film, occur mainly by means of gas dissolved in the soil solution. A gaseous phase is indispensable because it allows renewal of dissolved gases which have an especially important biological role. The solubility of gases varies: oxygen, for example, is nearly 40 times less soluble in water than carbon dioxide at a temperature of 20°C. Solubility increases as temperature decreases.

While there is no physical demarcation between the external terrestrial atmosphere and the soil atmosphere, their composition differs markedly due to the slowness of diffusion phenomena. Soil atmosphere contains only 15 to 18% oxygen on average. However, like the pH, this represents only an overall value. The local value can fall down to 2% in the vicinity of roots and is much lower in the cul de sacs filled with water or in the pockets inside aggregates. The average carbon dioxide content varies from 0.3 to 5%, which is 10 to 150 times higher than the values of the external atmosphere.

The gaseous phase of the soil also contains gases that are rare or absent in free air: nitrogen oxides, ammonia, ethylene, methane and traces of numerous compounds released by plant roots or microorganisms, most of which are still not known. At least some of these compounds, albeit present in very small quantities, seem to play an important role as inhibitors or transmitters of information; others act as stimulants.

Oxidoreduction (Redox) Potential

The metabolic reactions of all living beings require energy. This energy is provided by the cascade of oxidation and reduction reactions which transfer hydrogen or electrons from a reduced donor compound, which becomes oxidised, to an oxidised acceptor compound, which is thereby reduced. Each redox reaction is thus the final result of two simultaneous reactions:

$$1.\ \text{Red } 1 \rightarrow \text{Ox } 1 + n e^-$$
$$2.\ \text{Ox } 2 + n e^- \rightarrow \text{Red } 2$$

which gives

$$3.\ \text{Red } 1 + \text{Ox } 2 \rightarrow \text{Ox } 1 + \text{Red } 2$$

Each redox couple is characterised by an oxidoreduction potential, which can be measured using an electrode and expressed in millivolts, generally designated by the term E_h. For a redox system to oxidise another, it must have a higher oxidoreduction potential. For example, the redox potential of the Fe^{3+}/Fe^{2+} couple is higher than that of the I^-/I_3^- pair: ferrric iron can thus oxidise iodide by receiving electrons according to the reaction:

$$2Fe^{3+} + 3\ I^- \rightarrow 2\ Fe^{2+} + I_3^-.$$

Inside the living cell these transfers are effected by organic molecules, such as pyrimidines (NADH/NAD$^+$, NADPH/NADP$^+$) and flavoproteins for hydrogen, and cytochromes and ferridoxins for electrons. Passage of the electrons or hydrogen from one transporter to another (from NADH to flavoproteins for example), as in the case of mineral salts, occurs from pairs with low redox potential to pairs with a high potential.

In all aerobic organisms, oxygen is the final acceptor of electrons. However, in the absence of oxygen, a large number of normally aerobic bacteria are capable of using other terminal electron acceptors, such as nitrates and sulphates. They are known as facultative anaerobic bacteria. Strictly anaerobic bacteria constitute a third category of microorganisms, which do not tolerate the presence of oxygen.

HOW THE REDOX POTENTIAL OF A SOIL IS MEASURED

To compare the different systems of redox a frame of reference is required. This consists of a normal hydrogen electrode. This electrode consists of a platinum wire, immersed in an acid solution of pH = 1 (thus containing hydrogen ions) surrounded by gaseous hydrogen. This forms a redox system whose equilibrium can be written:

$$H^+ + e^- \rightleftharpoons 1/2H_2$$

In practice, the hydrogen electrode, not easy to handle, is replaced by a calomel electrode priorly standardised with reference to a normal electrode.

The redox potential value may also be found through computations using the Nernst formula

$$E_h = E_0 + \frac{RT}{nF} \log\frac{(Ox)}{(Red)} - 0.059 \frac{m}{n} pH$$

where E_0 is the normal redox potential, characteristic of the pair (by definition, $E_0 = 0$ for the $H^+ - H_2$ couple when pH = 0 under an H_2 pressure of 0.1 MPa); R the constant of perfect gases, T the absolute temperature and F the Faraday constant; n and m are the number of electrons and the number of H^+ ions exchanged respectively; (Ox) is the concentration of the oxidised form and (Red) the concentration of the reduced form of the couple.

Readings of this formula permit the observation that the redox potential varies slightly according to the pH. It decreases when the pH increases. At constant pH, the more reducing the environment, the lower the redox potential. The expression E_h is often replaced by the expression P_E modelled on the formulation of pH:

$$P_E = -\log (e^-)$$

P_E varies in the same manner as E_h. These two expressions are linked by the relation:

$$P_E = E_h/59$$

with E_h expressed in mV.

The degree of oxygenation of the soil can also be evaluated directly using a tensiometer. Once the water injected into the apparatus has attained equilibrium with the soil solution, the liquid contained in the porous plug is carefully removed and measured using an oxymeter (Vigouroux, 1997).

FOR FURTHER INFORMATION ON THE INERT FRACTION OF THE SOIL

Charman P.E.V. and Murphy B.W. (eds.). 2000. Soils. Their Properties and Management. Academic Press, NY (2nd ed.).

Hillel D. 1998. Environmental Soil Physics. Academic Press, NY.

Morel R. 1989. Les sols cultivés. Technique et Documentation. Lavoisier, Paris.

Sparks D.L. 1995. Environmental Soil Chemistry. Academic Press, NY.

2

Living Components of the Soil

The word 'soil' too often evokes that mineral substrate in which plants anchor their roots. It is difficult to imagine that soil constitutes a medium for life, similar to the oceans, and most people are surprised by the fact that earthworms can pass their entire life in such a habitat. Indeed, soil is currently considered a transient place: for example, the larvae of numerous insects sojourn there more or less a long time before escaping to reach free air. Many reveal their subterranean presence by the damage they cause in devouring the parenchyma of plant roots. Thanks to studies by Pasteur on sheep anthrax (caused by *Bacteridium anthracis*) and the epidemiology of infectious diseases, it is also known that soil can serve as a reservoir to a large number of pathogenic agents awaiting a new host. But only in a relatively recent epoch has soil begun to be considered a permanent habitat for a multitude of living beings whose number is increasingly recognised each year.

In this chapter we shall review the principal groups of soil inhabitants, most of which are microscopic and belong to the vast ambiguous world of **microorganisms.** Others, hardly visible to the naked eye, constitute the **microfauna.**

Since the nineteenth century it has been thought that the traditional separation of living beings into plants and animals does not make much sense, especially with reference to microorganisms. Indeed, one observes among them all the intermediates possible between immobile chlorophyllous organisms, comparable to green plants, and mobile organisms, which ingest their prey, similar to animals. Again, other microorganisms possess neither chlorophyll, as plants do, nor the movement and capacity for ingestion observed in animals. Consequently, taxonomists rather soon agreed to classify the microorganisms into a kingdom distinct from the animal kingdom and the plant kingdom. But this partition rapidly proved unsatisfactory.

In 1969, Whittaker proposed a more elaborate classification comprising five kingdoms (Fig. 2.1). In this system, the simplest microorganisms, possessing neither intracellular organelles nor well-differentiated nuclei, constitute the kingdom of **Prokaryota** (or Moneres). All the other microorganisms are termed **Eukaryota**, with well-differentiated nuclei and mitochondria in most cases and eventually other organelles as well.

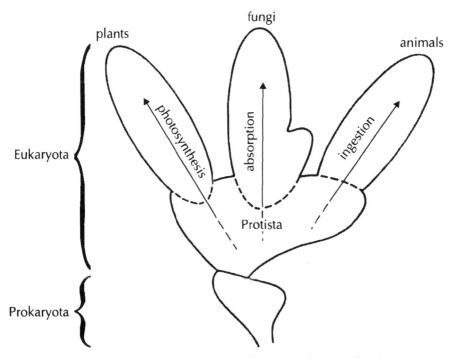

Fig. 2.1 Division of living beings into 5 kingdoms according to Whittaker (1969).

The kingdom Protista is large, comprising many phyla that include a wide variety of microorganisms which derive their nutrition from photosynthesis, absorption or ingestion. Whittaker visualised a trophic specialisation of these organisms, which helped in distinguishing plants, fungi and animal kingdoms.

Recent developments in molecular biology have caused a major upheaval in this concept of the living world (Woese et al., 1990). After comparing the ribosomal RNA sequences of the organisms known, 3 'domains' were distinguished: Bacteria, Archaea and Eucarya. They presumably originated from a common ancestor but these three groups of organisms apparently diverged very long ago, during the history of evolution. In this new light, each large differentiated phylum within a domain could itself constitute the equivalent of an ancient 'kingdom', so much differing genetically from the other phyla. The animal, plant and fungi kingdoms are now simply very close branches that appeared very late in the evolution of Eukaryota (Fig. 2.2).

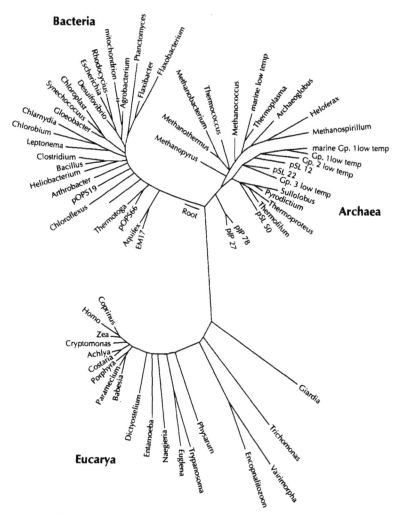

Fig. 2.2   Phylogenetic tree based on analysis of small subunit rRNA sequences of 64 representative organisms (Pace, 1997).

In practice, microorganisms (prokaryotes, monocellular eukaryotes and fungi) are often grouped under the name 'microbes'. Etymologically this word means 'very short-lived'. However, certain stages of microbes (endospores, cysts) can survive a very long time.

Besides microorganisms, the soil also harbours numerous representatives of the animal kingdom. Only those groups that intervene directly in the microbial equilibrium because of their size and activity are discussed in this chapter, namely, the nematodes, acarids and collemboles.

PROKARYOTES

These include the **bacteria** and the **archaea**. They are the smallest known living organisms whose dimensions, on average, vary from 1 to 2 × 1.5 to

4 μm with the difference between minimal and maximal values considerable. Thus, mycoplasma fall in the range of 0.1 μm in size while some filamentous bacteria can attain a length of 1 mm.

The role of prokaryotes in ecosystems, especially in the soil, is of paramount importance. So a few pages must be devoted to them even though paradoxically the advances in our knowledge prompt recognition of an even greater ignorance than thought a few years ago. Most of the information concerns the bacteria since the archaea have been studied only in the last few years.

Nuclear Apparatus

As in all living beings, all the information necessary for the functioning and multiplication of prokaryotes is coded by genes made up of deoxyribonucleic acid (DNA). These genes (a bacterium has nearly a thousand) are arranged on a double strand of curled DNA which upon unfolding, assumes the shape of a ring. This unique circular chromosome is not separated from the cytoplasm by a nuclear membrane, a characteristic which differentiates prokaryotes from the rest of living beings.

Besides the chromosome which ensures some sort of stability of character in the species, smaller molecules of DNA known as **plasmids** are frequently encountered; they are also circular and carry a limited number of genes (on average a hundred). The genes carried on the plasmids are not indispensable to cell functioning but very often confer selective advantages in hostile or inhospitable surroundings to the individuals possessing them, e.g. resistance to heavy metals and antibiotics, production of toxins, ability to degrade complex substrates, etc. Several plasmids can coexist in a single cell, some in several copies. In rhizobia, which contain very large plasmids, the extranuclear DNA can constitute up to 45% of the genome. Bacteria can quite easily lose or acquire such plasmids.

Bacteria reproduce by binary fission: after a growth period (which can last less than an hour under optimal conditions), the chromosome replicates into two identical chromosomes and a partition (septum) is formed in the middle of the cell, giving rise thereby to two identical daughter cells. Replication of the plasmids accompanies that of the chromosomes.

One phenomenon known in bacteria, namely **conjugation,** is considered by numerous scientists an indication of sexuality. Under the action of factor F coded by a plasmid, a donor cell can associate with a recipient cell and transfer either a part or all of its chromosome. The plasmids can also pass upon contact from a donor cell to a recipient one (of the same species, a related species or one of a different genus). Thus much genetic information can be transmitted from one individual to another in a single encounter and gradually to an entire population through the phenomenon of conjugation. The resultant variability is a formidable tool of adaptation for the microbial population, adding to the variability resulting from sporadic mutations leading to minor phenotypic modifications.

But that is not all! Prokaryotes are sometimes parasitised by viruses called phages, which usually destroy them but can, under certain circumstances, modify their behaviour either directly or indirectly. These phenomena are termed **conversion** and **transduction,** discussed under 'Viruses' (p. 63). The resultant phenotypic changes affect resistance to antibiotics, or production of toxins or antigens, or other metabolic activities.

Under certain conditions, by a process termed **transformation**, naked DNA fragments may be introduced into a prokaryote and subsequently integrated into its chromosome. Only some species of prokaryotes seem to be capable of this transformation; they have to be in a 'competent' phase which depends on their physiological state. The DNA molecule can remain very stable in soil. It is released into the soil either passively by the lysis of cells after death of a plant, animal or microorganism, or actively by certain bacteria. The phenomenon of transformation is rare but has nevertheless been demonstrated in a few experimental studies, thereby establishing that horizontal transfer of genes is theoretically possible in soil between genetically distant species.

All these mechanisms confer very great flexibility on bacterial populations, but this faculty for evolving constitutes an inconstancy for phenotypic characters at the scale of an individual. Exchanges in genetic material are possible however only between individuals in an active phase of development.

Envelopes

The cytoplasm of prokaryotes is delimited by a semi-permeable cellular membrane which allows the cell to exchange matter with the external medium and to maintain constant its internal pH and concentration of mineral salts and metabolites.

In most cases, a rigid wall is found outside the cellular membrane which protects the cell and gives it its characteristic shape. The absence or presence of this wall as well as its chemical composition are very important criteria in the classification of prokaryotes (Table 2.1). In motile bacteria, very fine filaments arising from the cytoplasm traverse the wall, viz. cilia or flagella, that assist movement by mechanisms not yet fully understood.

Table 2.1 Classification of prokaryotes according to the nature of their cell walls

Nature of cell wall	Type of organism
No cell wall	Mollicutes
Cell wall of peptidoglucan (N-acetylglucosamine + acetyl-muramic acid + diaminopimelic acid + other amino acids)	
—flexible cell wall	Myxobacteria
	Spirochaeta
—rigid cell wall	Rickettsia
	Chlamydia
	(no muramic acid)
	Eubacteria
	Cyanobacteria
Cell wall without peptidoglucan (many different types)	Archaea

Sometimes a more or less thick film of mucous polysaccharides which have a strong affinity for water surrounds the cell wall. This film is termed a **capsule** when its contours are well defined or a **mucous layer** when the contours are blurred. The capsule is not indispensable for normal functioning of bacteria but when present has a protective function: against desiccation, phages and even predators. In some species several hundred bacteria are sometimes enclosed inside a common capsule.

In some bacteria, such as *Bacillus* and *Clostridium*, under unfavourable environmental conditions an endospore may form inside a cell as a result of very complex metabolic processes. The endospore has a dense dehydrated cytoplasm surrounded by a thick wall comprising several superposed envelopes. These endospores show remarkable resistance to chemical (antibiotic, antiseptic) and physical agents (pressure, desiccation, UV, heat). For example, the endospores of *B. thermophilus* have to be heated for 4 hr at 100°C to kill them.

Nutrition

Bacterial cells require basic matter to build up their organic molecules. These basic materials include carbon, nitrogen, phosphorus, sulphur, various other mineral elements and eventually growth factors. Some bacteria, known as **autotrophs**, are capable of using carbon in the mineral form: atmospheric carbon dioxide or that dissolved in water is the element from which all their organic syntheses commence. A large number of bacteria require organic compounds already synthesised by other organisms. These are known as **heterotrophs.**

The energy needed for functioning of the bacterial machinery originates either from light or chemicals. In the former, energy is captured by pigments similar to chlorophyllous and carotenoid pigments. In the latter, energy comes from redox reactions. Species capable of utilising mineral or organic carbon are found in both categories.

Lastly, prokaryotes can be classified according to their energy and nutritional requirements (Table 2.2). Chemoorganotrophs are the better studied trophic group and many species have been identified. Recent works suggest that lithotrophic bacteria and archaea are definitely more abundant and diverse than previously thought.

Table 2.2 Classification of prokaryotes according to energy source and ability to use mineral carbon (CO_2)

			Energy source	
			Light	Chemicals
			Photosynthetic	Chemosynthetic
	CO_2	Autotrophic	Photolithotrophs	Chemolithotrophs
Carbon source	Organic compounds	Heterotrophic	Photoorganotrophs	Chemoorganotrophs

It is very important to note that bacteria have no possibility of ingesting the foods which nourish them. To ensure their utilisation, the nutrients must obligatorily cross the wall and then the cytoplasmic membrane. This is only possible for soluble and low molecular weight compounds. Nutrients with high molecular weight (cellulose or proteinic molecules for example) must first be broken down into smaller elements by the enzymes the bacteria excrete into the external medium. Hydrolysis of the molecules and diffusion of their fragments to the bacteria are impossible in a non-hydrated medium. Understandably, therefore, water is absolutely indispensable for bacterial development.

> Passage of nutrients into the cell is generally not the result of simple diffusion. It is assured by a particular category of enzymatic proteins, the permeases. They allow not only nutritive molecules to penetrate very rapidly into the cytoplasm, but also to accumulate in the cell at concentrations greatly exceeding those in the soil solution.

Chemotaxis

This is a property of motile cells that orients their movement in a gradient of chemical concentration. Chemotaxis is positive or negative depending on whether the bacteria are attracted or repelled. To accomplish this the cellular membrane must possess the corresponding receptors so that only a limited number of chemical compounds are attractive or repellent. A single receptor can generally recognise many analogous molecules.

Similarly, some bacteria are sensitive to gradients of oxygen or of temperature.

These phenomena imply that microorganisms are not randomly distributed in the soil. Rhizobia, for example, are selectively attracted to the extremities of roots of legumes.

Ecological Exigencies

The diversity of the innumerable species is so large that the prokaryotes, on the whole, seem to prosper in the most diverse environments: very low temperatures, hot springs, normal atmospheric or strictly anaerobic conditions and alkaline or highly acidic (pH near 0) media. However, this extraordinary versatility of microorganisms to colonise all these milieus does not mask the fact that a particular specialised species can generally develop only in very precise environmental conditions. In soil, mesophilic microorganisms (optimal temperature between 20 and 40°C) are usually found and they prefer neutral or slightly alkaline pH. Bacteria participating in the sulphur cycle, such as *Thiobacillus thiooxydans*, can withstand very acidic pH, however. Further, as mentioned above, anaerobic microorganisms can live side by side with strictly aerobic microorganisms.

Classification

The extreme diversity of the bacterial world and our as yet incomplete understanding of these organisms pose difficulties in developing a coherent classification of them, as has been done for the plant and animal kingdoms. The problem is further complicated by the challenge of precisely defining a prokaryotic species. A species is usually considered an ensemble of individuals whose mating or crossing results in the production of fertile offspring similar to the parents. But in the absence of true sexuality, this definition makes no sense. Furthermore, due to multiple possibilities of exchanges of portions of DNA between bacteria there exists a genetic continuum within species, even within genera. Therefore, among the prokaryotes, a species is only a somewhat arbitrary grouping of microorganisms that have a large number of characteristics in common.

Some criteria used for defining this group are very old, such as their morphology or staining by the Gram method. The Gram staining method, employed for more than 100 years, helps to differentiate bacteria which have only a glycoprotein wall (Gram positives) from those with a complex wall with an external phospholipid envelope (Gram negatives).

> The bacteria are spread on a glass slide and fixed with heat. After washing and drying, the preparation is stained with a solution of gentian violet, followed by a solution of Lugol iodo-iodurate. Excess dye is removed by washing the slide quickly under distilled water. This is followed by differentiation with alcohol. In the presence of ethanol, Gram-negative bacteria lose the violet coloration while Gram-positive ones retain it. The slide is rinsed again with distilled water and counterstained with basic fuchsin, whereupon Gram-negative bacteria turn pink. The slide preparation must be viewed directly under the microscope without a cover slip.
>
> The fact that the Gram-negative bacteria possess an aminopeptidase capable of hydrolysing L-alanine-4-nitroanilide can also be exploited (Cerny, 1976). The nitroaniline formed by the reaction stains the medium yellow in a few minutes; the Gram-positive bacteria do not react or very slightly.

Other criteria, more recently introduced, utilise biochemical properties. These are complemented by studies of proteins using electrophoresis or profiles of fatty acids obtained by gas-phase chromatography and by serological techniques. However, molecular biological techniques are more often used. These include, for example, determination of the ratio of guanine and cytosine with reference to adenine and thymine in chromosomal DNA; digestion of portions of DNA using restriction enzymes, followed by polymorphism studies of the size of the digested DNA fragments (RFLP); sequencing ribosomal RNA genes (obtained after cloning of DNA) and comparing their sequences with the equivalent sequences of known microorganisms; or evaluation of the degree of hybridisation of the DNA of an isolate with a reference strain.

The basic reference work for the study of prokaryote classification is Bergey's manual.

As indicated earlier, prokaryotes can be classified according to the nature of their walls:

— prokaryotes which lack a cell wall (Mollicutes) are not soil microorganisms but plant or animal parasites (*Spiroplasma*) whose DNA does

not hybridise with that of bacteria, indicating that spiroplasmids are genetically very distant from other prokaryotes;
— bacteria which have a two-layered wall, of which one layer consists of peptidoglucans (in general Gram negatives);
— bacteria with a thick wall containing peptidoglucans (generally Gram positives);
— archaea, which lack peptidoglucans and differ markedly from the preceding microorganisms.

The bacteria comprise the majority of the known species of prokaryotes and are often grouped as a function of their metabolic requirements. This classification is convenient but it is now known that the categories so defined contain species very distant from each other. A few examples:

Photolithotrophic bacteria
Green bacteria: Chlorobacteriaceae
Purple bacteria: Thiorhodaceae
Strict anaerobes, they oxidise hydrogen sulphide into sulphur in the presence of infrared or red radiation and do not produce oxygen:

$$CO_2 + 2H_2S \rightarrow (CH_2O) + H_2O + 2S$$

Photoorganotrophic bacteria
Non-sulphur purple bacteria: Athiorhodaceae

Chemolithotrophic bacteria
Nitrifying bacteria:

$$NH_4^+ \rightarrow NO_2^- \; \textit{Nitrosomonas}$$
$$NO_2^- \rightarrow NO_3^- \; \textit{Nitrobacter}$$

Acidifying bacteria of the sulphur cycle:

$$S^{--} \rightarrow S \rightarrow SO_4^{--} \; \textit{Thiobacillus}$$

Bacteria which oxidise iron and manganese: *Galionella, Leptothrix*
All these bacteria are aerobic and play an important role in mineral cycles.

Chemoorganotrophic soil bacteria
• *Motile bacteria without flagella (Myxobacteria)*
   Their supple walls allow them to move by gliding. They are active decomposers of organic matter and produce pigmented mucus. Cellulolytic bacteria: *Cytophaga*.
• *Aerobic Gram-negative bacilli and cocci*
— Pseudomonadaceae with polar flagella: *Pseudomonas, Burkholderia, Ralstonia* (Fig. 2.3).
— Azotobacteraceae with polar or peritrichous cilia. They fix atmospheric nitrogen: *Azotobacter, Beijerinckia, Azospirillum*.
— Rhizobiaceae with polar or subpolar flagella: *Rhizobium* and related genera, which fix atmospheric nitrogen when living in symbiotic association with leguminous plants; *Agrobacterium*, producers of galls or root proliferation.
• *Anaerobic facultative Gram-negative bacteria*
   Enterobacteriaceae, motile by use of peritrichous cilia; they can reduce nitrates to nitrites: *Erwinia*, plant parasites.
• *Gram-negative anaerobic bacilli*
Desulfovibrio: reduces sulphates to sulphides.

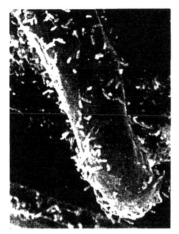

Fig. 2.3 Cells of *Ralstonia solanacearum* on the surface of an absorbent tomato hair (plate, J. Schmit, INRA).

- *Gram-positive sporogenous bacilli and cocci*
They can form endospores. Bacillaceae: *Bacillus* (aerobes) and *Clostridium* (anaerobes), *Thermoactinomyces*.
- *Coryneform bacteria*
Gram-positive bacteria of various forms. Corynebacteriaceae: *Clavibacter, Curtobacterium flaccumfaciens, Rhodococcus fascians,* plant parasites, *Cellulomonas,* cellulose decomposers, *Arthrobacter,* often involved in the degradation of industrial organic products.
- *Actinomycetes*
These Gram-positive bacteria form branching filaments and produce conidia and for this reason were long considered fungi. They play a major role in the decomposition of litter. Many synthesise antibiotics.
- — Actinomycetaceae: *Actinomyces,* produce little mycelium;
- — Streptomycetaceae: *Streptomyces,* morphologically very close to fungi;
- — Micromonosporaceae: *Micromonospora,* degraders of cellulose and chitin;
- — Frankiaceae: *Frankia.* These bacteria live in symbiosis with plants belonging to many tropical (Casuarinaceae) and temperate (Betulaceae, Eleagnaceae, Rhamnaceae) botanical families and form nodules on the roots, which fix atmospheric nitrogen.

Cyanobacteria, with chlorophyll-like pigments, come under these groups. They are present in small numbers in all soils and under certain circumstances play a very important role.

Cyanobacteria are prokaryotes which possess chlorophyll a (but not chloroplasts) and in the presence of light synthesise organic compounds using atmospheric CO_2. Like higher plants, they produce oxygen during this process.

Several species also have a nitrogenase which helps in fixing atmospheric nitrogen. Their nutritive requirements are thus reduced to a minimum: air, light and water. It must be added that their water requirement is very limited; certain forms living as crusts on the surface of rocks or desert soils are able to survive with water from morning dew.

Cyanobacteria live either singly or in multicellular filaments. Their walls are sometimes thick because of a polysaccharide capsule; or the walls may be thin and the cells move by gliding, similar to myxobacteria. In some groups, commencement of specialisation into vegetative cells, reproductive cells and nitrogen-fixing heterocysts has been observed.

In ecosystems such as rice fields, nitrogen-fixing cyanobacteria can contribute significantly to nitrogen fertilisation. Relevant species in this regard are *Anabaena azotica, Aulosira fertilissima, Tolypothrix tenuis.*

Archaea are very different from bacteria: the composition of their cell wall varies from genus to genus but is always devoid of peptidoglucans; their membrane and ribosomal RNA are also different. The electric neutrality of their DNA, as in eukaryotes, is due to histones and not polyamines; their RNA polymerases have points in common with eukaryotes and the regulation of their transcription mechanisms is also similar to that of eukaryotes.

As their name signifies, archaea have been considered primitive bacteria. The most recent studies do not confirm this hypothesis, however, suggesting contrarily that a branch common to archaea and eukarya diverged in the very early stages of evolution from bacteria.

All the archaea identified in the last few years appeared to be confined to 'extreme' environments: hot and acid sources (some enzymes of the thermoacidophiles function even at temperatures of 130°C!), brine (Halobacteriaceae) and environments totally deprived of oxygen (methanogens living in mud, rubbish and in the rumen). But studies of ribosomal DNA extracted from more usual environments show that archaea are doubtlessly present in large numbers in 'ordinary' habitats such as the soil or surface of oceans, where they coexist with bacteria. However, the method of culturing them is not yet known.

Isolation and Enumeration of Soil Prokaryotes

To obtain soil bacteria, a few grams of soil are suspended in water. After agitation and decantation, a few drops of the supernatant are spread on the surface of an appropriate gelified culture medium. Since the numbers of bacteria are considerable, dilutions of the initial suspension are always used for culturing. By this method, separate colonies are obtained, each arising in principle from a single bacterium.

ISOLATION OF SOIL BACTERIA

A good example of the difficulty of isolating soil bacteria is given by Steinberg (1987). A silty-clay soil sample was washed 20 times. After each wash an aliquot of the supernatant was taken, diluted and spread in petri dishes for enumeration.

The first wash yielded 6.88×10^7 bacteria per g dry soil. The number of desorbed bacteria decreased regularly until the 10th wash, then stabilised as shown in the graph below.

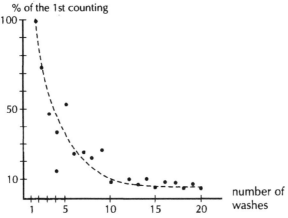

% of the 1st counting

number of washes

Fig. 2.3a Total microflora desorbed from the soil after 20 successive washes, expressed as percentage of the number obtained after the first wash.

The 20th wash still yielded 3.3×10^6 bacteria per g soil.

After the 20th wash, the soil was crushed and dilutions placed in culture. This operation again yielded 1.5×10^8 bacteria per g, which is still almost equivalent to all the combined yield of previous washes.

The total quantity of bacteria counted was 4.8×10^8 per g dry soil. One-third came from the internal compartment (crushed soil) and two-thirds from the external compartment (washes).

In this way it could be possible to obtain a clear representation of bacterial flora of a soil. But such a survey encounters many problems, for example:

(1) It is almost impossible to get all the bacteria present in a sample into suspension because most adhere strongly to the soil particles and settle with them, while many remain trapped inside the aggregates. To liberate such bound forms, the microaggregates must be finely crushed or pulverised ultrasonically, which engenders the risk of destroying at least part of the sample (see box above).

(2) Bacteria bound to soil particles form small masses of about ten cells, which are very difficult to separate. Under such conditions, an isolated colony does not really represent one single bacterium (Ozawa and Yamagushi, 1986).

(3) As already noted, anaerobic bacteria can coexist with aerobic bacteria in the same soil sample. It has also been shown that strictly anaerobic and aerobic bacteria can be actively present in the same particle (Marty and Bianchi, 1992). The conditions required for culturing them are not the same for each category, however.

(4) There is no culture medium that can simultaneously satisfy the nutritive requirements of all bacterial species. Moreover, some prokaryotes have special requirements which are so specific that they have never been grown on artificial media to date and hence their existence has long been ignored. An electron microscope revealed for the first time the presence of microscopic cells in the soil that had earlier been considered

dead; it was shown later that their genome was intact and functional (Bakken and Olsen, 1989). Studies of nucleic acid from soil extracts have confirmed the abundance and great diversity of these microorganisms which are mostly not known. It is estimated that a gram of soil can harbour many thousands of different species of prokaryotes.

Millar and Casida (1970) proposed that the bacterial biomass be evaluated by extracting muramic acid from soil, a characteristic constituent of cell walls. After alkaline hydrolysis, the muramic acid produces lactic acid, which can be measured. This method does not differentiate between the walls of living bacteria and those of dead ones, however. Moreover, Gram-positive bacteria contain ten times more muramic acid than do Gram-negative bacteria, which complicates interpretation of results. Lastly, the archaea, which do not possess muramic acid, are not accounted for by this method.

Direct methods can be used while counting microbial cells under the microscope after staining them on a slide. To count a specific group of microorganisms, serums combined with fluorescent dyes can be used. Using probes made of oligonucleotides coupled with fluorochromes allows very precise labelling and hence better targeting. With this technique, microorganisms can be identified in situ with no need for culturing them.

Thus, almost 150 years after the first works of Pasteur we still know very little about the soil prokaryotes. Microbiologists feel that 95 to 99% of the prokaryotes have still not been cultured and hence it is not easy to evaluate their number without a large margin of error. Most estimations yield a quantity between 10^7 and 10^{10} units per g soil in soil horizons rich in organic matter. In general, the density of the prokaryote population decreases as depth increases. Knowing the average dimensions and density of the bacterial cytoplasm, it is possible to deduce their fresh biomass. An extraordinarily high number (but probably underestimated) of 2.5 to 10 t ha^{-1} has been arrived at!

EUKARYOTES

General Characteristics

Whether they be unicellular microorganisms, mammals or higher plants, eukaryotes have characteristics in common, making them a much more homogeneous group than prokarytoes despite apparent diversity. We shall briefly recall these characteristics before studying some groups of soil eukaryotes.

Nuclear Apparatus

The nucleus, containing a nucleolus and DNA distributed in a determined number of chromosomes, is separated from the cytoplasm by a nuclear membrane. At the moment of cell division, the chromosomes separate out and undergo mitosis, which results is each daughter cell receiving an equal amount of chromosomal DNA.

Most eukaryotes are capable of sexual reproduction. Fusion of two gametes of sexually different polarities, containing a single copy of each chromosome, gives rise to an egg or zygote, which has two sets of chromosomes. Passage from the 2N-chromosome phase (diploid phase) to the N-chromosome phase (haploid phase) must obligatorily take place before fertilisation. This reduction is due to a particular type of nuclear division, **meiosis**. The cycle of all eukaryotic species is characterised by alternation of these haploid and diploid forms whose relative duration varies considerably depending on the organisms.

Principal Organelles

The cytoplasm of eukaryotes contains multiple demarcated elements whose number and appearance vary with cell age. Their complex structure and role can only be elucidated under an electron microscope.

Each organelle is delimited by a lipoproteinic membrane which isolates it from the cytoplasmic mass. The presence of organelles in the eukaryotic cell therefore results in a considerable augmentation of the membrane surfaces. The endoplasmic reticulum forms a highly dense network of flattened cavities, playing an important role in the production and transport of organic molecules and in the formation of membranes. Part of the endoplasmic reticulum is covered with ribosomes, small globules constituted of proteins and RNA which assure the assembly of amino acids into peptides. The endoplasmic reticulum is connected to the perinuclear envelope as well as the Golgi apparatus. The Golgi apparatus, formed by the interconnection of several groups of stacks of small cisterns, plays a role in the metabolic activity, especially in the synthesis of glycoproteins. This apparatus is not well differentiated in fungi. Lysosomes, vesicles containing hydrolases, assure the digestive process.

All eukaryotes except for a very few ancient microorganisms such as *Giardia* and *Trichomonas*, possess mitochondria, small elements derived from endosymbiotic bacteria which play a major role in energy metabolism (redox and phosphorylation reactions). Mitochondria contain specific DNA and divide autonomously.

The cytoplasm of photosynthetic eukaryotes also contains chloroplasts, flattened discs that contain chlorophyll a, b or c and various other pigments. These chloroplasts are the centres for transformation of light energy into chemical energy, a process that is the basis for synthesis of all organic compounds. Like mitochondria, chloroplasts have specific DNA and replicate independent of the cell. The chloroplasts originated from symbiotic cyanobacteria.

Bases for Classification

Among the eukaryotes the following can be distinguished:
— simple unicellular or multicellular organisms; in the latter, cells assemble in thalli but are not differentiated: fungi and protists;

— multicellular complexes with very specialised cells, which form tissues that differentiate into organs: plants and animals.

Fungi

Fungi can be defined as heterotrophic, filamentous, immobile micro-organisms. While they are usually heterotrophic, the filamentous characteristic is not an absolutely essential criterion. Many fungi of great practical importance are found in the unicellular form: these include yeasts and some other groups of lower fungi. While the absence of mobility is total in the so-called higher fungi, the flagellated motile zoospores aid in dispersal in certain other groups. Moreover, though all filamentous fungi and yeast possess a rigid cell wall, most other groups do not.

These few facts suffice to show that the fungal kingdom is in fact a very heterogeneous group and wrongly includes living organisms whose ancestors possessed almost no common traits.

Nevertheless, the numerical significance and the role played by the various groups are not at all the same and hence when speaking of fungi, we shall almost always mean the 'typical' filamentous organisms.

We must therefore say a few words about these filaments, more correctly termed hyphae.

Hyphae

These are more or less wide pipes (2 to 15 μm) containing cytoplasm, with a constant diameter for a given species. The rigidity of these hyphae is ensured by a wall made of chitin or cellulose fibres interspersed in a matrix of polymers of glucose (glucans) or mannose (mannans) and protein, occasionally covered with mucilage. The relative proportions of these components vary depending on the systematic position of the species. The walls are often impregnated with melanin, which gives them a brown colour. Melanins are complex molecules containing tyrosine and phenolic polymerised nuclei whose chemical structure is similar to lignin. After the fungi die, the melanins become part of the humus to whose composition they add the nitrogen from tyrosine. Melanised cell walls are more resistant to enzymatic hydrolysis than non-pigmented walls.

The hyphae are never partitioned in oomycetes and zygomycetes. In other classes of fungi, transverse septa of the wall are formed at irregular intervals and divide the hyphae into segments. There is no discontinuity in the cytoplasmic mass, however, because the septa are always incomplete and do not prevent exchanges between compartments.

Hyphae only develop through growth of their extremities and never by intercalary elongation. This apical elongation can be considered characteristic of fungi. Ramifications (branches) are formed behind the apex, which elongate and divide, forming thereby a **mycelium**. The location of the branches with reference to the apex and the angle they form with the primary hypha are species characteristics.

Cytoplasm and Nuclei

As in all eukaryotes, the cytoplasm contains organelles, and **cytoplasmic currents** flow towards the apical zones. This flux, readily observable in fungi whose hyphae lack septa, ensures a supply of nutrients and probably hormones to the growing parts. An inverse current, although much less intense, allows for flow of information towards the base and also helps in regulation of the main flux.

The cytoplasm is covered by a membrane made of phospholipids, proteins and sterols. Ergosterol is a characteristic component of the cytoplasmic membrane of fungi with partitioned mycelia. Oomycetes, non-partitioned, are incapable of synthesising this compound; we shall show later that these are not true fungi.

Genetic Aspects

Each bit of the mycelium contains several nuclei. If all the nuclei are identical, the fungal organism is homokaryotic and genetically homogeneous. In a dikaryotic mycelium, two haploid nuclei of different types coexist and divide at the same time, thereby ensuring phenotypic stability. However, two or several different nuclei may be present simultaneously and can divide independent of each other, in which event the mycelium is heterokaryotic. This can result in the relative proportions of the nuclei being modified and phenotypic fungal characters differing over a period of time. The heterokaryotic character is short lived because it is unstable. Heterokaryons can be formed either by mutation of some nuclei of a homokaryon or when hyphae of genetically different composition anastomose.

Anastomosis is a phenomenon characteristic of higher fungi: two hyphae move towards each other, their walls coalesce at this point of contact and the cytoplasms fuse, uniting the nuclear and cytoplasmic material (Fig. 2.4). This is generally regulated by several genes: for example, five different genes of compatibility are known in *Cryphonectria parasitica*, the causative

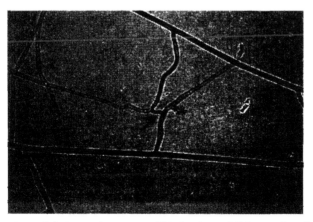

Fig. 2.4 Anastomosis between two compatible hyphae of *Rhizoctonia solani*. The arrow indicates the region where filaments have fused (plate, P. Davet, INRA).

agent of chestnut canker. When two isolates come into contact, it is imperative that they possess the same alleles on all the concerned loci for hyphal fusion. In such a situation, it can be said that the two isolates belong to the same anastomosing group. If one or several alleles differ, anastomosis is rare, even impossible, in which event it is considered that the isolates belong to different groups. Isolates in the same anastomosing group can therefore be considered genetically closer than those of the same species belonging to different anastomosing groups. This property is used in studies of population dynamics and in epidemiology. It also has practical consequences, an example of which is given in Chapter 9.

As in bacteria, genetic information can be carried by elements outside the nucleus. Plasmids, viruses and fragments of double-stranded RNA have been demonstrated in a few fungi. All these cytoplasmic genes are transmissible by anastomosis.

Fungi use various methods of sexual reproduction, all of them ending with the fusion of two haploid nuclei to produce a diploid zygote. The zygote is generally provided with thick walls which assure its survival during unfavourable conditions. The importance of the haploid and diploid phases varies considerably within groups but meiosis is an obligatory phase in sexual reproduction in which recombination of the genetic stock takes place.

Recombination of the genes after fusion of two haploid nuclei from a heterokaryon can also occur without fertilisation. This parasexuality is known to occur in certain deuteromycetes, a group of fungi which otherwise appear to be incapable of sexual reproduction.

Asexual Forms of Multiplication and Conservation

Sexual reproduction generally plays a minor role in the dissemination of fungi compared to the asexual method. Most species are capable of forming **spores**, either inside sporocysts (in lower fungi which have thalli without septa) or on differentiated branches of the mycelium (conidiophores). Though this method of dispersal is especially effective in fungi with aerial development (we are all familiar with the clouds of spores that a colony of *Botrytis cinerea* can release), many soil fungi also avail of this method of propagation. Dissemination of fungi in the soil is no longer ensured by the wind, but by microfauna, rain and irrigation (Fig. 2.5).

Under unfavourable conditions the cytoplasm of a compartment of a hypha can condense and cover itself with a very thick membrane, forming a **chlamydospore.** In some fungi, special organs called **sclerotia** are formed by intensive ramification of the mycelium. These sclerotia are often made up of a dense unpigmented central mass consisting of finely interlaced hyphae containing reserve substances and a melanised cortex, the latter playing a protective role, as in *Sclerotinia minor* and *Sclerotium rolfsii*. Often, these sclerotia have to pass through a maturation phase before they are ready for germination.

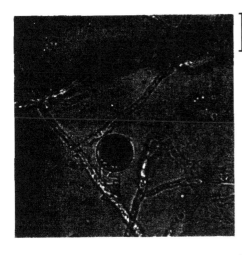

20 µm

Oospore and mycelium without
septum in *Pythium ultimum*
(plate, G. Forbes, INRA)

50 µm

Endoconids (hyaline, cylindrical)
and chlamydospores (brown,
multilocular) in *Chalara elegans*
(plate, M. Sailly, photo library, INRA)

20 µm

Croissant-shaped macroconids
of *Fusarium roseum* var *sambucinum*
(plate, B. Tivoli & E. Lemarchand,
photo library, INRA)

Fig. 2.5 A few examples of soil fungi.

Nutrition

Fungi are heterotrophs and use organic compounds as a source of carbon. As in bacteria, the digestion of large molecules starts in the external medium since only molecules of relatively small size can cross the cell walls to enter the cytoplasm. The panoply of enzymes produced by fungi is extremely rich, allowing them to use the most complex substrates even more effectively than bacteria. These substrates include cellulose, lignin, keratin, humic acids, etc. Several studies, recently summarised by Wainwright (1988), have also shown that they are capable of absorbing the smallest traces of organic carbon, uniquely present in the form of vapours in the atmosphere.

In some fungi, nutrients can be transported over a long distance by **rhizomorphs**, organs made of hyphae congregated parallel to each other and bound together by anastomoses and sometimes protected by a melanised cortical zone. Rhizomorphs are common in those fungi which decompose forest and fruit trees (*Armillaria mellea* and *A. ostoyae* are the more formidable representatives). With the help of rhizomorphs, starting from a stub that serves as a point of departure and as a food reserve, these fungi set out across areas devoid of nutritive resources in search of a new stump. Some rhizomorphs can reach several metres in length.

Under ordinary circumstances, oxidation of organic molecules provides the energy fungi need. But when these substrates are too few, many species can also derive their energy by the oxidation of ions or of minerals such as S, NH_4^+, H_2, Mn^{++}, and reserve the entire organic substrate as a supply of carbon. Under these conditions, it also seems that even a common fungus such as *Fusarium oxysporum* is capable of fixing atmospheric carbon dioxide, thus behaving like a chemiautotroph.

Chemotaxis and Chemotropism

In oomycetes and chytridiomycetes, chemotaxis is observed during a period in their cycle when they release flagellated zoospores and, in this respect, resemble bacteria and protozoans. For example, the zoospores of *Pythium* and *Phytophthora*, which are root parasites, are attracted by amino acids of the root exudates and by ethanol.

It seems that in strictly filamentous fungi apical growth of hyphae can be influenced by hormonal signals. That is why the reproductive organs of two zygomycetes of the same species and of sexually opposite polarities are mutually attracted and then unite; the hyphae of two compatible mycelia of ascomycetes or basidiomycetes move towards each other to anastomose. The response to trophic signals is still controversial. Fungi seem to grow systematically in the spaces around them rather than in specified directions. Chemotropism, however, seems to exist among the oomycetes whose hyphae are attracted by amino acids.

Ecological Exigencies

It was long believed that oxygen is indispensable for the development of fungi. Later observations showed that many common species are able to manage, at least temporarily, using other electron acceptors. Under such conditions, the organisms often require an external supply of vitamins. Soil micromycetes such as *Fusarium oxysporum*, *F. solani* or *Geotrichum candidum* can grow normally in a nitrogenous atmosphere in the presence of glucose and mineral nitrogen (Tabak and Cooke, 1968). Only in the last twenty-five years have strictly anaerobic fungi been discovered, first in the rumen of ruminants and then in the digestive tract of other herbivores. Being cellulolytic, they can probably be linked to chytridiomycetes.

Fungi need moisture but much less than that required by bacteria. That is why they are generally more numerous and more active than bacteria in the upper soil layers, which rapidly dry out. Fungi which have mycelia without septa are more sensitive to desiccation since their development stops when the hydric potential goes below −4 MPa whereas fungi which have mycelia with septa tolerate on average up to −10 MPa. However, *Aspergillus* and *Penicillium* species can generally grow in hydric potentials of the order of −20 MPa. The record is held by a species of *Xeromyces* which can sustain −65 MPa (note that the suction it has to exercise in this case to extract water from the surroundings is equal in absolute values to 650 times the atmospheric pressure!).

Most fungi can be considered mesophyllic because they develop at temperatures between 10 and 40°C. A few species, active in the primary stages of composting, develop at an optimal temperature of 40 to 45°C but cannot withstand temperatures greater than 60°C. Another very restricted group grows at low temperatures (between −5 and +10°C). Some parasites of cereals such as *Typhula* and *Gerlachia nivalis* can develop under snow.

Fungi are generally able to tolerate acidic pH and, in these conditions, are more competitive than bacteria for the substrate. Therefore, in acidic soils more fungi are found than bacteria. However, they cannot be considered acidophilic organisms since they are not more numerous in soils with low pH. As a matter of fact, these organisms can develop in a rather wide range of pH environments. Furthermore, many possess a range of enzymes capable of acting on the same substrates at different pH.

Classification

The classification still commonly used is based on morphological criteria and mode of reproduction. It is practical but not very satisfactory if recent phylogenetic studies are taken into account. Comparison of 18SrDNA gene sequences among increasing numbers of species has shown that the old group of Myxomycota (which comprises myxomycetes and plasmodiophoromycetes among others) appeared quite early during evolution and is actually very distantly related to fungi (Fig. 2.6). Similarly, the oomycetes, a very important group of soil microorganisms, are not really fungi because

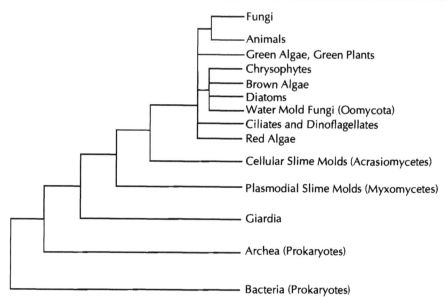

Fig. 2.6 Diagram of relationships of some organisms based on 18SrDNA molecular phylogenetic studies (from Taylor, 1995).

more closely related to brown algae. It can also be seen from Fig. 2.6 that according to the modern cladistic classification, fungi are closer to animals than to green algae and plants. At present, the true fungi include the following: chytridiomycetes, unicellular organisms or organisms consisting of short chains of cells (*Olpidium, Synchytrium*); zygomycetes (comprising the Mucorales), very close to the preceding group but having thalli with occasional septa; and ascomycetes and basidiomycetes which have mycelia with septa and are widespread and relatively well known compared to the other groups. Class Deuteromycetes is an entirely artificial group, unlike the other groups, because its species are arranged according to their mode of asexual reproduction. Some fungi are thus double-listed under different names—among Ascomycetes or Basidiomycetes and among Deuteromycetes depending on whether their sexual reproduction or asexual reproduction is taken into account. When both types of reproduction are known to occur in the fungal species, it is usually identified by the name used for its sexual form—but with many exceptions: *Fusarium solani* (a form classified under Deuteromycetes) continues to be referenced rather than *Hypomyces solani* (sexual form, Ascomycete) and *Rhizoctonia solani* (Deuteromycete) rather than *Thanathephorus cucumeris* (Basidiomycete).

Table 2.3 presents some distinctive characteristics of the fungi and organisms earlier included in this group. For the sake of convenience we shall continue to group Oomycetes and 'true' fungi under the name 'fungi' in this book.

Table 2.3 Main groups of microorganisms commonly referred to as 'fungi'

Cell wall	Septa	Zoospores	Sexual form	Name	
Absent	Absent	Variable	Variable	Myxomycetes (plasm-odial slime moulds)	
Cellulose + glucans	Absent	Present (2 flagella)	Oospore	Oomycetes (water moulds)	
Chitin + glucans	Absent	Present (1 flagellum)	Zygote	Chytridiomycetes	
Chitin + chitosans	Absent	Absent	Zygote	Zygomycetes	(fungi)
Chitin + glucans	Present	Absent	Ascus	Ascomycetes	
Chitin + glucans	Present	Absent	Basidium	Basidiomycetes	
Chitin + glucans	Present	Absent	Absent	Deuteromycetes	

Isolation and Enumeration of Soil Fungi

Specific isolation procedures are generally available for monitoring various fungal species. DNA can also be extracted from soil. If taxon-specific oligonucleotide primers are available, fungal DNA from soil can be amplified by polymerase chain reaction to identify the target species. To obtain an inventory of the different species in a soil sample, the suspension-dilution method intended for isolation of bacteria, can also be used for fungi. Successive dilutions in an agar medium, maintained in a molten state at 40°C, are distributed in petri dishes. To preclude concomitant development of bacteria, the medium is acidified or antibiotics are added. The suspension-dilution technique favours those fungi which show high sporulation and is not so suitable for species present only in the form of mycelium. This discrepancy can be overcome by directly culturing microaggregates after sieving the soil. The organic fragments retained on the sieve can, after washing, be incubated in a nutritive agar medium as well.

As always, this type of analysis gives only an approximate idea of the population of microorganisms which develop in an agar medium in the soil sample under study. The error can be not only qualitative since the nutritive substrate may not be suitable for some species of fungi, but also quantitative, as will be shown below.

Indeed, it is easy to define that which constitutes an individual in the case of a unicellular microorganism such as a bacterium or an amoeba and therefore (theoretically) enumerate them. But filamentous fungi pose a great problem. For example, if a soil sample contains a thallus A carrying a conidiophore with 10 conidia, after incubation there would be 10 colonies of fungus A. On the other hand, if a thallus B with an equivalent vegetative development, has a conidiophore producing 100 conidia, 100 colonies of fungus B would appear on the culture medium. Would it thus be correct to deduce that fungus B is 10 times more abundant than fungus A or its biomass in the sample 10 times higher? If on the other hand, the sample contains a sterile fungus C, the number of colonies of C could vary from 1 to n depending on the extent to which the fungal mycelium was fragmented during the course of soil preparation. Therefore, estimations of biomass arrived at by counting colonies on isolation media must be considered only approximate and interpreted with caution.

A more reliable method, requiring considerable patience, involves measurement of the length of the mycelial filaments present all around the aggregates or the organic debris. Such calculations have shown that the length of the hyphae is generally between 100 and 1000 m per g soil!

With perfection of analytical techniques, it is nevertheless possible to arrive at a chemical approach for evaluation of the fungal biomass by measuring the characteristic compounds such as chitin (after elimination of the microfauna) or ergosterol (knowing that in this case the oomycetes cannot be taken into account).

Despite all the problems listed above, most biomass estimations are concordant and yield values between 1 and 10 t ha^{-1}. So the biomass of soil fungi is as important as that of the bacteria.

Protists

This word is useful for encompassing all the eukaryotic microorganisms, but makes little sense from the taxonomic point of view because it groups together living organisms that differ markedly in genome. Likewise, among the protists it is convenient to distinguish the algae, which possess chloroplasts, and the protozoans, which lack chloroplasts and are motile and capable of phagocytosis. This separation, based on functional aspects, is actually artificial. In fact, there are numerous intermediate forms (such as the Euglenophytes for example) and each of the two groups per se comprises divergent branches.

Algae

GENERAL CHARACTERS

The fundamental difference between algae and protozoans (in spite of the previously listed reservations) is based on the presence of chloroplasts in algae which can use light energy to synthesise their cellular constituents using water and dissolved mineral elements. In common parlance, the term algae applies mainly to aquatic organisms, bringing to mind terrestrial plants with reference to their appearance and size since some algae can measure several tens of metres in length. Such remarkable individuals, however, should not make us forget that the majority of algal species are either unique cells or filaments comprising a few cells, and belong entirely to the microbial world, with average dimensions of a few μm.

The cell wall, when present, is supple and constituted of proteins (Euglenes) or rigid and made of cellulose, other polysaccharides, or pectin impregnated with silica (diatoms). Exocellular polysaccharide matrices generally envelop the cell walls. Most of the microscopic algae are motile, moving by changing the shape of their cells or by flagella. Some possess a photosensitive organelle that allows orientation to light.

REPRODUCTION

Unicellular algae usually multiply by binary fission, which can be very rapid when conditions are favourable. Under certain conditions, a haploid individual can transform entirely into a gamete and fuse with another apparently, identical cell of opposite polarity to form a zygote. After meiosis, and return to the haploid state, several asexual multiplication cycles can begin afresh and continue.

Many other forms of sexual reproduction are encountered in algae, some of which culminate in specialisation with the appearance of specialised organs, the gametangia, producing male motile gametes or immotile female gametes. However, gamete production occurs mostly in large multicellular marine forms such as the *Fucus*.

NUTRITION

Algae are photosynthetic organisms. They are therefore usually autotrophs albeit there are exceptions to this rule. For example, some species require vitamins or growth factors they cannot synthesise (auxotrophic forms). Others tend to behave as heterotrophs in the absence of light. In darkness, they can utilise simple organic molecules present in the environment (glucides, acetate, pyruvate) and could theoretically develop in deeper layers of the soil, although it is unlikely that they could survive the competition with other soil microorganisms under such conditions. Some algae lacking a cell wall are also capable of phagocytosis. It is estimated that 40 to 50% of soil algae are facultative heterotrophs.

ECOLOGICAL EXIGENCIES

Even though algae are typically aquatic and therefore play a significant role in fixation of atmospheric carbon dioxide in the oceans, they are nonetheless present in all soils. The microscopic terrestrial forms are particularly resistant to desiccation. They multiply rapidly on soil surfaces as soon as there is adequate moisture. For instance, they are readily observed around drippers in plantations where a trickle irrigation system is used. These algae are negligible in ordinary plantations, but can be problematical in hydroponic market gardening, especially with the nutrient film technique (NFT). They can block filters and injectors and compete with the main hydroponic cultures for mineral elements. The algal genera most commonly encountered under hydroponic conditions are *Ulotrix*, *Scenedesmus* and *Chlamydomonas* (as well as *Oscillatoria*, a cyanobacterium). It is very difficult to eliminate them as almost all algicides are phytotoxic (Coosemans, 1993). A better solution to this problem is to preclude algal growth insofar as possible by not exposing irrigation channels to light.

The polysaccharides that cover the cells are rather similar to those of the bacterial capsules. In addition to their probable function as reserves of water and eventually of carbon, it is thought that they also serve as floats to some extent by allowing the algae to remain fixed to particles of soil in the upper horizons, thereby preventing them from being washed into soil depths (Barclay and Lewin, 1985).

Algae can grow under low light intensity, which allows them to behave like active autotrophs several millimetres below the surface, especially in soils rich in translucent quartz particles (the light can diffuse even beyond 2 cm in certain soils). Under such conditions, the algae are well protected from desiccation.

The pH of the soil has an influence on the composition of the flora: diatoms predominate in neutral or alkaline soils while yellow-green algae and chlorophycophytes flourish in acidic soils.

CLASSIFICATION

This was based, on the one hand, on the nature of pigments other than chlorophyll a, b and c present in the cells giving them characteristic coloration (red algae, brown algae . . .) and, on the other hand, on the specific stages of the developmental cycles. Recent molecular phylogenetic studies based on comparison of 18S rRNA led to ample changes in classification. The ancient group of algae has disappeared. Algae are now distributed into nine very distinct phyla. Species adapted to a terrestrial life belong to the following branches:

— Chlorobionts: green algae, unicellular or filamentous terrestrial forms (*Chlamydomonas, Chlorella, Chlorococcum*);
— Stramenopileae: yellow-green algae (Xanthophyceae) and diatoms (Bacillariophyceae);
— Euglenobionts: Euglenophytes are motile cells with two flagella and a photosensitive organ; sexual reproduction is unknown (*Euglena*).

ISOLATION AND ENUMERATION OF SOIL ALGAE

Like fungi, algae can be observed and enumerated in situ on slides prepared as per the techniques of Rossi-Cholodny or Jones and Mollison.

The method of Rossi-Cholodny consists of digging the soil to introduce a glass plate for microscopic preparation. The surface of the plate is pressed firmly against the soil and kept in place for two to three weeks. It is then carefully retrieved and cleared of mineral and organic particles, fixed by heat treatment and stained with erythrosin in a phenolic solution before observation under the microscope.

In the Jones and Mollison method, a suspension of finely ground soil is incorporated in agar water. A drop of this gelified suspension is placed on a slide with a calibrated hole (a cell of a haemocytometer for example) and covered with a cover slip. After the agar solidifies, it gives a film of known dimensions and thickness which can be transposed onto an ordinary slide and microscopically observed after drying and staining.

In a heterogeneous suspension, algae can be distinguished because of chlorophyll which fluoresces red when illuminated with blue or ultraviolet light. The algae can also be separated from bacteria and concentrated in suspension by differential centrifugation. It is also possible to culture algae using soil dilutions. Sterile soil or a mineral solution with antibiotics, liquid or gelified, can be used as substrates (bacitracine eliminates the cyanobacteria).

The population size varies greatly from one soil to another and for the same soil seasonal variations in volume are common. For example, in temperate regions the maximum volume is attained at the beginning of spring and the minimum towards the end of autumn. Depending on the investigators, the numbers of algae vary between 10^2 and 10^9 per g soil, the most frequent reports on estimations being in the order of a few thousand cells g^{-1}.

One can deduce the biomass from the computations. Biomass estimations can also be made on quantity of chlorophyll extracted from the soil. However, it is not possible by this method to distinguish true algae from cyanobacteria. The algal biomass on average is 100 to 500 kg ha^{-1}.

Protozoans

These unicellular organisms are distinguishable from algae because they are heterotrophs and capable of movement. They vary in size from a few thousandths of an mm (some parasitic forms) to a few mm (Ciliae such as *Stentor*). Average dimensions lie in the range of tens of μm.

CYTOPLASM AND NUCLEI

The cytoplasm is surrounded by a lipoproteinic membrane and contains a considerable number of organelles performing very varied functions: respiratory (mitochondria), secretory (Golgi apparatus), muscular (contractile fibres), skeletal (microtubules), digestive and excretory (vacuoles) and organizational (centrosome). The cell is generally naked but can be protected by external sheaths (**test** or **theca**) and reinforced by an endoskeleton, often made of silica.

One individual normally contains one nucleus. In a few groups of protozoans, the nucleus can multiply whenever the cytoplasmic mass accumulates to form a **plasmodium** that is sometimes large: more than 1 cm in some foraminifers. All the nuclei are identical. At a particular moment in the cycle, the plasmodium fragments into uninuclear schizonts, each of which is capable of forming a new individual. In the Ciliae, each individual normally contains two non-identical nuclei. The larger nucleus, polyploid, takes control of the trophic functions while the other smaller nucleus is responsible for sexuality and transmission of genetic patrimony.

REPRODUCTION

Protozoans almost always follow an asexual method of reproduction, which most often involves binary fission that forms two daughter cells from the parent (amoebae, paramecia). As stated in the previous section, there can also be fragmentation of a plasmodium into several schizonts or, more rarely, budding from the mother cell, as in yeasts. When conditions are favourable, an amoeba or paramecium can divide every 6 to 8 hours.

Sexual reproduction is either absent (amoebae), occasional (flagellates) or systematic (sporozoids) depending on the groups. It results in the fusion of two gametes, which sometimes show marked sexual dimorphism. The relative importance of the haploid and diploid phases is extremely variable. Parasitic forms show very complex reproductive cycles.

FORMS OF CONSERVATION

When conditions are unfavourable, many protozoans are capable of encystment. After the lysis of cellular organelles and eventually one or several divisions of the nucleus, the cytoplasm synthesises reserve products, becomes dehydrated and envelops itself in a thick protective shell that is extremely resistant to external agents. These **cysts** play a role very similar to that of bacterial endospores. They are capable of withstanding prolonged dryness (several years), dilute acids (24 hours) or extreme temperatures (1 hour at 80 or 100°C, several hours at very low temperatures). The reversal of encystment is generally slow and seems to necessitate diffusion of products of bacterial metabolism across the wall.

NUTRITION

Protozoans consume organic matter but have very different diets. Most are capable of ingesting their nutrients. Some feed on organic substances present in solution in the external medium by pinocytosis: the membrane involutes and encircles a droplet of the solution in a vacuole which separates out and penetrates the cytoplasm. The vacuole then fuses with lysosomes which deliver their enzymatic content, becoming in this way a digestive vacuole. The elements necessary for microbial development are absorbed while the non-usable residue is jettisoned outside by a process mirroring ingestion. Other protozoans feed on organic debris or living microorganisms by an analogous mechanism known as phagocytosis. In some species, a permanent orifice plays the role of a buccal cavity. The prey may be bacteria, algae, other protozoans, yeasts or spores of fungi, or animal cells (the red blood corpuscles for example). The prey is sometimes larger than the predator (nematodes, mycelial hyphae) and then the fragments are apparently taken up by localised lysis, as in some amoebae which consume fungi. These amoebic mycophages clearly show trophic preferences in their diet.

Cells are capable of orienting movement according to stimuli that can be trophic or sexual (chemotaxis).

ECOLOGICAL EXIGENCIES

Most protozoans are aerobic but some are anaerobic (in the rumen for example). Protozoa are most widespread in the aquatic habitat as water (or an aqueous medium such as the tissues of the organisms they parasitise) is indispensable for them. However, they are not all that rare in the soil where representatives of most of the groups with free-living individuals are found, of which flagellates and rhizopods are the most common (Fig. 2.7). They move around in the water retained in micropores and on the surface of microaggregates. Their role in the regulation of microbial equilibria has been, and still is, mostly underestimated: the mass of bacteria ingested annually by soil protozoans lies in the range of 6 t ha^{-1} (Stout and Heal, 1967). The smallest forms such as *Colpoda* (ciliates) or *Naegleria* (amoebae)

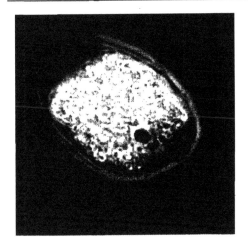

Thecamoeba granifera

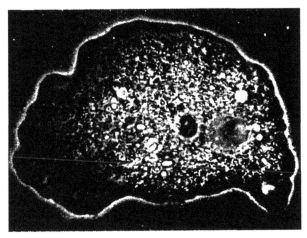

Cashia mycophaga

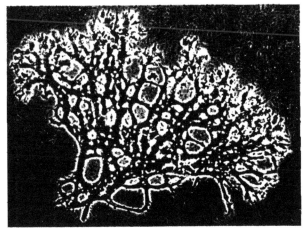

Leptomyxa reticulata

Fig. 2.7   Some soil protozoans (plates, M. Pussard, INRA).

are the most active because they can easily penetrate the soil pores where microbial colonies are found. Terrestrial forms are able to resist desiccation by rapid encystment. Moderate temperatures (10-30°C) are ideal for the development of protozoans. Several species of soil protozoans are very ubiquitous and have been found under quite different climatic conditions. Amoebae with thecae generally show a preference for acidic soils; some species, however, are exclusively calcicoles.

The genus *Phytomonas*, close to trypanosomes, has the peculiar habit of growing in the vascular system of certain tropical plants, which causes them to wilt.

CLASSIFICATION

Classification was formerly based on the type of locomotion, way of life (parasitic or non-parasitic) or the number of nuclei (Table 2.4) As for algae, a study of 16S/18S rRNA resulted in considerable modifications and the term 'protozoan' is henceforth no longer employed in systematics. Many phyla of eukaryotic microorganisms, sharing very few features, appeared at different periods along evolution. Some of these phyla amalgamate photo-synthetic and heterotrophic microorganisms.

Table 2.4 Simplified classification of protozoans

Main groups (phyla)	Characters	Examples
Rhizopoda	Locomotion by pseudopodia	Naked amoebae. Amoebae with thecae. Foraminifers
Labyrinthomorpha	Fusiform cells associated in a network	*Labyrinthula*
Actinopoda	Pseudopodia and retractile cellular expansions, central symmetry	Radiolarians
Mastigophora	Movement by flagella	Diatoms *Leptomonas* *Phytomonas* *Trypanosoma*
Sporozoa	Parasites with complex cycles	*Plasmodium*
Ciliophora	Presence of cilia, two types of nuclei	Ciliates *Paramecium* *Stentor* *Tetrahymena* *Vorticella*

ISOLATION AND ENUMERATION OF PROTOZOANS

The soil imprints obtained on glass slides using the technique of Rossi-Cholodny enable a qualitative in-situ study that reflects mainly the less motile forms. Amoebae with silicoorganic thecae (thecamoebians) can be specifically isolated using flotation techniques: after fixation of the sample and staining the cytoplasm to facilitate counting, the soil sample is suspended in water. Violent agitation (using bubbling for example) helps in dissociating the soil aggregates and separating the thecae. An aliquot of the suspension is then filtered on a membrane, clarified and used for counting under the microscope. The population counts of thecamoebians by this method in a superficial humic soil horizon (A_0) in a temperate forest contained 29,000 to 41,000 live individuals per g dry soil (Couteaux, 1967). It is also possible to culture them by techniques similar to those used in bacteriology or mycology by spreading soil particles in liquid media or on agar substrates. This method is especially convenient for studying bacteriophagous species. The substrate can also be made selective by the addition of a particular strain of bacterium.

But as a matter of fact, like for all the groups discussed above, none of the aforesaid techniques gives an exact idea of the actual protozoan populations in soils. It is possible, however, to propose estimations of the order of 10^8 to 10^9 individuals per m^2, and presume that a biomass of 150 to 700 kg ha^{-1} represents the average numbers. Naked amoebae and flagellates are the most abundant forms, of which amoebae alone may account for up to 90% of these populations.

Microfauna

Even though it has no taxonomic recognition, the term microfauna is convenient for referring to the smaller animals living in the soil. According to some researchers, this term refers only to animals smaller than 200 μm; others use the term to include a larger ensemble that covers animals of 1 to 2 mm in size. We use the word microfauna as signified in the latter category and hence include **nematodes** and very small **arthropods** as well as a few other animals of secondary importance.

Nematodes

GENERAL CHARACTERISTICS

Nematodes belong to the branch of Nemathelminthes, visually distinguishable from Plathelminthes by their spindly shape and circular aspect when viewed in cross-section. Both phyla are distinguished from the more evolved annelids by an embryonic mesodermal layer that is poorly developed, lacking a coelomic cavity, and giving rise only to a more or less lax mesenchymatic tissue. Nematodes, representing the most important class, possess a mouth with teeth, rasps or a stylet, extending into a digestive tube that terminates in an anus at the other end of the body. The excretory apparatus and nervous system are very rudimentary. Nematodes lack both a circulatory and respiratory system but the cuticle is permeable to gases and facilities exchanges between internal and external media. The cuticle is thick, consisting of several distinct layers and plays a very important protective role. Since it is not stretchable, four successive moults are necessary during growth of the animals.

Some nematodes which are animal parasites can attain a very great length; 30 cm in the female of *Ascaris megalocephala*, for example. The soil nematodes, of interest to us here, are always very small: 200 µm to 1 or 3 mm in length and 10 to 40 µm in diameter on average.

REPRODUCTION AND DEVELOPMENT

Reproduction is wholly by the sexual method. The sexes are separate. After copulation, the spermatozoids, non-flagellated, move up the oviducts by amoeboid movements and fertilise the ova. The eggs surround themselves with a double membrane and later are enveloped in a third membrane of maternal origin. Some forms are parthenogenetic. In some other species, the adults are hermaphrodites.

In small nematodes, cell division stops at the end of the embryonic stages. The number of cells which constitute each organ is therefore fixed very early and is constant for a given species. Further growth of the animal occurs by hypertrophy of the cells, as in rotifers. Therefore, in the event of a lesion, there is no scope for tissue regeneration. The nematode undergoes four larval stages and four moults successively before becoming an adult. The first two moults may occur before the egg hatches.

CONSERVATION FORMS

The egg, in its triple-layered covering, is well protected from external conditions. In Rhabditida, worms of the third larval stage are capable of encysting inside the envelopes of the preceding stage. In a dry atmosphere, the encysted nematodes show an astonishing survival capacity: some species can resist immersion for several hours in liquid helium (–272°C!) as well as temperatures of 80°C, and can also exist in a **resting stage** at ambient temperatures for many years. In *Heterodera*, encystment is a normal state during the biological cycle: the cyst wall is made up of the dry and tanned cuticle of dead females, which envelops the eggs (Fig. 2.8). Hatching of these parasite species is stimulated by root exudates of the host plant.

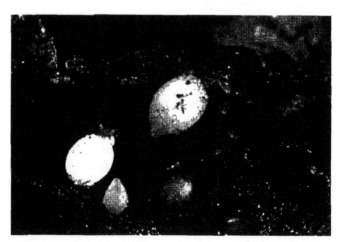

Fig. 2.8 Females (white) and cysts (grey or black) of *Heterodera schachtii* (plate, G. Caubel, photo library, INRA).

NUTRITION

Nematodes consume living organic matter. Many soil nematodes are plant parasites and cause serious damage to crops. Some of them puncture the cells with their stylet and ingest the contents. Others cause a dedifferentiation of the attacked cells, which transform into nurse tissue (Fig. 2.9). The free forms feed on algae, bacteria or spores of fungi. A single individual can consume several thousand bacteria in a minute. But part of the ingested bacteria or spores can traverse the digestive tract without being damaged and are ejected a little distance away from the place of their ingestion. These nematodes thus play a double role in regulation and dispersal of microbial populations. Other species live at the cost of fungal mycelia, sucking up the cellular contents of their prey. They often exhibit food preferences: *Aphelenchus avenae* is particularly fond of the parasite fungus *Rhizoctonia solani*. Lastly, other nematode species are predators of microfauna: protozoa, rotifers or other nematodes.

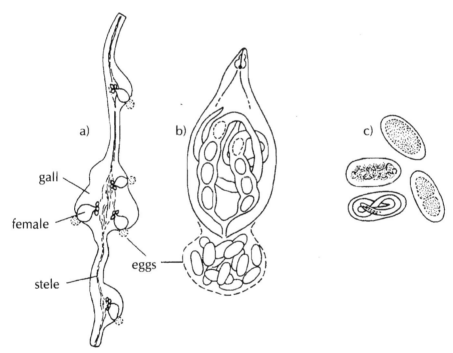

Fig. 2.9 Nematodes with galls (*Meloidogyne*), according to Messiaen et al., 1991: (a) radicle deformed by attack of *Meloidogyne* (longitudinal section); (b) adult female of *Meloidogyne* with mass of eggs; (c) eggs of *Meloidogyne* in various stages of embryological development.

ECOLOGICAL EXIGENCIES

Nematodes are found in all latitudes and all kinds of-environments, from glaciers to hot thermal sources. Like protozoans, they need free water to lead an active life and hence aquatic surroundings are their habitats of predilection. But they are equally very numerous in soil where they survive within water films retained by the clay-humus complex. They mainly occupy the top 10 to 20 cm of the soil but can migrate deeper in the course of a year depending on the degree of soil desiccation. Their ability to remain in a quiescent state allows them to overcome periods of drought and to colonise even desert soils. They are not very particular from the viewpoint of pH and can also withstand high variations in temperature. For a given species, the optimum temperature can vary depending on the stage of development.

CLASSIFICATION

The presence or absence of sensorial organs (phasmids for example), the anatomy of the excretory apparatus, morphology of the mouth parts and mode of life are the main criteria used to distinguish the classes and orders of nematodes. The most important groups are listed in Table 2.5.

Table 2.5 Some groups of nematodes of economic importance

Principal groups	Characters	Examples
Phasmidia	Phasmids present Lateral excretory canals Mouth with 3 or 6 lips Oesophagus non-cylindrical	
Tylenchida	Retractable stylet Carnivorous or phytophagous (in such cases, mainly sedentary endoparasites)	*Anguina* *Tylenchus* *Ditylenchus* *Pratylenchus* *Heterodera* *Meloidogyne* *Aphelenchus* *Aphelenchoides*
Rhabditida	No stylet	*Rhabditis*
Strongylida Ascaridida Spirurida	Animal parasites	strongylids ascarids and oxyurids filaria
Aphasmidia	No phasmids Excretory canals not lateral or absent Oesophagus not clubbed	
Dorylaimida	Plant parasites (mainly migrating ectoparasites)	*Xiphinema* *Trichodorus* *Longidorus*
Dioctophymatida	Animal parasites	

COLLECTION AND ENUMERATION OF SOIL NEMATODES

Various techniques have been used to isolate non-fixed forms of soil nematodes. The simplest and easiest method exploits the ability of these animals to move in water (Fig. 2.10). Other techniques consist of washing the soil on net sieves with progressively smaller meshes in the presence of

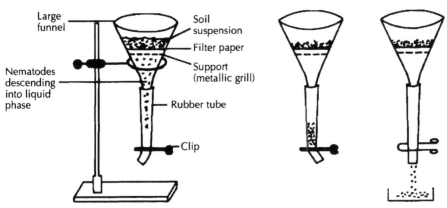

Fig. 2.10 Procedure for harvesting soil nematodes. The soil sample is suspended in water and poured into a funnel containing a metallic grill supporting a filter paper at the bottom. A piece of flexible tube, closed with a clip, is attached to the stem of the funnel (a). The nematodes swim and traverse the filter paper. After a few hours, most of them collect at the bottom of the tube (b). They can then be collected in a few ml of the suspension by loosening the clip (c).

countercurrents, which prevent sedimentation of the animals with mineral particles. Eventually the resultant suspension is centrifuged. For purposes of counting, 1 ml of a suspension is placed in a haemocytometer cell and observed under the microscope. A large number of samples must necessarily be studied to obtain a correct estimation because the distribution of nematodes in the soil is highly heterogeneous.

Most soil nematodes can be maintained artificially in the laboratory on agar media or on plant fragments.

Because of the diversity of species and number of individuals, nematodes are the most important constituents of the soil animal biomass. Under temperate conditions, their average biomass is in the range of 150 kg ha^{-1}. Their number can surpass 200 billion ha^{-1}, representing a biomass of 200 to 400 kg.

Earthworms are annelids, constituting a biomass considerably greater than nematodes. They are agronomically very important and have a role in pedogenesis.

However, they are not discussed here because they belong neither to microfauna nor to mesofauna given their large size. Unlike nematodes, the specific diversity of earthworms is very limited, especially in northern regions.

Microarthropods: Collemboles and Acarids

Because of their size, microarthropods constitute the limit between microfauna sensu stricto and the group commonly called mesofauna. These small animals, barely visible to the naked eye, range in size from 0.2 to a few mm. Their body is protected by an external skeleton made up of chitinous plaques. For growth, they must renew their carapace after moulting.

The two groups of microarthropods better represented in the soil are the **Collembola** and the **Acaridida.**

COLLEMBOLES: GENERAL CHARACTERISTICS

Collemboles are primitive **insects** deprived of wings (apterygotes) that grow to adult size without undergoing metamorphosis (ametaboles). Fossils, which probably represent the most ancient forms of insects known, and which are close to the present-day forms, have been found in sandstone of the Devonian era.

Collemboles possess three pairs of long legs attached to the thoracic segments, six abdominal segments and a sort of crutch (furca) dependent under the fourth segment. Developed in those species which live on the soil surface, the furca, on releasing, enables them to make jumps of several centimetres. Respiratory gas exchanges occur across the cuticle. Some species possess rudimentary tracheae.

In the event of aggression, collemboles can secrete a repellent liquid from rudimentary or well-formed pores as an act of self-defence.

REPRODUCTION AND DEVELOPMENT OF COLLEMBOLES

Reproduction occurs without copulation. Fertilisation of females takes place by rubbing the abdomen on spermatophores priorly deposited on the soil by males. After the eggs have hatched, the emergent larvae undergo a variable number of moults depending on the species but their morphology changes very little upto the adult stage. Instances of parthenogenesis are also known.

NUTRITION OF COLLEMBOLES

These insects feed mainly on bacteria, algae, filaments and spores of fungi. Their nutritional preferences are generally eclectic but some species can be very specialised. Thus, *Proisotoma minuta* and *Onychiurus eucarpatus* consume *Rhizoctonia solani* in vitro but not *Laetisaria arvalis* (Lartey et al., 1989) while *Trichoderma harzianum* and *T. virens* seem to exercise a toxic effect on them similar to that of fly agaric or death cup on human consumers. Collemboles also consume plant debris that accumulates on the soil surface, preferring the fragments degraded by microflora. They play an important role in organic matter decomposition but cannot digest cellulose and lignin. There are also some carnivorous species which feed on nematodes, eggs of various animals and even other collemboles.

HABITAT AND ECOLOGICAL EXIGENCIES OF COLLEMBOLES

Some species are better adapted to life on the soil surface than others. They are highly pigmented and hairy, with a well-developed saltatory organ, long antennae and compound eyes. Wholly terrestrial species live at a depth of a few centimetres and are well represented up to a depth of 15 cm. These insects are small, slightly pigmented and less hairy; the antennae are short, the furca and eyes poorly developed, but they possess complex sensory organs.

In general, collemboles prefer moist habitats and do not tolerate dry heat. In summer, they burrow into the upper soil layers to avoid desiccation, rapidly returning to the top layer after a thunderstorm to feed on the plentiful microorganisms. They generally prefer moderate temperatures (10 to 30°C) and thus in temperate regions maximal activity is seen in spring and autumn. Contrarily, some species are well adapted to cold and can live on the surface of glaciers even in polar regions.

In case of temporary submersion of the soil, collemboles are able to carry on respiratory exchanges for several hours because of the hair that covers their body. This hair cover retains air and the animal is thus enveloped in a kind of bubble in which oxygen can diffuse from the surrounding liquid phase.

CLASSIFICATION OF COLLEMBOLES

Two suborders can be distinguished:
— Arthropleona, characterised by an elongated body with visible segmentation.
  Most of the soil species belong to this group (genera *Onychiurus, Hypogastrura, Isotomiella, Folsomia, Tomocerus*).
— Symphypleona, characterised by a globular body and indistinct segmentation.
  Mainly epigeal species are included in this group (genera *Sminthurus, Dicyrtoma, Megalothorax*).

ACARIDS: GENERAL CHARACTERISTICS

Unlike crustaceans, myriapods and insects, the **Arachnida** to which acarids belong, lack antennae but do have chelicerae, organs in the form of claws or pincers (Fig. 2.11) on the anterior part of the body. In acarids, the posterior part of the body is not segmented and is continuous with the anterior part. There are four pairs of legs. Respiration generally occurs through the trachea. The body bears various sensory organs whose functions are not yet entirely clear.

REPRODUCTION AND DEVELOPMENT OF ACARIDS

Copulation is not the only mode of fertilisation. Like the collemboles, many males deposit their sperm in spermatophores. The eggs are rarely abundant. Five moults are generally undergone to attain adulthood, requiring several weeks, sometimes several months. In the nymphal stage, some species possess the capability for attaching themselves to the surface of larger organisms (insects, myriapods . . .) by a suction pad or adhesive secretions, enabling their transport from one place to another. This phenomenon, termed **phoresy,** helps in the dispersal of populations.

Some species conserve shreds of the exuvia of the previous moult attached to their new test. Soil particles slowly accumulate on these shreds and give the animal a very strange appearance.

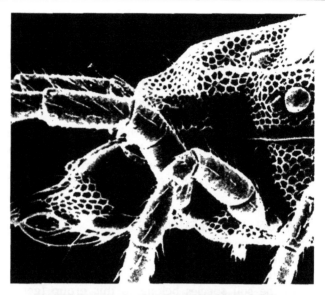

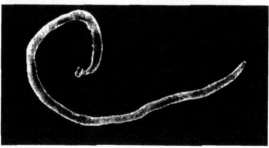

Fig. 2.11 Soil acarids (plates, R. Cléva and Y. Coineau, Museum of Natural History).
Top: Anterior part of *Labidostoma luteum* viewed through a scanning electron microscope, showing the chelicerae used for capturing prey.
Bottom: Example of adaptation to surroundings. The acarid *Gordialycus tuzetae*, which lives in sand in littoral sites is Languedoc, is filiform.

NUTRITION OF ACARIDS

Soil acarids are very active consumers of plant debris: the daily ration of litter consumed by oribatids can reach one-fifth of their weight. Because of this characteristic, acarids play an important role in the early stages of decomposition of plant matter. Other species prefer to feed on algae or fungi. Still others are predators and feed on nematodes, collemboles, larvae of other insects, or even other acarids.

HABITAT AND ECOLOGICAL EXIGENCIES OF ACARIDS

Acarids are plentiful in the superficial soil layers (between 0 and 5 cm) but some can live much deeper in well-structured soil: adaptation to deeper habitats involves reduction in girth accompanied by elongation of the body (Fig. 2.11), characteristics that allow them to move through the soil pores. They are able to withstand aridity much better than collemboles and remain active in summer. Among the oribatids, phthiracarids are able to coil themselves completely just like woodlice, a mechanism which protects them from desiccation and also from dangerous predators.

Acarids are found in all types of soil. However, some species are particularly abundant in acid soils.

CLASSIFICATION OF ACARIDS

Acarids (Fig. 2.12) constitute a subclass of Arachnida. They are divided into two superorders, based on the presence or absence of actinopilin in their tegumentary hair.
— Anactinotrichids (actinopilin absent). Predatory gamasids are the main representatives along with ixodes (which include the ticks).
— Actinotrichids (actinopilin present). Among these the oribatids represent about 70% of the soil acarids. The other orders are Actinedida (or prostigmates) and the Acaridida (or astigmates).

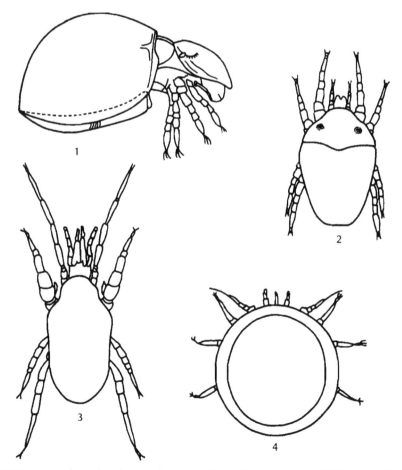

Fig. 2.12 Examples of soil acarids (after Owen Evans et al., 1961). 1: oribatid; 2: actinedid; 3: gamasid; 4: mesostigmate.

COLLECTION AND ENUMERATION OF MICROARTHROPODS

The techniques are the same for all small arthropods. Isolation can be done by the wet method, as in collecting nematodes, or by the dry method. The principle of these methods, though there are several variations, is given below.

Isolation by the dry method: The soil sample is spread 2.5 to 4 cm thick on a sieve with meshes about 1 mm and the sieve placed in a large funnel (Fig. 2.13). The tube of the funnel is immersed in a test-tube containing 70% ethanol (or simply a sticky film if the animals are to be collected live). As the moisture decreases on the surface of the sample, the arthropods penetrate deeper into the soil and finally fall into the test-tube. Desiccation of the soil can be accelerated by placing a source of mild heat slightly above the funnel. Only some specimens are recovered, and these even after waiting several days, while the inactive forms (eggs, larvae) remain in the soil sample. This method, devised by Berlese at the beginning of the twentieth century, and its numerous variations subsequently, enables collection of arthropods at different intervals over a period of time. Let it also be noted that acarids fall into the tube later than collemboles, which confirms their better adaptation to dry environments.

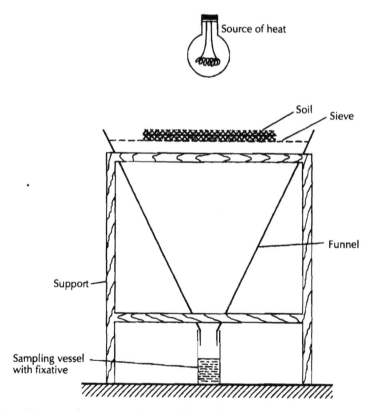

Fig. 2.13   Diagram of apparatus designed by Berlese to isolate small soil arthropods by the dry method.

Isolation by washing: The soil colloids are dispersed using sodium pyrophosphate or hexametaphosphate. The sample is then passed over sieves with smaller and smaller meshes. The deposit left on each sieve is again put into suspension: the arthropods that are non-wettable float on the surface. Water can be replaced by solutions whose density can be adjusted to facilitate separation of the animals from the sediments. With this method, it is possible to collect active as well as immobile forms.

The animals can then be identified and counted under a stereomicroscope. Identification of the species often requires microscopic preparations.

Acarids are the most numerous of the soil arthropods. They are generally more abundant in forest soils than in grassland soils. Their population in uncultivated soils varies from 50,000 to 500,000 per m^2; however, their numbers are lower in cultivated soils, about 30,000 m^2, which represents a biomass of nearly 3 kg ha^{-1}.

Collemboles are the next most abundant group of arthropods. They are better able to withstand soil tillage and, generally speaking, environmental disturbances than acarids—to such an extent in fact that the ratio of number of acarids to number of collemboles serves as an indicator of degree of degradation of the biocoenose. The number of collemboles lies in the range of 10,000 to 200,000 m^{-2}, a biomass of less than 2 kg ha^{-1}. They live in groups as colonies, which makes sampling a tricky.

Other Components of Microfauna

The soil microfauna includes animals other than nematodes and small arthropods, but their number and role may be considered negligible. The main groups are as follows:

—**Rotifers:** 0.2 to 1 mm in length, they constitute a separate phylum. They are extremely resistant to desiccation and, once dehydrated, can withstand very hostile surroundings. They feed on bacteria, protists and nematodes. Their population is between 5×10^4 and 10^6 m^{-2} in the upper soil layers, which represents an average biomass of 2.5 kg ha^{-1}.

—**Tardigrads:** Close to arthropods but nevertheless different, they live mainly in moist litter. Their size is less than 1 mm. Like rotifers, tardigrads are highly resistant to desiccation and also withstand extreme temperatures and X-rays. They feed on plant cell contents or prey on protozoans, rotifers or nematodes. They number is about 10^4 to 2×10^5 m^{-2}, representing an average biomass of 1 kg ha^{-1}.

—**Turbellarias:** These are small flat worms of the phylum Plathelminthes about 1 mm in length. They are also called geoplanairs. They prey on protists, rotifers and nematodes. Their population in the soil is much less than that of the two preceding groups.

Despite the difficulty and imprecision of enumeration techniques, biomass estimations provide some idea of the importance of soil microorganisms and help in comparing the size of the main groups. In an arable soil with a pH close to neutral, it can be estimated that the distribution of microorganisms is close to that indicated in Fig. 2.14. In an acidic soil, fungi would predominate.

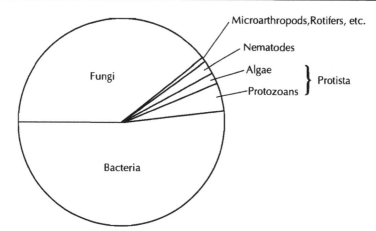

Fig. 2.14   Relative importance of biomasses of groups of microorganisms in a neutral soil. (Diagram depicts the most widely accepted average estimations.)

VIRUSES

Definition

The question still unanswered is whether viruses are portions of genetic information that have been deregulated and escaped from more complex nuclear entities or whether they are specific organisms that have attained a degree of absolute parasitism. Viruses are always obligate parasites and intracellular; they lack metabolism and can neither grow nor replicate except by utilising the biomolecular machinery of their host. Viruses are made up of a single type of nucleic acid (DNA or RNA) surrounded by a proteinic coat, the **capsid**, which is sometimes enclosed in a lipoproteinic envelope, of which at least a part comes from the host cell membrane. The ensemble is extremely small (on average less than 0.1 µm in diameter) and easily traverses filters capable of retaining bacteria—for which characteristic the name 'filterable virus' was formerly assigned. Most living beings—bacteria, archaea, fungi, plants and animals— can be parasitised by viruses.

Multiplication of Viruses

The mechanisms of penetration, replication and dispersal have been specially well studied in the viruses of bacteria (or **bacteriophages**). Penetration is preceded by a phase of adsorption that involves electrostatic attraction. It requires the presence of specific lipopolysaccharidic receptors situated on the bacterial membrane or on its appendages (flagella, pili). Specificity in the organ of attachment of the viral particle corresponds to the specificity of the host cell receptor. Once fixed on the bacterium, the phage 'injects' its nucleic acid inside the cell while its proteinic capsid remains outside. The functioning of the bacterium is then entirely diverted by the parasite for its benefit. Proteins and nucleic acid are synthesised in bulk, then assembled during a maturation process to form new bacteriophages, liberated by the hundreds after rupture of the bacterial wall. The entire operation lasts no more than some tens of minutes.

In plants, viral infection is only possible if a tissue has just been injured: the cells are apparently protected as long as their cellulosic walls remain intact. Even a very small lesion suffices, which in most cases is caused by bites of insects, acarids, or nematodes and sometimes even by fungi. The steps involved in penetration are still not well understood. Synthesis of viral nucleic acid begins after decapsidation. For each category of virus, replication is associated with a specific cellular structure: cytoplasmic membrane, mitochondria, chloroplasts, etc. The complete cycle of plant-infecting viruses spans more time than that of the bacteriophages (several hours) but each infected cell produces many more infective particles (on average 10^6).

Lysogeny

In some cases the bacteriophage-infected bacteria are not killed and continue to grow and multiply normally. Such bacteriophages are termed 'temperate'. The nucleic acid, instead of replicating autonomously, integrates into the bacterial chromosome and multiplies concomitantly as though part of the genetic patrimony of the cell. The virus is then termed a **prophage** (Fig. 2.15). The bacterium-prophage equilibrium can last indefinitely, from generation to generation, but this relationship can also snap spontaneously when a large number of viral particles are formed and lyse the host cell. Because of this potential for lysis, bacteria carrying prophages are called lysogenic. The frequency of occurrence of lysogeny is very low: the other bacteria of the population are not simultaneously affected and continue to transmit the prophage to their descendants.

Transduction and Conversion

At the time of maturation of the phages, bacterial DNA fragments can be included with part of the viral genes in the proteinic capsids. After lysis of the cell they are able to penetrate other bacteria and insert themselves into their genome, thus conferring characters of the donor cell on the receptor cells. This phenomenon is called **transduction** (Fig. 2.15)

Conversion involves the lysogenic bacteria, which can acquire new properties as a result of their infection by a temperate phage. In this instance, it is the genome of the prophage itself which transmits new characters to the lysogenic bacteria. Cases of this kind of conversion have been observed in pseudomonads and xanthomonads.

Ecology of Viruses

Free viruses placed in light outside a living cell are rapidly inactivated. But in the soil, once adsorbed, they can conserve their infectivity for a long time. Too large to intercalate between the clay layers, the viruses attach to the edges or the surface of colloidal clay particles with certain specificity. They can also be present in infected or in dead tissue. Even after six months of burial, the roots of tobacco contaminated with the tobacco mosaic virus could still serve as source of infection for susceptible plants. Even a very labile virus such as the cucumber mosaic virus can remain viable for a long time in infected plant roots left behind in the soil and be transmitted to the next crop (Pares and Gunn, 1989).

Some parasitic viruses of plants are transmitted by **vectors**, which often also play a role in virus conservation (see Chapter 7). For example, nematodes of the group Trichodorida can conserve and transmit the tobacco rattle virus for many months; the filamentous viruses of Gramineae can remain viable in the encysted spores of *Polymyxa graminis* for at least 2 years. This is also true of the potato mop top virus which remains in the cysts of *Spongospora subterranea.*

UNDERGROUND ORGANS OF PLANTS

Introduction

Soil not only supports an intense microbial life. It also serves as the medium in which seeds of the greater majority of plants germinate. Soil is the substrate in which plants anchor their roots, through which they absorb the water and mineral salts required. In this way, the roots of plants are fully soil inhabitants. They play a major role in microbial equilibria—a fact that cannot be oversighted by the microbiologist. Similarly, the botanist or agronomist cannot ignore the effects exercised by microorganisms on plants through the mediation of roots.

Overview of Root Anatomy

At the tip of the roots there is a cap surrounded by a polysaccharidic mucilage (Fig. 2.16). The cap and mucilage respectively serve as shield and lubricant, thereby allowing penetration of the root into the soil without damage to successive tissues. These tissues are especially precious because they comprise the meristems, which represent the sources for all new root cells.

The root cap meristem produces the cells of the cap and the **border cells**, which form a kind of sheath extending the cap. The border cells are distributed in one or several layers and their number is characteristic of the botanical family to which the plant belongs: the daily production of these cells varies from zero to several thousands per root apex depending on the plant species. The cells are not attached to each other (the middle lamella is absent) but as long as there is no free water, are maintained around the root

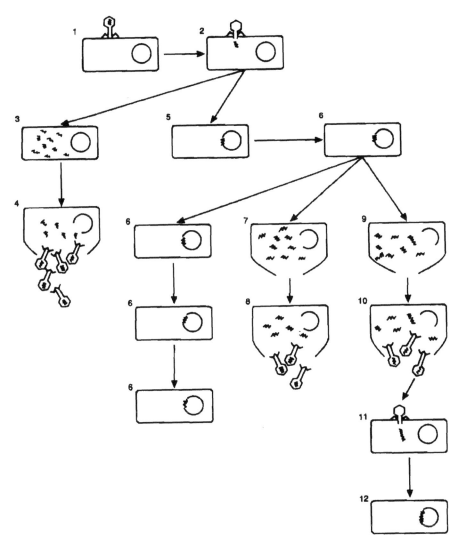

Fig. 2.15   Diagram illustrating the various changes that can occur after infection of a bacterium by a bacteriophage.

After adsorption of a phage (1) and penetration of the viral nucleic acid into the cell (2), multiplication occurs if the bacteriophage is virulent (3) and liberation of new virions occurs after lysis of the infected cell (4). If the phage is temperate, its genome can integrate into the bacterial cellular genome (5). The virus remains in the form of a prophage in the descendants of the bacterium (6). There, at any given time, the phage can cause induction of a multiplication cycle (7), leading to liberation of bacteriophages (8). The proviral DNA is not always excised properly and sometimes carries with it a fragment of the bacterial DNA (9). The resultant phages (10) introduce in their new bacterial hosts this foreign DNA (11) which can eventually be expressed there: this results in transduction (12).

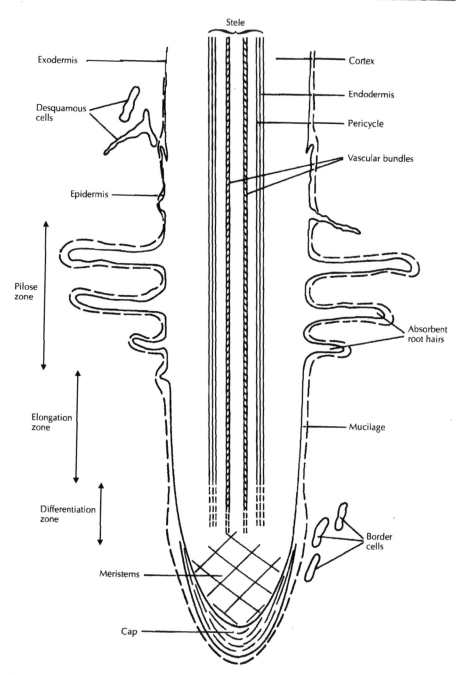

Fig. 2.16 Diagrammatic representation of a longitudinal section of a root (not to scale).

by a mucilaginous matrix, resembling the epidermis. When water is present, this matrix swells and the cells break up rapidly. Since they were observed in the vicinity of the roots, such cells were long regarded as dead or senescent, desquamous debris. However, recent studies have shown that the border cells are indeed alive. They maintain a metabolic activity for several days and even possess some genes that are specifically expressed once they are detached from the root apex. These cells could be regarded as a tissue capable of modifying the environment ahead of the root (Hawes et al., 1998). For instance, they can synthesise phosphatases, phytoalexins or the *nod* gene-inducing signals, the latter only in legumes.

The meristematic cells situated away from the apex give birth to the future tissues. After a large number of divisions, the initial cells elongate and differentiate. Elongation is preceded by an excretion of protons in such a manner that the pH is clearly much lower all around the elongation zone than in nearby sites: the difference can exceed one unit (Pilet et al., 1983). The waterproofness of this rapidly growing region is difficult of ensure and the fluids arising from the interior of the root diffuse into the external medium, constituting the root exudates. Behind the elongation zone is the pilose zone, formed by extension of certain cells of the external cellular layer (epidermis). The very fine wall of the absorbent root hairs and the large surface the hairs occupy make them a privileged (but not exclusive) site for the absorption of nutritive elements. They are generally covered with a mucilaginous sheath and their walls are often relatively rich in **lectins**. Lectins are glycoproteins which play, as we shall see, an important role in the recognition phenomena between organisms: lectins specifically interact with some sugar residues. The absorbent root hairs live only a few days so that the pilose zone maintains almost constant dimensions and moves in the soil at the same rate as the root apex. The absorbent root hairs and desquamous epidermis are replaced by an underlying layer of cells whose walls are impregnated with impermeable suberin.

The root thus comprises two zones (Fig. 2.17): a central cylinder containing the xylem and phloem vessels, parenchymatous tissue, meristems and eventually the sclerified support tissue forming the interior, and the exterior comprising a cortical zone with parenchymatous tissue protected by a suberised exodermis. This parenchyma generally serves to accumulate plant reserves. Its texture is sometimes rather lax and significant gaps between the cells allow gaseous exchanges with the exterior. In aquatic plants, such as rice, a true conductor tissue, the aerenchyma, ensures diffusion of oxygen from the atmosphere to the root extremities. The innermost cortical belt close to the central cylinder is the **endodermis** (Fig. 2.17). A thick layer of suberin is deposited on the radial walls of the endodermal cells (Casparian strip), followed by the tangential walls. Some cells however escape this tangential deposit, thus acting as intermediaries for the passage of metabolites between the central cylinder and the cortical zone.

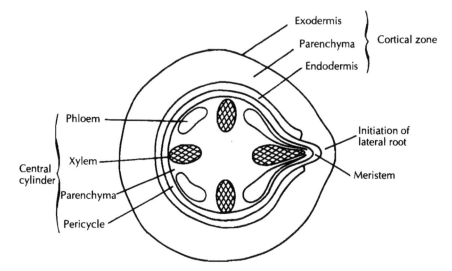

Fig. 2.17 Cross-section of a young root in the act of forming a lateral root.

Development usually stops at this stage in annual plants. In perennials, the root continues to grow in thickness because of the meristems situated in the central cylinder. The primary cortical zone tears and disappears. It is replaced by a secondary bark, the periderm, which is partly suberised. Thus roots, like shoots, lengthen only by apical growth but thicken all along their length. Lateral roots arise from the meristematic belt situated in the periphery of the central cylinder, the pericycle. The root profile develops from inside outwards, traversing the endodermis, cortical parenchyma and exodermis, thereby rupturing the continuity of the protective covering (Fig. 2.17). Each expansion of a lateral root thus results in a scar, a source for abundant exudation and a preferential site for entry of parasitic microorganisms.

Root Functions

Aside from their role in anchoring the plant and in absorption of nutrients, roots also play a role in storage. Nutritive reserves accumulated in the root parenchyma during the vegetative phase are mobilised en masse during the reproductive phase. In certain plants, this results in major modifications in root functioning, which can greatly change the equilibrium maintained with the microorganisms surrounding them. Examples of this phenomenon are given in Chapter 7.

Root Renewal

The root system is perpetually changing. In a young plant, it elongates and branches rapidly. Elongation of the primary root occurs at the rate of a few millimetres to a few centimetres per day and can even reach 9 cm day^{-1} in wheat or maize. Secondary roots grow at a slower rate. In sunflower, the average elongation of the root system in the top 10 cm of soil is 72 cm day^{-1} dm^{-3} during the initial weeks of plant life (Maertens and Bosc, 1981).

The roots are constantly renewed. In monocotyledons, the radicle emerging from the seed has a very limited life. In cereals, for example, the radicle is rapidly replaced by adventitious roots which first appear on the primary nodes, at the base of the seedling, and subsequent nodes in the tillered section (Fig. 2.18). Generally speaking, in all plants most of the radicles die a few days after their formation. It is not clear whether this disappearance is due to a physically hostile environment, parasitic attack or natural senescence. Nevertheless it seems that their life is ephemeral and unfavourable conditions only hasten an already programmed death. For example, Huisman (1982) showed in cotton that 50% of the root tissue disappeared in 100 days, and that during the last two-thirds of development time, there was an equilibrium between the formation of new roots and the disappearance of old ones. When a plant approaches ripening, production of new roots decreases, then stops completely, followed by rapid dimunition of the root mass (Fig. 2.19).

As a result of this manner of development of the root system, the plants, though immobile, are in fact capable of exploring the soil to a greater extent than more active microorganisms. This allows renewal of their nutritional sources while constantly exploring new surroundings. But this phenomenon has two other important consequences: this incessant prospection significantly augments the chances of interaction between roots and microorganisms and secondly, dead root cells serve as precious nutritive resources for the soil habitants.

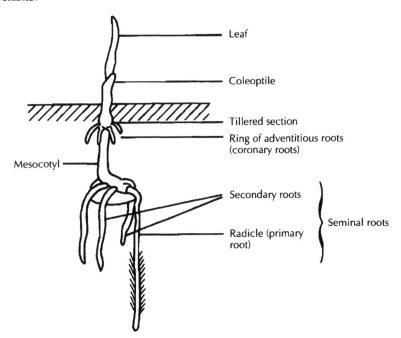

Fig. 2.18   Diagram showing development of a root system during germination of a maize grain (monocotyledon). Seminal roots disappear at the stage of 5–6 leaves and plant nutrition is then taken over by the coronary roots.

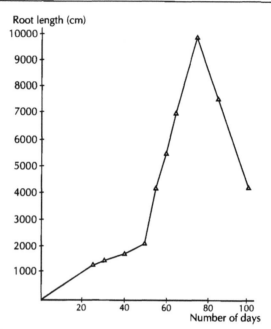

Root length (cm)

Fig. 2.19   Development with age of the plants of root length measured in a column of 1 dm² section, dug up under a row of sunflowers (after Maertens and Bosc, 1981).

Studies on root development are beset with major technical problems and hence are quite rare. Despite this drawback, however, estimations of root biomass under diverse ecosystems are available. In a temperate forest, the dry root mass lies, for example, in the range of 2 t ha^{-1}; in a grassland with humid soil, it is nearly 12 t ha^{-1}.

Other Underground Organs

Tubercles are modified underground stems. They store reserve material and in temperate and cold regions ensure conservation of the plant during winter. Rhizomes are also stems but morphologically resemble roots that develop horizontally. Their function is the same as that of the tubercles. Rhizomes also participate in anchoring the plant in the soil. Bulbs are constituted of modified leaves, borne by a flat part which is the equivalent of a rhizome. These specialised thick leaves contain no chlorophyll, are reserve organs and play a role in conserving the plant during cold or dry seasons.

These underground organs never have an absorption function. This role is fulfilled by the adventitious roots (generally deprived of absorbent root hairs) they are carrying.

For further Information on the Living Components of Soil
- *General*
Prescott L.M., Harley J.P. and Klein D.A. 1999. Microbiology. McGraw-Hill Publ. Co., NY (4th ed.)
- *Bacteria*
Goto M. 1992. Fundamentals of Bacterial Plant Pathology. Academic Press, London.
- *Fungi*
Alexopoulos C.J., Mims C.W. and Blackwell M. 1996. Introductory Mycology. John Wiley and Sons, NY (4th ed.)
Carlile M.J. and Watkinson S.C. 1994. The Fungi. Academic Press, London.
Deacon J.W. 1980. Introduction to Modern Mycology. Blackwell Scientific Publ., London.
Jennings D.H. and Lysek G. 1996. Fungal Biology: Understanding the Fungal Lifestyle. Bios Scientific Publ., Oxford, UK.
Microscopic Algae
Borowitzka M.A. and Borowitzka L.J. (eds.) 1988. Micro-algal Biotechnology. Cambridge Univ. Press, Cambridge.
Kojima H. and Lee Y.K. (eds.). 2001. Photosynthetic Microorganisms in Environmental Biotechnology. Springer, NY.
Metting B. 1981. The systematics and ecology of soil algae. Bot. Rev. 47: 195-312.
- *Protozoans*
Darbyshire J.F. (ed.). 1994. Soil Protozoa. C.A.B. Intn'l., Wallingford.
- *Microfauna*
—Nematodes
Dropkin V.H. 1980. Introduction to Plant Nematology. John Wiley and Sons, NY.
Freckman D.W. and Caswell E.P. 1985. The Nematodes in Agroecosystems. Annu. Rev. Phytopathol. 23: 275-296.
Veech J.A. and Dickson D.W. (eds.) 1987. Vistas on Nematology. Soc. Nematologists, Hyattsville, MD.
—Arthropods
Bachelier G. 1978. La faune des sols. Son écologie et son action. ORSTOM, Paris.
Butcher J.W., Snider R. and Snider R.J. 1971. Bioecology of edaphic Collembola and Acarina. Annu. Rev. Entomol. 16: 249-288.
Eisenbeis G. and Wichard W. 1987. Atlas on the Biology of Soil Arthropods. Springer-Verlag, Berlin.
- *Viruses*
Cornuet P. 1987. Eléments de virologie végétale. INRA, Paris.
Girard M. and Hirth L. 1989. Virologie moléculaire. Doin, Paris.
- *Roots*
Feldman L.J. 1984. Regulation of root development. Annu. Rev. Plant Physiol. 35: 223-242.

Foster R.C., Rovira A.D. and Cock T.W. 1983. Ultrastructure of the Root-Soil Interface. Amer. Phytopath. Soc., St. Paul, MN.

Gregory P.J., Lake J.V. and Rose D.A. (eds.). 1987. Root Development and Function. Cambridge Univ. Press, Cambridge.

Waisel Y., Eshel A. and Kafkafi U. (eds.) 1991. Plant Roots: the Hidden Half. Marcel Dekker Inc., NY.

3

Biological Activity of the Soil

BIOMASS

Introduction

The quantity of living matter present in the soil is the **biomass**. The relative stability of this biomass is the result of dynamic equilibria: at any given instant organisms are born, multiply and die. Microorganisms degrade substrates, synthesise new organic matter and then are subsequently lysed. Their total number varies more or less within wide limits, depending on environmental conditions and all the external interventions. The qualitative and quantitative composition of the biomass is one of the characteristics that contribute to defining a soil but, on the other hand, this composition is partly the result of the physicochemical nature and structure of that soil. Maintenance of the biomass is ensured by the nutrients supplied by plants. It depends to a great extent on numerous metabolic reactions effected by **enzymes**. Some of these processes, such as redox, are common to all organisms. Other enzymatic systems come into play only in specific groups— cellulolytic organisms, nitrogen-fixators for example.

Evaluation of the biomass or measurement of enzymatic activity at a given moment and, more importantly, their variation over time, can give valuable information on the functioning of the ecosystem. It is therefore important to understand how the biological activity of a soil can be measured and translated in terms of biomass or enzymatic activity.

Direct Methods of Biomass Evaluation

The underlying principle is very simple: it suffices to observe soil samples under a microscope, count all the bacterial cells and measure all the fungal mycelial fragments and spores. This operation, repeated a sufficient number of times, enables determining the volumes of bacteria and fungi (principal hosts of the soil) and, knowing their density, computing the corresponding masses. However, implementation of this operation is often complicated. The soil is an opaque and congested medium in which observations are difficult. Besides being tedious, this procedure is not conducive to distinguishing between living and dead organisms. The problem can be overcome in principle by using fluorescent dyes, such as isothiocyanate and fluorescein diacetate or europium chelate, which emit light in living cells upon excitation by an ultraviolet source. But interpretation of observations is not always easy.

Methods based on enumeration of colonies after plating diluted soil suspension are related to direct methods since they purportedly estimate the number of individuals present in the sample (see Chapter 2). Such methods are quite inaccurate, however, and seldom used for computation of the biomass.

Indirect Methods of Biomass Evaluation

These methods do not attempt to directly ascertain the number of microorganisms in a soil sample but to evaluate a basic activity presumed common to all. They therefore have the advantage of being concerned only with living organisms. It is implicitly postulated that the greater the intensity of activity measured in the soil, the more numerous the organisms responsible for that activity, albeit this cannot be considered a rigid rule. Before validation, however, these methods have to be standardised by comparison with other classic methods of direct determination or after incorporation of known quantities of microorganisms in the soil. Results have shown that the initial postulation can be confirmed in most cases.

The results of biomass evaluations by indirect methods are almost always expressed as values of organic carbon mass.

Measurement of Mineralised Carbon

Carbon dioxide represents the final stage of oxidation of organic substrates and the older and still more commonly used method for estimation of biomass consists of measuring the carbon dioxide released consequent to microbial activity. This measurement, earlier done using chemical reactions, is currently executed directly by infrared spectrophotometry, which has considerably simplified the procedures. However, since microorganisms are not the only sources of carbon dioxide in the soil, it is important to take into account respiration of the roots and, if required, also the abiotic origin from carbonate (calcareous) rocks.

In the **fumigation-incubation** method (Jenkinson and Powlson, 1976), the microorganisms in a soil sample are first killed by fumigation with chloroform. This biocide does not modify the solubility of the non-microbial organic matter, but destroys the selective permeability of the cell membranes, thereby providing access to the cell contents. After fumigation, reseeding with microorganisms is done using a very small quantity of the original soil. Recolonisation of the sterilised sample is very rapid. Microbial development results in an intense release of carbon dioxide, much more in volume than that released by an identical non-fumigated sample. This increase in CO_2 production (F) is attributable to utilisation by the reseeded microorganisms of the biomass killed by fumigation (B). Only a part of the carbon of biomass B is mineralised and converted to carbon dioxide, the rest being incorporated in the newly formed biomass.

Thus, $F = KB$, where $B = F/K$.

The value of K has been determined experimentally several times: it is of the order of 0.4 when incubation occurs at 22°C. This method is fairly drawn out as the release of carbon dioxide must be measured over a period of 10 days.

The **fumigation-extraction** method yields results much more quickly. It consists of measuring the organic carbon released from the lysed microbial biomass after chloroform fumigation. Extraction and measurement of carbon can be automated, which allows a large number of samples to be processed (Wu et al., 1990). The biomass $B = k (C_1 - C_0)$. C_1 represents the organic carbon of the chloroform-fumigated soil and C_0 the organic carbon of a control sample of the same mass. The fumigation-extraction method also enables measurement of nitrogen in the biomass.

Another method consists of measuring the increase in release of carbon dioxide induced by the incorporation of an organic substrate into a soil sample. This method is termed **rapid respiration**. A substrate that can be readily metabolised is selected, for example glucose. The maximal quantities of carbon dioxide released at the beginning of the test are a linear function of the soil biomass (Anderson and Domsch, 1978).

Measurement of ATP

Adenosine 5'triphosphate (ATP) plays a role in the transfer of energy inside cells of all types of organisms. Rapidly hydrolysed when the cells die, ATP is a good marker of living organisms. Its measurement uses the same bioluminescence reaction observed in fireflies. Bioluminescence requires energy and results from the enzymatic action of luciferase on a substrate, luciferin, after its activation by ATP. The luminous emission, measurable with a photometer, is proportional to the quantity of ATP. After measuring the ATP, the biomass is computed using a formula of the type: biomass $B = K (ATP)$. The coefficient K of this equation is obviously not of the same value as those shown in the preceding section. The ATP must be extracted rapidly before it can be hydrolysed or before it gets fixed on the adsorbent soil complex. It is possible to surmount the methodological problems encountered on employing the luciferin-luciferase couple while directly measuring the ATP by HPLC (Prévost et al., 1991).

Ammonification of Arginine

Most microorganisms are capable of using amino acids as sources of carbon liberating ammonia. Standard tests have shown that ammonification is only achieved by metabolically active microorganisms and that, among the various amino acids, arginine gives the best results. It therefore suffices for an estimation of the biomass to measure the quantity of ammonia released after introduction of a known quantity of arginine into the soil (Alef and Kleiner, 1986).

Measurement of Activity of Electron Transporter Systems (Dehydrogenase Activity)

General biological activity is based on a cascade of redox reactions. In aerobic systems, oxygen represents the final acceptor. However, it is possible to replace oxygen with an organic oxidant such as triphenyltetrazolium chloride that can be readily reduced by the microorganisms. Once reduced, this soluble, colourless compound becomes an insoluble red-coloured triphenylformazan. It is easy to carry out a colorimetric estimation of this compound after extraction into methanol. Iodophenyl-nitrophenyl-phenyltetrazolium chloride, even more sensitive, may be used in the presence of oxygen (Trevors, 1984) as well as under anaerobic conditions.

Similarly, a very large number of soil microorganisms can utilise dimethyl sulphoxide (DMSO) as an electron acceptor and reduce it to dimethyl sulphide (DMS). The greater the biomass, the greater the DMS released under trial conditions. The sample must be incubated in the absence of oxygen, in a nitrogenous atmosphere. The volatile DMS is measured by gas-phase chromatography. This very sensitive method can be used even with very small samples or in soils of little biomass (Sparling and Searle, 1993), irrespective of root activity.

Measurement of Phospholipids

Phospholipids enter into the composition of all cellular membranes and are rapidly degraded after cell mortality. After extraction using a mixture of chloroform and ethanol buffered at pH = 4, the phospholipids can be readily estimated by a simple enough colorimetric method. The results correlate very well with the biomass values estimated by ATP measurement (Hill et al., 1993). This method appears promising for evaluation of biomass of particular microbial groups: indeed, it is possible to selectively evaluate certain fatty acids that are biological markers by using chromatography to separate the lipid components extracted from the soil (Zelles et al., 1992).

Other Techniques

Other techniques have also been considered, such as microcalorimetry or measurement of nucleic acids or certain nucleotides. But these techniques do not allow distinction between the DNA of living cells and that of dead ones. Other methods, specific to particular microbial groups, have also been considered in the preceding chapter: measurement of muramic acid, chitin, ergosterol, etc.

The great diversity of the methods proposed clearly demonstrates that none is totally satisfactory under all conditions. Depending on the type of soil and level of microfloral activity, the various estimations of biomass may be well, moderately or totally not correlated and compatible with one another. The choice of method must therefore be decided relevant to experimental conditions and the objective of the research.

Specific Techniques for Observation of a Specified Organism

Instead of following the evolution of the entire biomass, one may wish to study the behaviour of one species or one particular strain.

One of the simplest techniques consists of using a strain resistant to a toxic product (most often a fungicide or an antibiotic). Enumeration is carried out after isolation on a culture medium rendered selective by incorporation of the biocide.

Observations can also be done in situ using immunofluorescence. In this method a serum is prepared by injecting a rabbit with antigens specific to the microorganism under study. After purification, the antibodies are coupled with fluorescein isothiocyanate. Observed under ultraviolet light, the cells recognised by this preparation emit a characteristic fluorescence. Immuno-enzymatic techniques (ELISA) can also be used by coupling the antibody of the foregoing prepared serum to an enzyme. Acid phosphatase is generally used which hydrolyses the paranitrophenylphosphate, yielding a yellow-brown compound. Only the wells containing the marked bacteria are coloured on the microtritration plate.

It is also possible to use marker genes introduced into the microorganisms using appropriate vectors. The most common are the *Gus* and *lac* Z genes, which direct the synthesis of a β-glucuronidase and a β-galactosidase respectively. In the presence of an appropriate substrate, the appearance of a coloured reaction signals their presence and hence that of the microorganisms. Gene coding for the bioluminescence has also been used sometimes.

Techniques of DNA hybridisation after amplification enable very sensitive and very specific detection, provided the probes are carefully chosen. But they do not assist in determining whether the nucleic acid detected has come from living or dead cells.

Measurement of Specific Enzymatic Activity

When the objective is no longer estimation of the overall biological activity of the soil, but rather the activity of a particular category of microorganisms, the characteristic enzymatic activities of that **functional group** are utilised. For this purpose, generally a known quantity of substrate is introduced into the soil and under precise conditions of incubation, the disappearance of the substrate or the appearance of products of degradation is measured (such as ammonia for example). By this method, the activity of pectinolytic, cellulolytic or nitrogen fixators, ammonification or nitrification, sulphooxidation or sulphato-reduction reactions, and biodegradation of a pesticide etc. can be assessed. Use of substrates marked with ^{14}C can help in precising the metabolic pathways. The protocols can be as diverse as the enzymatic systems studied. The one general rule in manipulations and interpretation of results, is to take into account the possible activity of soil enzymes.

SOIL ENZYMES

Introduction

In addition to the syntheses and degradations resulting from the activity of microorganisms, flora and fauna, the soil has its own enzymatic activity that is not directly related to metabolic reactions. This activity, suspected since the end of the nineteenth century, can be demonstrated after soil sterilisation, which has to be sufficiently harsh to eliminate all biological activity, yet concomitantly mild enough not to denature the enzymatic proteins. Heat is therefore excluded. The antiseptic employed most often is toluene. Other agents, such as chloroform, sodium azide and gamma-ray irradiation, are also used sometimes. The choice of sterilisation method depends on the category of enzymes to be studied since each method has drawbacks.

Origin of Soil Enzymes

Soil enzymes are derived in part from plants and animals but it is common knowledge that they are mostly of microbial origin. As a matter of fact, as already mentioned, digestion in microorganisms starts outside the cell through the action of enzymes that traverse the cell membranes and are thus released in the extracellular surroundings. Other enzymes, especially those of higher molecular weight, remain inside the cells as long as they live, or bind to their membranes. After the cells die, lysis of the cell walls brings these enzymes into contact with the external milieu. A part of them remains attached to the membrane fragments. When a soil sample is prepared for microscopic examination, specific enzymatic activities can be demonstrated in these phantom cells using histochemical techniques. The remaining enzymes pass directly into the soil solution where, in fact, they do not remain free for long. Like all organic molecules, they may serve as substrates for microorganisms and then disappear, or be degraded by soil proteases, or subject to physicochemical denaturation. They may also be adsorbed on the clay-humus complex, in which case these enzymes may be trapped inside the clay layers, or be covered with organic matter, or even affixed on the external surfaces of mineral or organic particles (Fig. 3.1).

The main enzymes identified in the soil to date are these:
— *oxidoreductases*: dehydrogenases, glucose-oxidase, aldehyde-oxidase, uricase, phenol-oxidases, ascorbate-oxidase, catalase, peroxidase, nitrate-reductase.
— *transferases*: dextran-sucrase, levan-sucrase, aminotransferase, rhodanase.
— *hydrolases*: carboxylesterase, arylesterase, lipase, phosphatase, nucleases and nucleotidases, phytase, arylsulphatase, amylase, cellulase,

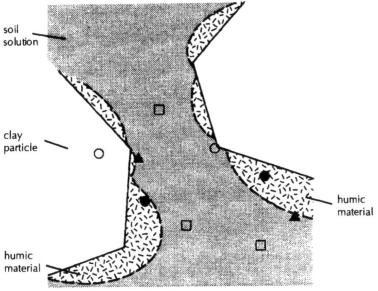

soil
solution

clay
particle

humic
material

humic
material

Fig. 3.1 Probable localisation of soil enzymes (after Burns, 1978). ● enzyme covered with humic matter; O enzyme adsorbed on clay particle or trapped inside the layers; ▲ enzyme retained on surface of humic colloid film; □ enzyme free in soil solution and of limited lifetime.

laminarinase, inulase, xylanase, dextranase, levanase, polygalacturonase, maltase, α- and β-galactosidases, α- and β-glucosidases, invertase, proteinases and peptidases, asparaginase, glutaminase, urease, pyrophosphatase.
— *lyases*: decarboxylases.

Consequences of Enzyme Adsorption

The phenomena of adsorption, which vary as a function of numerous parameters (nature of the clay-humus colloids, structure and molecular weight of the enzyme, pH, ionic composition of the medium...), have important consequences:
— *protection against degradation:* The activity of adsorbed enzymes is not attenuated by drastic changes in environmental conditions, which otherwise would significantly reduce the activity of free enzymes (Lähdesmäki and Piispanen, 1992), and furthermore their life is considerably lengthened. Urease is a well-known example; it resists denaturation by heat as well as by proteolysis when inserted in the organic fraction of the soil. Urease-like activity has been demonstrated in Alaskan soils frozen for 9000 years and hence deprived of all life (Skujins and McLaren, 1969)! Phosphatase also persists in soil for a long time. The protection of enzymatic molecules could be due to physical causes such as inaccessibility or modification of protein structure. The inhibitory effect of aromatic compounds (tanin, humic acids) present in the organic adsorbent could be another cause. In

summary, the detection of an enzymatic activity in a soil is by no means proof of a concomitant biological activity.

Several studies on urease have shown that a soil can only keep a certain quantity of enzyme in reserve. Beyond a **threshold** (whose value depends on the type of soil), excess urease is not stored and is rapidly degraded (Zantua and Bremner, 1977). The capacity to stock urease increases grosso modo with organic matter content.

— *modifications of activity*: Protection of enzymes against biodegradation is often accompanied by a reduction in their activity because they are beyond reach of their substrate or because they themselves are inhibited by organic compounds. The inhibition is generally reversible and varies with the cation composition of the soil solution. On the other hand, cases wherein adsorption stimulates catalytic activity are also known: we have already cited the example of catalase which, adsorbed on montmorillonite, becomes four times more active than free catalase (Stotzky, 1974).

Various Reasons for Disappearance of an Organic Compound from the Soil

An old experiment of Durand (1966) illustrates well the diversity of mechanisms that play a role in the disappearance of even a simple organic substrate from the soil.

Three samples of the same soil are compared. One serves as control and the other two are subjected to one of the following treatments:
— sterilisation using heat, which kills the microorganisms and denatures the enzymes;
— sterilisation by the addition of toluene, which inhibits the microorganisms without denaturing the enzymes.

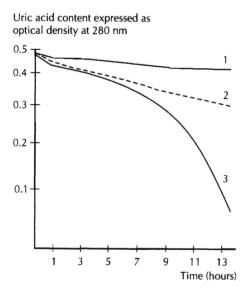

Uric acid content expressed as optical density at 280 nm

Time (hours)

An equal quantity of uric acid is added to each sample, which is incubated under standard conditions and the compound measured at regular intervals. It is noted that in the absence of any enzymatic activity of both biotic and abiotic origin, part of the uric acid disappears under the effect of physicochemical processes (curve 1). When only the microorganisms are inhibited, the effect of hydrolysis by soil uricase is added to the physicochemical phenomena (curve 2). Curve 3 represents the overall result due to all processes of degradation.

Ecological Importance

Soil microorganisms are confronted with a dual problem: nutrients are scarce and are often large insoluble molecules (cellulose for example) which can only traverse the membranes and enter the cytoplasm after being cut into smaller molecules. To accomplish this cleavage, the organism must synthesise and secrete extracellular enzymes, which supposes the presence of appropriate substrates.

But the microorganisms have no means of sensing such substrates. In a way, they have to cope with the following dilemma: either they 'wager' that an adequate substrate is, or will be shortly, available and they secrete enzymes at the cost of depleting their reserve materials if the bet is lost; or they economise these reserves and lead a quiescent life at the risk of missing

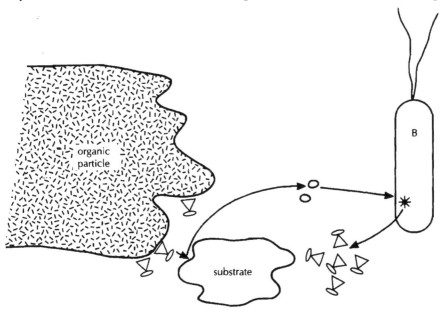

Fig. 3.2 The enzymes (⋎) which are adsorbed at the surface of an organic particle take off a few molecules of a soluble oligomere (O) from the insoluble substrate (cellobiose, for instance, from cellulose pieces). These soluble molecules can diffuse into the bacterial cell (B) and lead to elaboration of the enzymes that are necessary for degrading the substrate (cellulases in this example). (After Burns, 1982).

an opportunity when a consumable substrate becomes available. According to Burns (1982), the soil enzymes offer the possibility of overcoming both these impasses. Adsorbed on the organic polymers in the external environment, protected from degradation while still active, they can act on a suitable substrate as soon as it comes and initiate its cleaving, releasing by this way a few small soluble molecules into the aqueous phase of the soil. These molecules diffuse to the cell, pass through the walls and serve as a signal for induction of enzyme synthesis. The reaction thus initiated continues for the benefit of the cell.

If this hypothesis is correct, the enzymes immobilised on the organomineral complex in soil would have a very important ecological role because they would allow the microorganisms to considerably economise on energy. As per this scheme, the notion of an overall metabolic activity of the soil can be introduced. It would comprise the sum of highly integrated multiple activities arising from living microbial cells, free and adsorbed extracellular enzymes and physicochemical processes.

For Further Information on the Biological Activity of the Soil

Burns R.G. (ed.). 1978. Soil Enzymes. Academic Press, London.

Burns R.G. and Slater H. 1982. Experimental Microbial Ecology. Blackwell Sci. Publ., Oxford.

Jenkinson D.S. and Ladd J.N. 1981. Microbial biomass in soil: measurement and turnover. In: Soil Biochemistry, vol. 5, pp. 415-471. E.A. Paul and J.N. Ladd (eds.). Marcel Dekker, NY.

Sparling G.P. 1985. The soil biomass. Dev. Plant Soil Sci. 16: 223-262.

Effect of Environment on Microorganisms

AVAILABILITY OF WATER

Water is an indispensable chemical for the development of metabolic processes. It also has multiple mechanical or physical effects. When abundant, water augments the availability of soluble elements and facilitates the moving of mobile microorganisms in search of substrates or prey; when scarce, it enables better gaseous exchanges with the external atmosphere and more rapid renewal of oxygen. The thermal inertia of water also has a very important buffering effect on high temperature variations. Five times more calories are required to increase the temperature of 1 g of H_2O by 1°C than are needed for 1 g of dry soil.

Effects of Desiccation

Reduction in Populations

Severe desiccation always results in the death of a large number of individuals. The intensity of the phenomenon augments with the value of the hydric potential attained, duration of period of drought and temperature. The effect is attenuated in soils rich in humus and in swollen clays such as montmorillonite. The survival rate of microorganisms depends largely on their physiological state. The youngest and most active fraction of the biomass, which attained its exponential phase of growth at the time of dehydration, is the most vulnerable, while the fraction reaching a dormant or stationary stage is least affected. The former corresponds to the microflora of the biotope termed by us the external compartment, while the latter essentially refers to the internal compartment. In semi-arid regions where the soil is regularly subjected to prolonged and intense drought, the rapidly growing microflora of the external compartment seems to have acquired some resistance to desiccation; hence it cannot be much more affected than microflora which develops slowly (Van Gestel et al., 1993).

Microfauna does not take well to drought but each group shows some adaptive strategies: collemboles burrow into the soil in search of moisture, nematodes encyst, while acarids envelop themselves in protective sheaths.

Reduction of Microbial Activity

Microbial activity steadily diminishes with availability of water (Fig. 4.1). Bacteria are on average more sensitive than fungi. Bacterial activity is low around the wilting point (–1.5 MPa) and is arrested around –6 to –8 MPa, but there are significant variations depending on the taxonomic groups. Gram-negative bacteria, which have thin walls, are by and large less resistant than Gram-positive bacteria, enclosed in thick walls. The thinness of the wall can be compensated in some cases by the presence of a protective polysaccharidic capsule. The capsule of *Xanthomonas campestris*, for example, consists of a three-dimensional network of xanthan capable of retaining water 180 times its weight. In a dry atmosphere, the network gradually dehydrates and flattens (Robert and Schmit, 1982). Actinomycetes and cyanobacteria are more resistant than the average. *Azotobacter* spp., on the other hand, are very sensitive. In general, atmospheric nitrogen fixation is

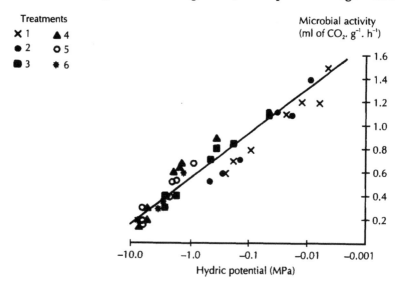

Fig. 4.1 Relationship between hydric potential and respiration. The soil was divided into six lots and the potentials adjusted prior to the experiment to –0.005, –0.01, –0.05 –0.41, –1.0 and –1.5 MPa (treatments 1 to 6 respectively). Measurements were taken simultaneously during the 37-day period of incubation at 25°C, of respiration (by gas phase chromatography) and the hydric potential (by thermocouple hygrometer) (after Orchard and Cook, 1983).

arrested as soon as the potential falls below −1 MPa while nitrification diminishes below this threshold and ceases around −5 MPa. The microorganisms responsible for ammonification are more resistant, continuing activity to about −6 MPa and sometimes well beyond (Dommergues and Mangenot, 1970).

The threshold limit of activity for fungi is of the order −20 MPa, but here also large variations are observed depending on the species (Griffin, 1972). *Aspergillus* spp. and *Penicillium* spp., very common in soils, are very resistant to drought: their spores are capable of germination at hydric potentials of −30 MPa. Some fungi are still more resistant: the champion seems to be *Xeromyces bisporus*, capable of development at a tension of −65 to −69 MPa!

Consequences of Soil Desiccation

Nitrifying activity is arrested prior to cessation of formation of ammoniacal nitrogen from organic matter. This results in accumulation of ammoniacal nitrogen in the soil, which under certain conditions can lead to loss of nitrogen in the form of gaseous ammonia. This phenomenon is aggravated in regions of strong sunlight because infrared rays induce non-biological ammonification by destroying the bonds between the NH_4^+ ion and organic complexes (Dommergues, 1962). A concomitant increase is also observed in assimilable forms of potassium, manganese, phosphorus and sulphur (Dommergues and Mangenot, 1970), for which there is no clear explanation as yet. The rupture of aggregates caused by the mechanics of desiccation could be the cause for the temporarily sequestered nutritive elements coming back into circulation.

Effects of Rehydration

Rewetting a dry soil induced very rapid revival of microbial activity which, for a few days, was much higher than that of a sample of the same soil kept constantly wet. Emission of carbon dioxide increased considerably, reaching up to 40 times the initial level (Orchard and Cook, 1983), resulting from intense mineralisation of organic carbon (a part of the organic carbon is also mineralised by the non-biological path). Increase in mineralisation of organic nitrogen paralleled that of carbon (Van Gestel et al., 1993). Denitrification can lead to an overall loss of nitrogen. The activity decreased progressively with depletion of resources and, after 2 to 3 weeks, fell to a level equal to or lower than that of non-dried (control) soil.

This intensification of microbial activity in a soil rewetted after drying is due to consumption of substances which were not available before desiccation: substrates liberated by fragmentation of aggregates, desorbed organic molecules and microbial cells which were killed by drought. A succession of several cycles of drought followed by rewetting (as often happens in Mediterranean climate) results in depletion of nitrogen reserves and diminution of the quality of organic matter that ensures soil stability.

Effects of Submersion

Generalities

Prolonged submersion results in severe disturbance of biological equilibria. Fungal populations are considerably reduced and replaced by bacteria; terrestrial arthropods are eliminated. (This effect has been used on several occasions to eradicate parasites. Stover (1962) inundated banana plantations for 6 months before planting and successfully curtailed vascular disease of banana caused by *Fusarium oxysporum* f. sp. *cubense*. In France, even today vineyards are inundated in winter in the plains of Narbonne to preclude attacks by *Phylloxera* on the roots of ungrafted plants.) In the liquid phase, algae develop and zooplankton appear. Some nematodes can survive in hydromorphic soils.

Under such conditions, the microflora is initially mainly aerobic. However, oxygen diffuses slowly in the water and development of microbial communities depends on the degree of oxygenation of the medium or, in other words, the redox potential. Understanding the succession of equilibria that occur as the environment progresses towards anaerobic conditions will be facilitated by taking a rice field as an example. Similar situations are likewise encountered in soils accidentally inundated or deficient in oxygen consequent to too strong compaction or the addition of excessive organic matter. At a different level, such equilibria also occur in the interior of aggregates.

Prolonged Submersion: Example of a Rice Field

The following report is largely based on the synthesis of Watanabe and Furusaka (1980).

LIQUID LAYER

The fields are submerged after the rice has been planted (Fig. 4.2). Bacteria, which consume the organic particles detached from the soil surface, and especially algae grow in the water and serve as nourishment for the zooplankton. Filamentous cyanobacteria, followed by other unicellular forms, progressively replace the initial flora comprising members of Chlorophyceae and diatoms, as and when sunlight decreases with crop growth (Roger and Reynaud, 1976). Due to the photosynthetic activity of micro- and macroflora, the quantity of oxygen dissolved in the water increases during sunlight hours while the concentration of carbon dioxide decreases, resulting in an elevation of pH. The opposite occurs at night: oxygen is consumed and carbon dioxide released in respiration, thereby lowering the pH. Several species of cyanobacteria are capable of using atmospheric nitrogen but the quantity fixed by natural populations is highly variable, fluctuating between 0.4 to over 40 kg nitrogen ha^{-1} and, on average, amounting to 30 kg ha^{-1}. This non-negligible supply of nitrogen explains the maintenance of high productivity in traditional monocultures of rice. The yield of this natural

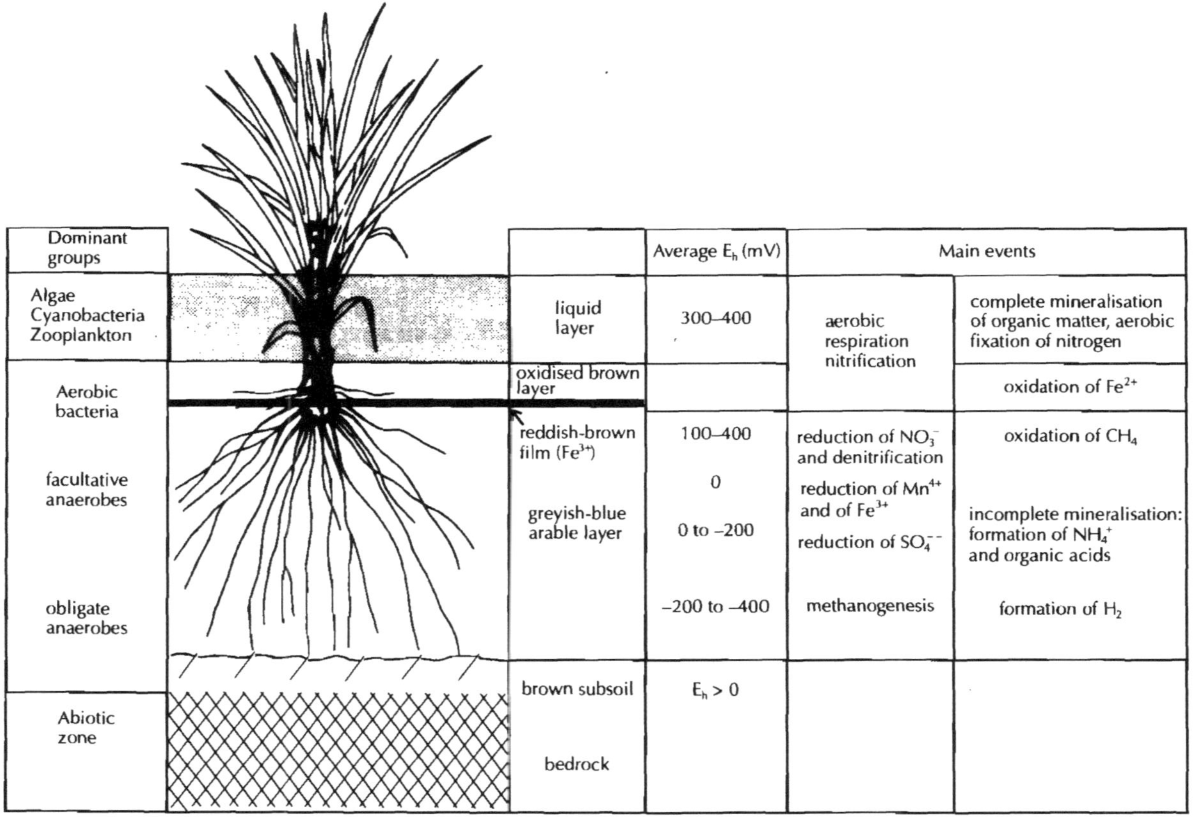

Dominant groups		Soil layer	Average E_h (mV)		Main events
Algae Cyanobacteria Zooplankton		liquid layer	300–400	aerobic respiration nitrification	complete mineralisation of organic matter, aerobic fixation of nitrogen
		oxidised brown layer			oxidation of Fe^{2+}
Aerobic bacteria		reddish-brown film (Fe^{3+})	100–400	reduction of NO_3^- and denitrification	oxidation of CH_4
facultative anaerobes		greyish-blue arable layer	0	reduction of Mn^{4+} and of Fe^{3+}	incomplete mineralisation: formation of NH_4^+ and organic acids
			0 to –200	reduction of SO_4^{2-}	
obligate anaerobes		brown subsoil	–200 to –400	methanogenesis	formation of H_2
Abiotic zone		bedrock	$E_h > 0$		

Fig. 4.2 Profile of soil in a rice field and principal biological and physicochemical characteristics of the different horizons (after Watanabe and Furusaka, 1980).

fertilisation can be enhanced by seeding the rice field with selected strains of cyanobacteria. (See page 134 and 278)

ARABLE LAYER

The roots of a rice plant are submerged in mud where oxygen diffusion becomes more and more difficult with depth. The presence of a special lacunar tissue, the **aerenchyma**, in this particular plant permits direct gaseous exchanges between the roots and the atmosphere, thereby preventing asphyxiation of the roots and facilitating maintenance of relatively aerobic conditions in their immediate vicinity. It should be mentioned that gaseous exchanges can take place as well from the soil towards the atmosphere and by this path the gases reduced in the anaerobic zone are preferentially released (Buresh et al., 1993).

Anaerobiosis is gradually augmented with passage of time and since submergence of rice and with distance from the soil surface. The intensity of anaerobiosis is reflected by the value of the redox potential (E_h): as E_h decreases, the environment becomes more and more reducing and the bacteria have greater difficulty in finding electron acceptors.

First of all, nitrates are reduced. This is accomplished in the first step by a large number of bacteria, leading to the formation of nitrites. The other steps of denitrification—production of nitrogen oxides and afterwards gaseous nitrogen—are carried out by fewer species. If the environment becomes more reducing, the loss due to denitrification can be compensated by nitrogen fixation by such anaerobic bacteria as *Clostridium* spp., abundant on the organic debris in rice soils.

At lower values of redox potential, reduction of iron and manganese is observed. Ferric salts are enzymatically reduced to soluble ferrous salts that impart a characteristic greyish-blue colour to the mud. Common facultative anaerobic bacteria (*Pseudomonas*, *Bacillus* and *Enterobacter*) as well as strict anaerobes (*Clostridium*) effect this reduction. The biological nature and practical importance of reduction of manganic to manganous salts are debated by soil scientists.

If the value of the redox potential continues to fall and becomes negative, the sulphates act as electron acceptors. Reduction of sulphates is effected by a small number of specialised bacteria (*Desulfovibrio*, *Desulfotomaculum*). The hydrogen sulphide formed by this reduction can damage the roots of rice. Fortunately, in most cases it combines with reduced iron to form non-toxic ferrous sulphide which precipitates.

Redox potential values lower than –200 mV are conducive to the growth of archaea which reduce organic acids and carbon dioxide in the process of producing methane. As the methane rises to the surface water layer, it is oxidised by aerobic bacteria in such a way that the release of gases is generally low, except when temperature and degree of anaerobiosis are elevated.

SUBSOIL

Under the thin arable layer, rich in organic matter, inundated and reducing, lies a subsoil generally untouched by tillage. Being very coarse, it is neither saturated with water nor reducing. To the contrary it is generally rich in insoluble salts of metallic oxides. Strictly and facultative anaerobic bacteria of the upper layer slowly attack the subsoil zone. As a result of this bacterial activity, the reduced iron and manganese, more soluble, gradually migrate to the upper soil horizons.

OXIDISED LAYER

The superficial horizon of the rice field constitutes a transition zone between the well-oxygenated liquid layer and the reducing arable layer. In this region of a few millimetres in thickness where active aerobic microflora live, nitrification (by oxidation of ammonia arising from mineralisation of organic deposits) overtakes denitrification. Nitrogen-fixing bacteria (*Azotobacter, Beijerinckia*) can thrive in this region. On the soil surface, exposed to light, unique associations of aerobic heterotrophic bacteria and anaerobic photosynthetic bacteria form.

The reduced organic and mineral compounds reaching this zone are reoxidised. Metallic hydroxides precipitate in the vicinity of the surface: a reddish-brown pellicle forms due to ferric hydroxides, which contrasts with the bluish colour of the subjacent soil.

Effects of Temporary Soil Flooding

In soils rich in sulphates, as encountered in Tunisia for example, a sudden decline in crops is sometimes observed when hot weather follows a major storm or an irrigation. Dommergues et al. (1969) studied this problem experimentally and offered an explanation. In the inundated soil the redox potential drops dramatically. The combination of heat, water and an unlimited quantity of electron acceptor sulphates provokes a very rapid and considerable increase in number and activity of sulphato-reducing bacteria. This phenomenon manifests only in the rhizosphere where immediately utilisable organic substrates abound. This results in production of toxic hydrogen sulphide and consequent deterioration of the plants.

Such situations do not develop in an 'ordinary' soil. However, temporary soil flooding does disturb respiration of plant parts. Due to partial asphyxiation, roots and germinating seeds excrete ethanol. It has been experimentally demonstrated that zoospores of several parasitic oomycetes (*Pythium, Phytophthora*) are selectively attracted by ethanol: exposed to a gradient of ethanol, they orient towards the highest concentrations (Allen and Newhook, 1973). Moreover, germination of zoosporangia, which release zoospores, is effected in the presence of water, so it is readily understandable why the risks of attack by oomycetes are much greater in flooded soils than in well-drained ones.

SOIL pH

Though it is easy to measure a pH, interpreting these measurements can be dicey. As already mentioned, the pH of a soil represents only a rough mean

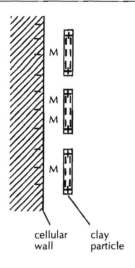

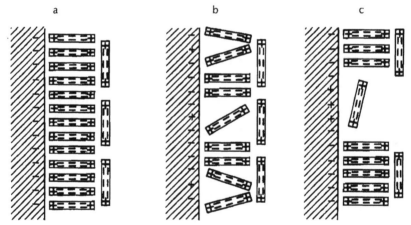

a   b   c

Fig. 4.3   When the cellular walls carry negative charges (pH > pI), they can adsorb clay
particles either by intermediate bridges formed by metallic ions M (above:
bonds formed face to face) or by direct bonds between the negative charges
of the carboxyl groups and the positively charged extremities of the particles
(a: bonds formed side to face). Due to the presence of amine functions that
carry a positive charge, the disposition is most likely of type b (amine groups
scattered) or of type c (amine functions grouped) (after Marshall, 1968).

COMPOSITION OF SOIL ATMOSPHERE

Given their volatile nature, the active gases in the soil atmosphere do not
require the presence of a continuous water film to diffuse and hence can
reach considerable distances at the microbial scale, say 5 to 6 cm or more.
Only the most common and the most concentrated of these gases have been
identified and studied to date, with a few rare exceptions. A very large
number of active and volatile compounds in infinitesimal concentrations

(such as pheromones in insects) very likely play a significant role that as yet is not known.

Only volatile products of physicochemical or plant origin are considered in this chapter. Compounds of microbial origin are discussed in Chapter 5.

Effect of Oxygen Content

Oxygen plays an essential role in the manifestation of microbial activities. Unlike all the volatile products we shall speak of later, O_2 is exogenous in origin. Renewal of the oxygen consumed in the soil during the course of oxidative processes is ensured by its constant diffusion from the external atmosphere. This diffusion depends on the porosity and the hydric state of the soil, which is why this topic was covered on page 86. It should be remembered that a large number of aerobic bacteria and fungi are capable of developing in the absence of oxygen, provided the redox potential is maintained at a sufficient level and that other electron acceptors are available. For a large number of archaea, the absence of oxygen is a necessity.

Effect of Carbon Dioxide

The proportion of carbon dioxide in the soil atmosphere is of the order of 0.3 to 5% but can reach and even exceed 20% when the biological activity is intense, which is the case in the rhizosphere or near a significant source of organic matter. (In the external atmosphere, the carbon dioxide content is 'only' 0.037%.) The response of microorganisms to carbon dioxide is highly variable. Bacterial fixators of nitrogen function poorly when the carbon dioxide content increases. Nitrifying bacteria prefer a concentration around 0.15%. As for fungi, Louvet and Bulit (1964) have shown that in vitro the growth of *Sclerotinia minor* decreases very rapidly as the concentration of carbon dioxide increases: it diminishes by half when the atmospheric content of CO_2 rises from 0.03 to 5% (Fig. 4.4). On the contrary, *Fusarium oxysporum* f. sp. *melonis* is able to handle concentrations above 20% with no problem. More recently, Dijst (1990) proposed that the carbon dioxide released due to respiration of the underground organs of potato could inhibit formation of sclerotia on the tubers caused by the fungal parasite *Rhizoctonia solani* as long as the plant is living. After leaf cutting, respiration diminishes considerably and the tubers become covered with sclerotia. Moreover, carbon dioxide has a stimulatory or morphogenic effect on several fungi, which in some cases is even indispensable: for example, absence of CO_2 impedes development of *Verticillium albo-atrum* when the culture medium contains glucose as a carbon source. Utilisation of $^{14}CO_2$ led to demonstrate that CO_2 was used for the synthesis of C_4 amino acids (Hartman et al., 1972).

Effect of Ammonia

Like carbon dioxide, ammonia is a universal component of the soil atmosphere. The hydrolysis of urea, proteins and nucleic acids from organic debris is the main source of ammonia, but it can also derive from the

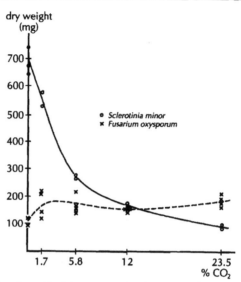

Fig. 4.4   Effect of carbon dioxide content in the atmosphere on growth by weight of *Sclerotinia minor* and *Fusarium oxysporum* in vitro (Louvet and Bulit, 1964).

reduction of nitrates if the melieu is anaerobic. Partly retained as NH_4^+ ions on the adsorbent complex, ammonia can be released into the soil with no microbial interference under certain circumstances. The amount of ammonia in the soil is highly variable, from very low in a 'resting' soil to very high after the addition of nitrogenous organic matter.

Bacteria are in general not very sensitive to ammonia and their growth may even be stimulated sometimes. However, *Nitrobacter*, responsible for the transformation of nitrites to nitrates, is inhibited by high concentrations of ammonia. This situation can temporarily exist in alkaline soils after the addition of organic matter or a high dose of urea, which then leads to the accumulation of phytotoxic nitrites. Some fungi such as *Fusarium solani*, *Gliocladium roseum*, *Trichoderma harzianum* for example, seem to be unaffected (Schippers et al., 1982). In the case of *F. oxysporum*, formation of chlamydospores is stimulated up to concentrations of 15 µL L^{-1} air (Löffler et al., 1986). Most fungi seem to be sensitive to ammonia, however. Even a concentration as low as 1 µg g^{-1} air suffices to prevent germination of the conidia of *Botrytis cinerea* or *Penicillium chrysogenum*. A field trial showed that after the addition of *Vicia villosa* green manure, the ammonia content of the soil atmosphere reached 0.14 ppm after one week, which sufficed to significantly reduce germination of the chlamydospores of *Chalara elegans* and hence reduce attacks on the cotton crop (Candole and Rothrock, 1997). In vitro, chlamydospores of *C. elegans* are killed by ammonia above concentrations of 0.4 ppm. Prolongation of the antifungal effect of soil disinfection can be explained in part by the high ammoniacal concentrations resulting from mineralisation of the biomass killed by these treatments. Some alkaline soils manifest high fungistatic power when moistened, an effect of physicochemical origin due apparently to sorting out significant quantities of ammonia in the presence of water (Ko et al., 1974). The aridity of the regions in which these soils have been studied explains why the ammoniacal complex has not already been exhausted.

Effect of Volatile Compounds Released by Roots

All of us who have smelled the fragrance of a flower know that plants represent an important source of volatile products. This emission is not limited to just the aerial parts, however; it also arises from the roots.

The concentration of carbon dioxide is higher in the rhizosphere than in the rest of the soil because of respiration. In addition to carbon dioxide, acetaldehyde, ethanol, formic acid and traces of many more specific as yet unidentified gases are found in the vicinity of the roots. For example, it has recently come to light that the roots of leguminous plants emit species-specific flavonoids which "inform" the rhizobia of the nearness of a compatible partner. Similarly, plant roots of various botanical families send signals (not yet identified) to their endosymbiotic fungi which stimulate spore germination and orientation of mycelial growth in their direction (Koske, 1982). Crucifers are among the rare plants that lack endomycorrhizae. This absence could be explained by the presence in their near atmosphere of toxic isothiocyanates derived from the sulphur-glucosinolates contained in the roots (Vierheilig and Ocampo, 1990). Ligneous plants and ectomycorrhizal fungi likewise exchange volatile signals. Once the symbiosis is established, the mycorrhizal roots excrete gaseous substances (mainly terpenes) which are generally not produced by either of the two partners when present singly. It has also been demonstrated in vitro that these volatile compounds exert a fungistatic effect on various root parasites (Krupa and Nylund, 1972) and could significantly augment development of bacterial populations (Schisler and Linderman, 1989).

Effect of Volatile Compounds Released by Germinating Seeds

Germinating seeds themselves also constitute important sources of exudates, a great part of which is volatile. Emission is maximum during the initial 2 days after seed swelling, followed by a decline and complete stoppage by the end of the 4th day. The intensity of volatile compounds production increases with the quantity of reserve substances stored in the seeds. Dry seeds or moistened but dead seeds do not release such substances.

A large number of products have been identified in the atmosphere encompassing seeds during germination. These are essentially aldehydes (acetaldehyde, propionaldehyde, formaldehyde), alcohols (methanol and ethanol), acetone, formic acid, ethylene and propylene.

Generally, these compounds have a stimulatory effect on the soil microflora. In an in-vitro trial, they allowed growth of 6 out of 8 bacteria and 4 out of 6 fungi which were tested experimentally on a mineral culture medium deprived of carbon substrate (Schenck and Stotzky, 1975). The aldehydes constituted the principal sources of carbon; alcohols were little or not utilised. These molecules also stimulated germination of conidia and sclerotia in several fungal species, while others were unaffected; it was observed that the mycelium preferentially orients towards the emitting source. There seemed to be no marked qualitative differences in the emission of the volatile compounds among plant species. On the contrary, quantitative differences of genetic or physiological origin, can have significant consequences. For example, germination of the conidia of *Fusarium oxysporum* was inhibited by the presence of seeds of a lentil cultivar slightly sensitive to fusariosis but was stimulated by the presence of a susceptible plant cultivar. The susceptible cultivar secreted more ethanol, methanol and acetaldehyde than the resistant cultivar (Čatska and Vančurā, 1980). The age of the seeds can lead to physiological differences between two lots of genetically identical composition. With ageing, peroxidation of the lipid complexes present in the seed reserves liberates fatty acids which, at the moment of germination, are transformed by lipolysis into alcohols, aldehydes and ketones of lower molecular weight. For example, old pea seeds released 16 to 20 times more aldehydes than young seeds of the same cultivar, inducing a much more intense stimulation of the soil microflora (Harman et al., 1978). The concentrations of aldehydes sufficient to evoke such a 'stimulatory' response remained extremely low, of the order of 10^{-4} M.

Microbiological equilibria of the soil are therefore profoundly modified in the proximity of germinating seeds. The epidemiological consequences of these phenomena vis-à-vis damping off agents depend on the relative sensitivity of the species to the stimuli on the one hand, and to competition on the other. Thus, populations of *Pythium* were stimulated by the volatile exudates of pea sown in a sterile soil whereas in the same but non-sterile soil, such stimulation was counterbalanced by the rapid development of bacteria (especially pseudomonads) competing with *Pythium* (Norton and Harman, 1984).

Simultaneous with stimulating spore germination, these volatile compounds generally inhibit sporulation of the thalli. This double action could confer on sensitive organisms an ecological advantage: on the one hand, response to stimulation guarantees germination of spores only when a potential host is present and, on the other, inhibition (temporary) of sporulation allows dedication of all their energy to colonisation of their new substrate (Harman et al., 1980).

Effect of Volatile Compounds Emitted by Plant Debris

Decomposing plant debris, much less rich in nitrogen than organic animal waste, generally does not constitute an important source of ammonia, but does produce large quantities of alcohols (methanol and ethanol) and aldehydes (acetaldehyde, isobutyraldehyde, isovaleraldehyde, valeraldehyde, 2-methylbutanal). Pulverised and hydrated alfalfa is most commonly used in trials, all of which have shown its stimulatory effect on microflora, evidenced by rapid increase in respiration and number of bacterial and fungal populations (Linderman and Gilbert, 1973). This overall effect can be effaced, in some plants, by toxic compounds. The tissues of crucifers, as already noted, are rich in sulphur and while undergoing decomposition rapidly release various methyl sulphides and very likely isothiocyanates. Within 2 days these compounds attained sufficient concentrations in the soil atmosphere to irreversibly inhibit a fungal parasite such as *Aphanomyces euteiches* (Lewis and Papavizas, 1971).

Dry wood fragments also produce as yet mostly unidentified volatile products, which can distantly attract or repulse fungi (Mowe et al., 1983). For example, nonanal strongly stimulates the growth of wood decomposer basidiomycetes (Fries, 1973).

CLAYS

Because of their properties, discussed on page 7, (structural effect, buffering effect, storing effect), clays play a very important regulatory role in microbial life. The greater their exchange capacity, the greater their surface for interaction and ability to swell in the presence of water, and hence the greater their role in microbial life. This explains why a clay such as montmorillonite has a biological activity superior to that of kaolinite. The biological effects of clays can be highlighted by comparing soils of different texture or by experimenting with sandy soil gradually enriched with clay. Let us study a few examples.

Effect on Populations

The biomass of a sandy soil enriched with clay increases significantly, which enhances respiratory activity (Fig. 4.5). This augmentation occurs in both indigenous as well as experimentally introduced bacterial and fungal populations. For example, 225 days after the addition of an inoculum of *Fusarium oxysporum* f. sp. *lini* into a sandy soil, only 0.4% of the initial population could be found whereas in a similar sample enriched with montmorillonite, the remaining population was 46% (Amir and Alabouvette, 1993). When two previously sterilised soils of different textures were inoculated with 8 strains of fluorescent pseudomonads, it was observed that the density and structure of the bacterial populations differed significantly in the two soils after a stabilisation period of 3 weeks; the total bacterial population was 2.3 times greater in the soil containing more clay (Latour et al., 1999).

At equilibrium, clayey soils retain abundant reserves of organic substrates. It is also possible that the clay stimulates certain aspects of microbial metabolism. Furthermore, in a non-sterile soil, clay increases the survival level of bacterial populations. This protective effect is exercised in different ways. Clay can improve resistance to desiccation by ensuring more steady dehydration of the cell contents (Bushby and Marshall, 1977). It probably reduces diffusion and hence the inhibitory effect of toxins and antibiotics by

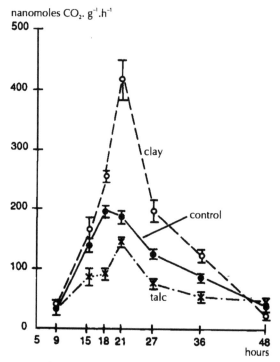

nanomoles $CO_2 . g^{-1}.h^{-1}$

Fig. 4.5 Modification in the kinetics of release of carbon dioxide when 25% montmorillonite is added to a sandy soil. The release of carbon dioxide was measured after the introduction of 1 mg glucose per g soil (Amir and Alabouvette, 1993).

adsorbing them (Campbell and Ephgrave, 1983). The presence of clay also results in the formation of a large number of microhabitats with access passages in the range of 3 to 6 μm, sufficient for the entry of bacteria but too narrow for the passage of protozoans, their primary predators in soil (Heijnen and Van Venn, 1991). The protective effect of clays against predators seems to be confirmed by the fact that development of a population of *Rhizobium leguminosarum* introduced into a sterile sandy soil was not at all modified by the addition of bentonite. But when protozoans were also introduced into the sterile samples, results similar to those observed in a non-disinfected sample were obtained: the bacteria decreased significantly in the control (sandy) group whereas the addition of clay allowed for their maintenance and, as a consequence, severe regression in the populations of amoebae and ciliates occurred (Fig. 4.6). Smectite (bentonite) provides better protection to bacteria than does non-swelling clay (kaolinite): once water is imbibed, bentonite offers more protective microsites than kaolinite.

Effect on Infectivity

A certain number of diseases, especially diseases of the vascular system caused by specialised forms of *Fusarium oxysporum*, are rarer or less serious

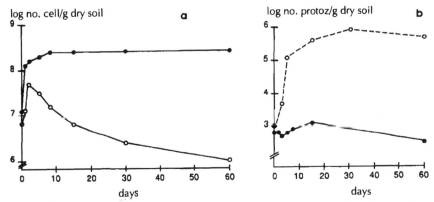

Fig. 4.6 Dynamics of populations of rhizobia and protozoans introduced as a mixture into a previously sterilised silty-sandy soil. Incubation at 15°C and 18% water content (after Heijnen et al., 1988).
(a) Development of population of *R. leguminosarum* biovar *trifolii* in an unchanged soil (O) and in a soil to which 10% bentonite (●) was added. Counting was done after tagging the bacteria using immunofluorescence.
(b) Development of population of flagellates and amoebae introduced along with rhizobia in sterile unchanged soil (O) and in a soil to which 10% bentonite (●) was added.

in soils rich in clays of the smectite family. It has been shown in most such cases that the effect is indirect, and due to the action of clays on the biological equilibria. These aspects are covered in the discussion of resistant soils in Chapter 6.

In bacterial wilting of tomato and egg-plant caused by *Ralstonia solanacearum*, the reduction in severity is apparently due to a direct effect on the pathogen. A series of tests done in the French Antilles established that severity of the disease, very variable depending on type of soil, was not related to the pH, but strongly dependent on the nature of the clays. Bacterial wilting was severe in soils rich in kaolinite but benign in vertisols containing more than 50% montmorillonite. The phenomenon was not due to biological equilibria unfavourable to the bacterial pathogen because it still manifested after autoclaving the soil 3 times at 120°C for 30 min at 24-h intervals. It is possible that montmorillonite traps some of the bacteria between its layers when swollen in the presence of water (Fig. 4.7). If moistening is followed by a period of dryness (hydric potential between –1 and –2.5 MPa), the bacteria apparently do not survive contraction of the layers. The soil is thus naturally purged during the dry season. Sometimes, however, through intensification of cultivation, irrigation is provided which maintains soil moisture even during the dry season, thus vitiating the curative effect and allowing re-establishment of *R. solanacearum* (Schmit et al., 1990).

Effect on Vertical Descent

Rainfall and irrigation would rapidly result in microorganisms moving deeper into the soil if there were no mechanisms to prevent it. Bacteria have

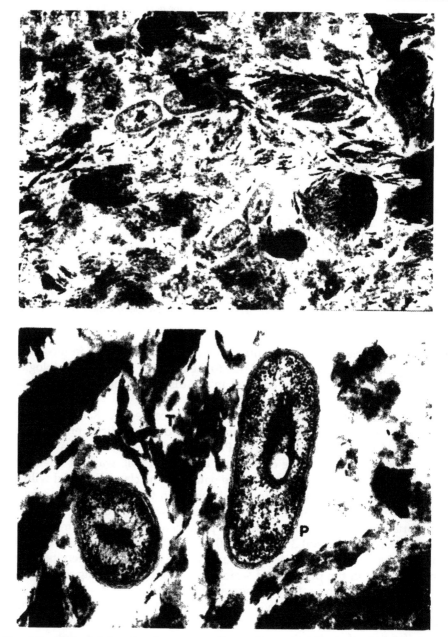

Fig. 4.7 Cells of *Ralstonia solanacearum* observed in montmorillonite extract of a soil from Guadeloupe, maintained at a hydric potential of −10 kPa (plates, J. Schmit, INRA).

Top: general view of bacteria in clay matrix.

Bottom: crystallites of montmorillonite associate to form deformable structures, the tactoids (T), which delimit the pores (P) in which the bacteria are surrounded. The network of tactoids narrows down or contracts when the clay dries up.

a great capacity for adherence to soil particles, including quartz grains and leaching is thus restricted, even in sandy soil. However, downward movement into deeper horizons is always less when the soil is enriched with clay since the possibility of adherence is increased and the diameter of the soil pores reduced where the cells are retained (Huysman and Verstraete, 1993).

TEMPERATURE

Temperature is a principal determinant of the speed of chemical reactions. It plays a role in the fluidity of the cell membranes and cell contents. At high temperatures proteins are denatured.

Soil temperature depends on the intensity of absorbed sun radiation. Temperature rises more or less rapidly depending on the degree of humidity and the nature of the plant cover. The amplitude of diurnal variations of temperature, which can be considerable in summer near the soil surface, rapidly diminishes with soil depth. But seasonal variations are perceptible at a great depth.

Direct Effect of Temperature on Microorganisms

Each organism is characterised by a minimum and a maximum temperature beyond which its development is no longer possible. Between these limits lies an optimal temperature at which growth is maximal. Therefore a certain geographic distribution of species is observed, determined by the balance between cardinal temperature and climate. For example, *Ralstonia solanacearum*, a bacterial parasite of the vascular system of the tomato plant in tropical and equatorial zones, is replaced by *Clavibacter michiganensis* in the temperate zone. Similarly, *Microdochium nivale*, the causative agent of winter rot in cereals exists (as denoted by its name) only in countries where winters are cold while *Sclerotium rolfsii*, parasite of a large spectrum of hosts, is apparently unable to withstand cold. Nevertheless, a large number of studies carried out in very different regions have led to the conclusion that the habitants of soil have very few climatic exigencies and their ubiquity is more the rule than the exception. To illustrate: specialised forms of one and the same fungus, *Fusarium oxysporum*, can parasitise the flax plant in northern Europe or the banana plant in Central America. A certain differentiation can occur in strains within one species depending on the climatic zone. One finds, for example, 'hot' and 'tepid' strains of *Pyrenochaeta lycopersici* (optimum 24 to 28°C) in the eastern Mediterranean while the majority of 'cool' strains of the same fungus (optimum 22°C) are found in temperate Europe (Clerjeau, 1976).

These observations lead to the conclusion that temperature, an essential component of climate, is not a factor likely to indefinitely conflict with the introduction of a microorganism into a new region. Thus, *Bradyrhizobium japonicum*, a symbiotic bacterium of soybean of subtropical origin, maintains itself perfectly in the Alsatian plains where the climate is continental.

Similarly, nothing is going to gainsay *R. solanacearum*, present in Morocco, finding its way into Europe via the areas of vegetable production located in southern Spain.

Microorganisms are able to subsist well beyond the extreme temperatures required for their growth even though there are lower and higher **lethal temperatures** beyond which they cannot survive. Most bacteria are generally killed around 90°C and fungi 65°C. These are only average values since the damage due to heat is cumulative, depending on both the temperature attained and the duration of exposure. Beyond a certain threshold, which varies according to the species, the organisms cannot recuperate. It is thus experimentally possible to construct curves for prediction of the survival level of the propagules of a fungus, for example, maintained for a specified period at a predetermined temperature (Fig. 4.8).

These results could go wrong, however, if the lethal temperatures are not attained rapidly. In fact, exposure of microorganisms to temperatures slightly less than lethal for some tens of minutes results in the appearance of **thermal shock proteins** whose synthesis is induced by special genes which seem to have been conserved in all the eukaryotes during evolution of the species. The resistance of microorganisms to lethal temperatures is closely associated with the synthesis of these proteins and does not take place if protein synthesis is prevented (Plesofsky-Vig and Brambl, 1985). The genes that determine synthesis of the thermal shock proteins also respond to stress other than heat. Thus the proteins produced in response to starvation also confer protection against elevated temperatures in the cells (Jouper-Jaan et al., 1992), the younger cells responding better than the older. In a population, only a few individuals seem capable of benefiting from pretreatment with sublethal temperatures. There is an exposure time for which their number attains maximum (Table 4.1).

The lethal temperature for active nematodes is in the range of 50–55°C (but encysted nematodes can survive exposure up to 80°C). Hatching of eggs, activity of juvenile stages and rate of reproduction depend largely on

Table 4.1 Influence of exposure time to a sublethal temperature on the survival level of propagules of *Fusarium oxysporum* f. sp. *dianthi* collected from cultures aged either 23 or 75 days (after Castejon-Munoz and Bollen, 1993)

Age of culture (in days)	Time of previous exposure at 45°C (in min)	Proportion of individuals surviving a 30-min exposure at 55°C
23	0	0.01
	30	0.14
	60	0.73
	90	0.58
75	0	0.01
	30	0.12
	60	0.15
	90	0.14

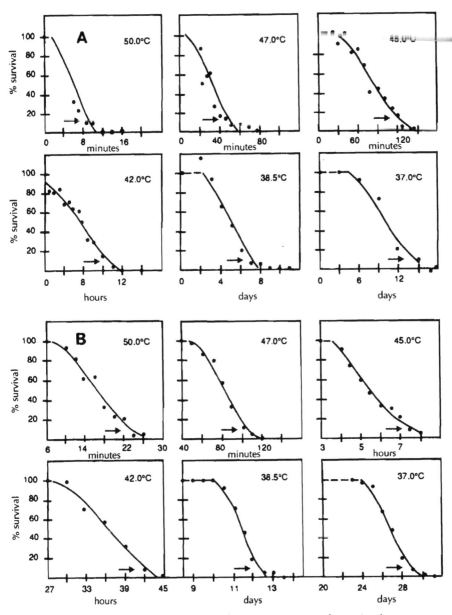

Fig. 4.8 Relationship between duration of exposure to a predetermined temperature and level of survival of the fungus *Verticillium dahliae* in a natural soil wetted to retention capacity (A) and in a petri dish in a nutritive agar medium (B) (Pullman et al., 1981). The arrow indicates the exposure duration required to kill 90% of the propagules: depending on the temperature, this duration can be expressed in minutes, hours or days.

temperature. In parasitic forms, the duration of larval survival before attachement to a host, an indispensable step for completion of their life cycle, depends on the reserves they possess. The rate of consumption of reserves increases with temperature; for example, the energy resources of infective larvae of *Meloidogyne javanica* remained quite significant after 16 days at 15°C whereas these reserves were exhausted at 30°C (Van Gundy et al., 1967).

Movement of motile microorganisms is often oriented to the temperature gradient. Protozoans and myxomycetes respond positively to variations of a few hundredths of a degree per centimetre. The nematode *Meloidogyne incognita* is sensitive to a temperature gradient of less than 10^{-3}°C cm^{-1} (Pline et al., 1988).

Effect on Microbial Equilibria

The antagonistic, competitive or mutually helpful relationships established between microbial populations and which condition their equilibrium, are closely dependent on temperature. Simply speaking, the result of competition between two species for a given substrate present in a limited quantity, at a given temperature, is dependent on their relative rate of growth at that temperature. A variation of a few degrees can completely reverse the equilibrium. For example, it was observed during in-vitro colonisation of root fragments of tomato with a mixture of *Fusarium oxysporum* (optimum: 28°C) and *F. solani* (optimum: 31°C) that colonisation by *F. oxysporum* predominated at 20°C and *F. solani* at 28°C (Table 4.2).

Table 4.2 Percentage of root-trap fragments colonised by a mixture of two species of *Fusarium* after 3 days of incubation in a sterile soil. Many fragments contained both species, which explains the total for each temperature being more than 100. Figures in parentheses are the limits of intervals of confidence at 5% (after Davet, 1976 a)

Incubation temperature	Percentage of root fragments colonised by	
	F. oxysporum	*F. solani*
20°C	100	57.4 (43–70)
28°C	75.9 (62–86)	92.5 (83–99)

The situation is obviously more complex under natural conditions as the success of a microorganism in competition depends not only on its rate of growth, but also on other factors, such as sensitivity to antibiotics or the ability to produce them. Moreover, the optimal temperature for synthesis of an antibiotic (or for any other function) may not necessarily coincide with the optimal temperature for growth.

This influence of temperature on the equilibria between microbial populations helps us understand why, within the same place and on the same substrate, the order of microbial successions can differ depending on the season in which the substrate is available. An example of analysis of such successions is given in the box below:

Seasonal Changes in Soil Temperature Regulates
Succession of Fungi on a Substrate

Roots of tomato cultivated in open air on the Lebanese coast show lesions of different aspect depending on whether the plants are cultivated in autumn or in spring. Irrespective of the season, the following fungi are readily isolated from these lesions: *Pyrenochaeta lycopersici, Colletotrichum coccodes, Rhizoctonia solani, Fusarium oxysporum* and *F. solani*. An experimental study of the relations among these fungi, and between this parasitic complex, the host plant and the general soil microflora, showed that equilibria vary depending on the temperature gradient corresponding to the period of tomato plant cultivation.

In spring, the soil is cool at the time of planting (16 to 18°C). At these temperatures *P. lycopersici*, whose competitive ability is very low, encounters little competition from *F. oxysporum, F. solani* and *Rhizoctonia solani*, all three of which require higher temperatures. Stimulated by root exudates, the propagules of *P. lycopersici* germinate rapidly and the fungus invades the roots. As the soil gradually warms up, the activity of the other three species increases. As a result, penetration of *P. lycopersici* into newly formed roots becomes more difficult and therefore more recent roots are attacked less. However, the already existing lesions, abundant on older, larger roots, are invaded by the fusaria and then by *R. solani*, which continue to cause damage. By the end of cultivation these roots show severe but atypical necrosis. Similar phenomena unfold with the other primary parasite, *Colletotrichum coccodes*, which begins to manifest at slightly higher temperatures.

At the beginning of autumn, on the other hand, the soil is warm (26°C) at planting time. *F. oxysporum, F. solani* and *R. solani* are abundant and active on the root surface but, less aggressive, do not penetrate them. Their presence impedes penetration of *P. lycopersici* and *C. coccodes*. Hence few lesions are observed at commencement of the crop. But as the soil gets colder, the phenomena of competition attenuate and the new roots are less protected. *P. lycopersici* then becomes dominant in the tissues and the delimited brown rot characteristic of this fungus appears on the young roots. *C. coccodes*, less active at temperatures below 20–22°C, remains discrete in this season despite lessening of competition (Davet, 1976b).

Effect on Equilibrium among Microbes and Plants

Analysis of the effect of temperature on the infectivity of a pathogen or a symbiont is dicey: it may be a direct effect on the development of the organism or, as seen above, be an indirect effect on the microbial equilibria outside the host plant; it can also influence the plant-parasite relationship. Thus, even though the optimal temperature of *Verticillium dahliae* is high and close to 28°C, all the host cultivars of cotton become resistant at 32°C, even those lacking the genes for resistance. Contrarily, even resistant cultivars are susceptible at 22°C. The reason is that production of phytoalexins by the cotton plant, very weak at 22°C, augments with rise in temperature to about 32°C, and then effectively counters vascular invasion by *V. dahliae* (Bell and Presley, 1969). Similarly, the 'Gros Michel' variety of banana susceptible to fusariosis, becomes resistant around 34°C because the speed of its reaction to the invasion increases and becomes sufficient to arrest development of the pathogen (Beckman et al., 1962).

In plants which possess a gene responsible for hypersensitivity reaction, an increase in temperature often, contrarily, provokes loss of resistance. The critical temperature is lower for heterozygotes than for homozygotes.

It happens therefore that temperature affects the plant-microorganism relations in a manner not directly forseeable when the behaviour of each partner is considered discretely.

XENOBIOTIC COMPOUNDS

Generalities

Industrial countries manufacture and utilise greater and greater quantities of synthetic organic compounds that accidentally or intentionally are emptied into the soil or into water courses. These products, which do not exist in a natural state, are called **xenobiotics**. If they remain long enough on the soil surface, a part may be lost through volatilisation or by photochemical degradation. The remainder may be carried away by rain or irrigation water. But these molecules, usually large in size and carrying strongly reactive chemical groups, are not readily washed away: the soil behaves towards them somewhat like a chromatographic column and more or less adsorbs them onto the humus-clay complex. Among the pesticides for example, benomyl migrates very slowly and stays in the top few centimetres of soil; organo-mercuric compounds contrarily traverse the soil profile rapidly, while iprodione and vinclozolin are intermediate in behaviour. The distribution of these products in the soil profile is thus far from homogeneous. It depends largely on the nature of the molecules and the adsorption capacity of the soil.

Maximum Depressive Effect and Recovery Time

Introduction of a xenobiotic compound into the soil can have important ecological consequences. The effects on the microorganisms can be estimated by counting the populations, or measuring a particular activity, which can be general such as respiration, or more or less specialised such as nitrification or cellulose degradation. If the compound leads to disappearance of all the microbial populations under study or cessation of all activity, it is easy to draw conclusions regarding its toxicity! This situation is nevertheless very rare and it is important that observations be carried out over a sufficiently long period, minimally 2 months. Usually a reduction in population members or in the specific activity selected as a criterion is observed. If progress of the phenomenon is studied over time, it can be established that the depressive effect reaches maximum, then gradually declines. After a more or less long period of time, a return to the initial situation is observed. Far more than the maximal depressive effect (which can be considerable even in perfectly natural situations: after a period of drought for example), it is the **recovery time** that is important in judging the degree of harm done by a xenobiotic compound to the microbial groups under consideration (Fig. 4.9).

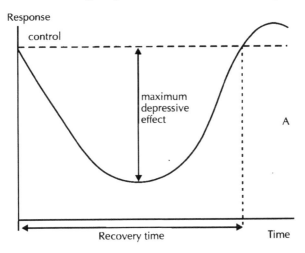

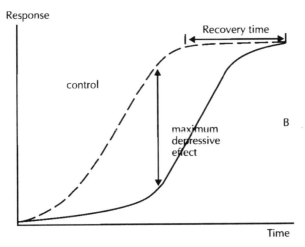

Fig. 4.9   Types of response to the stress exerted by the environment. In the case of curve A it could be a variation in the biomass. In curve B the variable measured could be the consumption of a substrate or the synthesis of a metabolite. Knowing the response of a control not subjected to stress, it is possible in both cases to define the maximum depressive effect, enabling assessment of the amplitude of disturbance; and a recovery time, which allows a more precise estimation of its gravity (Domsch et al., 1983).

With respect to depressive effects of natural origin, if the recovery time is less than 1 month, the consequences for the ecosystem can be considered negligible; between 1 and 2 months, tolerable; and above 2 months, definitely toxic (Domsch, 1985). The experimental conditions must certainly be taken into account while interpreting the results: the pH, temperature and water content of the soil have to allow normal microbial activity. It is particularly important to ensure that the recuperative capability of the populations is not limited by insufficient availability of metabolisable organic substrates.

Effect of Pesticides

Pesticides represent, by far, xenobiotic compounds that are the most systematically introduced into the environment and the most extensively used in a variety of cultures. Herbicides, nematicides, insecticides and fungicides are either buried directly in the soil or spread on its surface, or sprayed on the leaves, of which at least some part falls back immediately on the soil or after rainfall. The quantities frequently used for aerial treatment range from 0.5 to 5 kg active ingredient per hectare. Assuming that the product is carried downwards and is uniformly distributed in the ten top centimetres of soil (as already mentioned, this is only an approximation), this represents concentrations of 0.5 to 5 mg per litre of soil, corresponding approximately to the range of activity of the preparations.

Despite the significant quantities used, very little is known about the secondary effects of pesticides on soil microorganisms. Only the impact of very old products and those of large-scale usage has been studied and given the great diversity of commercial products, it is difficult to draw general conclusions about their effects on soil. Therefore just a few examples are presented here.

The more generalised the nature of the biological activity, the less it seems disturbed. In fact, the disappearance of certain species of microorganisms is rapidly compensated by the development of others not sensitive to treatment in such a way that the overall effect of the pesticides on such phenomena as absorption of oxygen or the release of carbon dioxide becomes insignificant. The effect of treatments on the decomposition of straw is likewise generally of little importance because a large number of different microorganisms possess cellulolytic enzymes.

On the other hand, measurements of specific activities relevant to a particular flora represented by a small number of species reveal many noteworthy variations. Numerous products have been shown to be toxic for mycorrhizal fungi and rhizobial nitrogen fixators associated with the roots of leguminous plants; it is important to bear this in mind when selecting seed treatments. Nitrification is disrupted, sometimes strongly, by most pesticides (except herbicides). Ammonification contrarily is generally stimulated. This could be the indirect effect of addition to the soil of fresh organic matter of plant (herbicide), animal (insecticide and nematicide) or fungal origin (fungicide). Use of pesticide can also have consequences for the development of diseases, probably an indirect effect (change in microbial equilibria, action on host plants) rather than a direct one on the parasites. For example, overuse of benomyl to eradicate grey rot (*Botrytis cinerea*) in a sheltered culture would lead to concomitant elimination of antagonistic flora and thereby augment attacks by *Pythium* and *Phytophthora* which are not sensitive to this fungicide. Trifluraline, a popular herbicide, augments attacks by *Rhizoctonia solani* on kidney-bean, due not to stimulation of the fungus, but rather inhibition of phytoalexin production by the plant (Romig and Sasser, 1972).

OTHER FACTORS

Metallic Ions

Metallic ions present in excess can have an inhibitory effect on some microorganisms, especially in acid soils where their solubility is higher. For example, in volcanic soils of the Hawaiian islands or Japan in which the pH lies between 4.2 and 6, exchangeable aluminium represents up to 10 meq per 100 g soil, whereas in soils with a pH between 6 and 7 this element is present only in traces. The presence of Al^{3+} ions can strongly inhibit germination and mycelial growth of certain fungal pathogens, for example *Rhizoctonia solani* (Kobayashi and Ko, 1985), *Chalara elegans* (Meyer et al., 1994) and *Fusarium solani* f. sp. *phaseoli* (Furuya et al., 1999). Elevation of the pH attenuates the phenomenon by making part of the toxic ions insoluble, which then precipitate in the form of hydroxides. The toxicity of heavy metals is particularly well known, due largely to their affinity for the –SH groups present on enzymes, which results in inactivation of the enzymes binding to metallic ions. The consequences of heavy metals for microbial life in soil are still controversial, however. Several studies reveal the serious inhibitory effect of heavy metals vis-à-vis mycorrhizal fungi (for example, Hepper and Smith, 1976) or rhizobia: in soils contaminated for several successive years by waste slush used as an organic amendment, it was found that the genetic diversity, host spectrum and efficiency of *R. leguminosarum* populations had markedly reduced (Hirsch et al., 1993). However, due to the formation of very stable complexes between the humic substances and the heavy metals, it was very difficult to ascertain the 'biologically active' concentration of metals in the medium. On the other hand, it can be said that the microorganisms, especially the bacteria, exhibit remarkable possibilities of adaptation. For example, *Sinorhizobium meliloti* populations isolated from a highly contaminated area in the proximity of a zinc mine were able to tolerate concentrations of zinc, copper, nickel and cadmium 1000 times greater than concentrations effectively present in the soil solution (El Aziz et al., 1991). Tolerance to heavy metals seems in some cases to pre-exist selection pressure. Such tolerance may be due to the impermeability of the bacterial polysaccharide capsules which currently surround bacteria. Chelation of cations by soil organic acids (citric, oxalic . . .)

or by specific proteins inside cells are other possible mechanisms. Some bacteria and yeasts detoxify mercuric salts by reducing the metallic cation to the elementary state, followed by disappearance of mercury from the environs through evaporation (Gadd and Griffiths, 1978). The rapidity with which resistance is acquired by populations of Gram-negative bacteria is an argument in favour of the intervention of plasmids. Such plasmids, which are very bulky, have been demonstrated in *Alcaligenes eutrophus*, capable of withstanding various heavy metal concentrations of several millimoles (Lattudy, 1990).

Examples of resistance to heavy metals are also known in fungi. For example, *Oidiodendrum maius*, a fungus that forms ericoid mycorrhizae (see page 236) with bilberry (*Vaccinium myrtillus*) can grow even in terrains severely polluted with industrial residues of zinc and cadmium. Unlike the strains present in healthy soil, which are inhibited in polluted soils, the mycelia of isolates from toxic soil are covered with very abundant, very strongly pigmented mucilage (Martino et al., 2000). The mucilage and pigments could contribute to detoxification of the medium by chelating the heavy metals. Melanin has many active sites capable of fixing cations. Rizzo et al. (1992) showed that melanised rhizomorphs of many species of *Armillaria* were surficially covered with almost a sheath of metallic ions (including lead, copper or zinc) adsorbed at concentrations 4 to 96 times higher than that in neighbouring soil, and remained unaffected. This covering, strictly external, may protect the rhizomorphs and eventually other melanised structures too, such as the sclerotia or certain chlamydospores, from microbial competitors.

At non-inhibiting doses, heavy metals can also affect the equilibria among microorganisms. For example, the auxiliary bacterium *Pseudomonas fluorescens* CHAO is normally not very effective against *Fusarium oxysporum* f. sp. *radicis lycopersici*, a pathogen of tomato collar and roots. The fusaric acid secreted by the fungus, while not affecting growth of the bacterium, does inhibit production of 2,4-diacetylphloroglucinol, an antibiotic. Small concentrations of zinc (from 10 µg ml^{-1}) can significantly increase the antagonism of *Pseudomonas* against *Fusarium* indirectly: zinc suppresses the secretion of fusaric acid, thereby facilitating synthesis by the bacterium of 2,4-diacetylphloroglucinol, itself an inhibitor of the fungus (Duffy and Défago, 1997).

Pressure

One factor rarely taken into account is the pressure exerted by the thickness of the soil layer. Punja and Jenkins (1984) carried out studies by placing adjustable weights on the sclerotia of *Sclerotium rolfsii* in such a way that a pressure equivalent to that of a column of soil a few centimetres in height was exerted. Under these conditions the sclerotia exuded significant quantities of hydrates of carbon and amino acids. Having thereby lost their nutritive reserves, the sclerotia were no longer capable of germination (in the absence of an exogenous energy supply), which could explain why their spontaneous germination under natural conditions is possible only in superficial soil layers.

For Further Information on the Effect of Environment on Microorganisms

Atlas R.M. and Bartha R. 1998. Microbial Ecology: Fundamentals and Applications. Benjamin Cummings, Redwood City, CA.

Coleman D.C. and Crossley Jr. D.A. 1996. Fundamentals of Soil Ecology. Academic Press, NY.

Dix N.J. and Webster. J. 1994. Fungal Ecology. Chapman & Hall, London, UK.

Durrieu G. 1993. Ecologie des Champignons. Masson, Paris.

Griffin D.M. 1972. Ecology of Soil Fungi. Chapman & Hall, London, UK

Hattori T. and Hattori R. 1976. The physical environment in soil microbiology: an attempt to extend principles of microbiology to soil microorganisms. CRC Crit. Rev. Microbiol. 4: 423-461.

Lynch J.M. and Hobbie J.E. (eds.). 1988. Micro-organisms in Action: Concepts and Applications in Microbial Ecology. Blackwell Sci. Publ., Oxford, UK.

Marshall K.C. 1975. Clay mineralogy in relation to survival of soil bacteria. Annu. Rev. Phytopathol. 13: 357–373.

Stotzky G. 1980. Surface interactions between clay minerals and microbes, viruses and soluble organics, and the probable importance of the interactions to the ecology of microbes in soil. In: Microbial Adhesion to Surfaces, pp. 231–249. R.C.W. Berkeley, J.M Lynch, J. Melling, P.R. Rutter and B. Vincent (eds.). Ellis, Horwood, Chichester, England.

Vyas S.C. 1988. Nontarget Effects of Agricultural Fungicides. CRC Press, Boca Raton, FL.

Watanabe I. and Furusaka C. 1980. Microbial ecology of flooded rice soils. In: Advances in Microbial Ecology, vol. 4, pp. 125-158. M. Alexander (ed.). Plenum Press, NY.

Part II
Effects of Microorganisms

5

Modifications of Physico-chemical Characteristics of the Environment due to the Effect of Microorganisms

EFFECTS ON pH

Due to their action on mineral or organic compounds, microorganisms can modify the pH of the soil in a localised way or, on the contrary, in such a manner that the consequences can be significant for the entire ecosystem.

Acidification of Environment

Sulpho-oxidation

The sulphides and reduced sulphur of organic compounds are oxidised in the soil into polythionates and then into sulphates by the action of bacteria and fungi, and the resultant SO_4^{--} ions cause a local lowering of the pH. The organisms most active in this process are the widely distributed aerobic chemolithotrophic bacteria, the *Thiobacillus*. In a soil in equilibrium, these sulphates are in turn reduced and incorporated into organic compounds by plants and microorganisms (see page 150). This could, however, result in the accumulation of SO_4^{--} ions and thereby cause severe acidification of the environment under two particular situations:

— *Massive addition of sulphur*: These additions could be intentional, for the purpose of lowering the pH (see page 293). But accumulation of sulphur in the soil could also be the unwanted result of spraying done to protect the aerial parts of plants against oidia. Use of sulphur was drastically reduced once organic fungicides were developed. Meanwhile, since no other remedy was available, significant quantities were used, year after year, especially in the vineyards of southern France. It is thus known that between 1950–1960, some fields were so replete with sulphur that the soil became uncultivable, with pH values sometimes falling below 3.

— *Draining of hydromorphic soils*: As already mentioned (chapter 4), hydromorphic soils rich in organic matter contain significant quantities of sulphides. When these soils are drained for cultivation, spontaneous and biological oxidation of sulphides leads to their rapid and significant acidification. Polders, old rice fields or mangroves can only be cultivated after the addition of limestone, which increases the pH.

Nitrification

Biological oxidation of ammoniacal salts results in the formation of nitrates. Resultant reduction in pH is only transient, however, because these nitrates are themselves very rapidly absorbed by the plants, reduced by denitrifying microflora or leached.

Synthesis of Organic Acids

A large variety of organic acids appear during the course of mineralisation of organic matter. Many, produced during hydrolysis of lignin, contain aromatic nuclei and are formed under aerobic conditions. Others, such as fatty acids, appear mainly under anaerobic conditions. Because of the acidity of the COOH groups formed in a reducing medium, peat cannot be utilised as a substrate for cultivation until subjected to drastic neutralisation measures.

Some fungal parasites have very high acidifying power: *Sclerotium rolfsii* or *Sclerotinia minor* for example, synthesise oxalic acid which, in addition to having a toxic effect on the host plant, also lowers the pH to a value favourable to the action of their hydrolases (Bateman and Beer, 1965).

Alkalinisation of the Environment

Hydrolysis of proteins and, more generally, of nitrogenous organic compounds (urea for example) results in the formation of ammonia, which raises the ambient pH. Addition of urea can increase the pH locally to 8 or 9. This elevation is normally counterbalanced by the activity of nitrifying flora, but this flora being more susceptible to unfavourable conditions than ammonifying flora, reestablishment of the initial equilibrium is sometimes slow.

EFFECTS ON SOIL STRUCTURE

As already seen (page 10), the quality of soil structure depends on the stability of its aggregates. This stability itself is largely the result of the activity of microorganisms.

Binders of Microbial Origin

A large number of bacteria and unicellular algae are encircled by a thick capsule consisting of glycoproteins and polysaccharides. A mucous layer similar in nature often covers the fungal hyphae also. The role of these polymers in the constitution of microaggregates, long suspected, has now been confirmed.

The degree of aggregation of a soil increases as soon as cultures of bacteria or yeasts are introduced, which is proportional to the quantity of cells added (Lynch, 1981). Similarly, very significant improvement of the soil structure is noticed one week after the addition of glucose. Marking the glucose with ^{14}C helped demonstrate that the compounds responsible for the increment in level of stable aggregates are essentially the polysaccharide chains newly synthesised by the microorganisms (Guckert et al., 1975). A similar stabilising effect is obtained even if only polysaccharides extracted from microbial cultures are added to the soil. If the soil is treated with periodic acid (which opens the cyclic bonds in the sugars), followed by alkalinisation, fragmentation of the polysaccharides into simpler molecules results and simultaneously the stability of the aggregates is destroyed. Lastly, observations using an electron microscope have shown that elementary particles of clay attach to the mucus present on the surface of the hyphae and on the bacterial colonies. The bond between these clay particles and the mucilage seems, at least in part, assured by cations because the effect of the addition of polysaccharide extracts is more prolonged in the presence of Al^{+++} and Ca^{++} ions (Tisdall, 1991).

The polysaccharide glues play a role in the initial stages of aggregate formation. After this phase of aggregation there is a stabilisation phase during which part of the synthesised compounds is biodegraded while another part is incorporated into much more stable humic compounds. The binders formed from the readily usable substrates, such as glucose, are shortlived compared to the cements synthesised from substrates difficult to degrade (lignocellulose plant fragments for example). A third stage involves the collection of several microaggregates into one macroaggregate. The very fine roots and filamentous fungi play a significant role at this stage.

Role of Mycelial Hyphae

Observations made using optic and electron microscopes have shown that the elementary particles can be trapped by mycelial hyphae as though caught in a net (Fig. 5.1). Studies carried out using soil from the great Canadian prairie indicate a clear relation between aggregate size and the fungal biomass

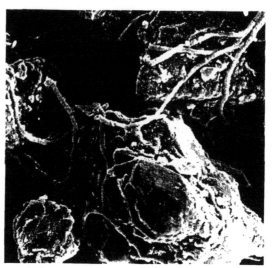

Fig. 5.1   Network of mycelial filaments clustering mineral and organic particles (plate, B
Tivoli and E. Lemarchand, photo library, INRA)

associated with the aggregates: the maximum biomass is associated with
macroaggregates between 0.25 and 1 mm in diameter, whereas microaggregates carry very few filaments (Gupta and Germida, 1988). When land
is cultivated, a marked decrease in number of macroaggregates occurs, which
is attributable largely to destruction of the network of mycelial filaments by
ploughing.

These filaments belong to saprophytic fungi proliferating on organic
fragments. But these substrates are rapidly used up and it appears almost
certain that most of the mycelial clusters effecting cohesion of the
macroaggregates are produced by endomycorrhizal symbionts (see page
233). This could explain why the macroaggregates are much more abundant
in the proximity of the roots than in the rest of the soil. It could also explain
why the structure of a soil continuously maintained under grass is generally
superior to that of a cultivated soil (alternately covered and barren), which
per se contains more macroaggregates than soil maintained fallow (Tisdall,
1991). On average, the mycelium of an endomycorrhizal fungus extends 6
to 9 cm from the roots. But the significance of the mycelial network depends
on the type of plant: endomycorrhizal fungi of C_4 plants have better
developed filaments than C_3 plants and provide better stability to the soil
(Miller and Jastrow, 1990).

Biological Ameliorators of Structure

The growing importance of the phenomenon of erosion, related to
degradation of the structure of cultivated soils, has led to research on soil
restructuring agents. Effective synthetic soil-improving agents are generally
derivatives of petroleum products (soil conditioners) but their high cost
rarely justifies utilisation of them in open fields. Therefore research is oriented

wards discovery of biological ameliorators that could improve soil
ructure.

Among bacteria, the number of possible candidates is large. For example,
ytophaga of Myxobacteriales transform cellulose waste into a polysaccharide
lly within about 10 days, which confers lasting stability to the soil. Some
nicellular algae seem to be of much greater interest, however.

Some chlorophycophytes (*Asterococcus* and *Chlamydomonas*) are
urrounded by a mucilaginous capsule constituted of polysaccharides of
ery high molecular weight (greater than 20,000), whose mass could represent
p to 75% of their total dry mass. Compared to most other bacteria, these
nlorophycophytes have the advantage of being less susceptible to acidity
nd multiply readily at pH in the range of 6 to 8 (Barclay and Lewin, 1985).
Infortunately, algae have a very high requirement for moisture. They cannot
nultiply properly unless the water content of the soil is close to its water-
etention capacity. As they only develop on the surface, they are especially
usceptible to desiccation. It is therefore impossible to think of their being
sed except in highly densely sown irrigated crops whereby the soil is
rotected against very severe variations in water content. A few trials have
een promising, however. For example, two species, *Chlamydomonas mexicana*
nd *C. sajao*, were spread surficially on maize fields at the rate of 5×10^{11}
ells per hectare (i.e., 7.8 kg ha^{-1} dry mass). The dose was applied in the
orm of two sprayings—beginning of spring and during spring—for three
uccessive years. A significant improvement in the resistance of aggregates
o dispersion was observed from the very first year. At the end of three
ears, soil cohesion had increased by more than 20% at the surface and 10 to
2% in the top 30 centimetres (Metting, 1987).

Undoubtedly, a better knowledge of the ecology of endomycorrhizal
ymbiotic organisms could, in the near future, lead to better utilisation of
hese organisms for consolidation of aggregates: conceivably, plants could
e grown along with strains which could develop into the densest mycelial
networks.

MINERAL CYCLES

Microbial communities play an absolutely major role in the recycling of
organic matter. They ensure renewal of the supply of most of the mineral
ons of the soil. But this function is not without vested interest as the
microorganisms generally begin by using the minerals for themselves. While
storing part of the mineral elements for their own requirement, they can
enter into competition with the plants. This immobilisation is nonetheless
always temporary. In a well-equilibrated soil, the microbial biomass behaves
ike a **storehouse of mineral elements**: it retains them in the upper soil
ayers, protects them from leaching and gradually supplies them to the
plants. A study of each of the major mineral cycles could constitute a large
reatise. Obviously, it is not possible to dedicate more than a few pages in
his book to these mineral cycles and only the essential aspects are covered.

Carbon Cycle

The natural supply of fresh organic matter of animal origin (excreta an cadavers) represents less than 10% of the total residues accumulated on th surface of the soil and these wastes, relatively poor in carbon, contribut only a small part of the supply of organic carbon. Mineralisation of organi carbon therefore essentially involves substrates of **plant origin** (Fig. 5.2). is estimated that the quantity fixed annually by photosynthesis on the surfac of the Earth is 70 billion tons of carbon per year (Paul and Clark, 1989 Immobilisation and mineralisation of carbon were in equilibrium until mid 19th century. Around 1850, the carbon dioxide content of the atmospher was around 280 ppm. It has crossed 367 ppm at the present time, whic represents an increment of more than 25% in less than 150 years. Thi significant increase is due to industrialisation and combustion of carbo fossil reserves' which liberate 6 billion tons of carbon per year. In the las few years, deforestation of intertropical regions (17 million hectares pe year) represents an annual supplementary supply of 1.7 billion ton (Goudriaan, 1992) and this number does not take into account the majo fires that ravaged the forests of Indonesia in 1999, Canada in 2000 and th USA in 2002.

The carbon from organic matter transformed by microbial activity ca follow three different paths. It can go back into the atmosphere in the forn of carbon dioxide and, to a much lesser extent as methane, or be assimilatec and transformed into biomass, or incorporated into humic substances. W shall briefly study these three pathways.

Atmospheric Methane

The atmosphere contains nearly 1.7 ppm methane—nearly 200 times less than the carbon dioxide content. But these traces of gas are of considerable importance since it is estimated that methane contributes 10 to 15% to the greenhouse effect. Just like the carbon dioxide content, the methane content of the atmosphere has steadily risen in the last few decades. The increase has been 1% per annum over the last ten years (Blake and Rowland, 1988) and the present concentration represents more than twice that in air bubbles trapped 200 years ago in polar ice. This increment in methane level could be due to multiplication of herds of ruminants (the rumen is a very important source of methane), intensification of rice cultivation and increment in number of rubbish dumps. Systematic addition of nitrogenous mineral manure to cultivated land could also entail a lessening of soil capacity to biologically oxidise methane (Hütsch et al., 1993).

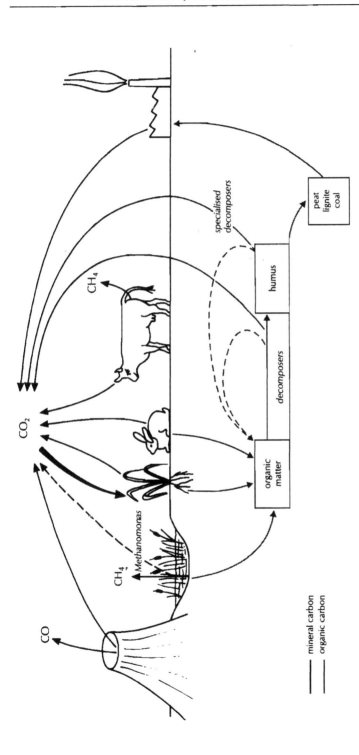

Fig. 5.2 The carbon cycle. This diagram does not take into account the oceans, which play a significant role in the balance of carbon dioxide, and which produce methane.

The Greenhouse Effect

The main gases present in the terrestrial atmosphere responsible for the greenhouse effect are carbon dioxide and methane. The oxides of nitrogen and chlorofluorocarbons are also contributors. These gases absorb part of the solar radiations that the soil deflects back into space. Without them, the average temperature of the Earth would be –18°C. Therefore, these gases are necessary for maintaining life. But the increase in concentration of all of them over the last 150 years is worrisome because the atmosphere has heated up, which could lead to grave consequences. This increment is clearly related to industrial activity and intensification of agriculture. The Kyoto protocol of 1997 stipulated that the major industrial countries should engage in checking and then reducing their emission of gases responsible for the greenhouse effect: between 1990 and 2000, however, these emissions were not reduced in the European Union, and even rose in the United States by 11%!

Decomposition and Mineralisation of Organic Substrates

Plant material (already partly colonised by epiphytic microflora) is rapidly invaded as soon as it arrives on soil. Soluble substances and those of lower molecular weights (Table 5.1) are consumed first and rapidly mineralised by a very varied flora, mainly constituted of consumers of simple sugars whose population declines as soon as these substrates are exhausted. Hemicelluloses, and especially cellulose, the major constituent of plant cells, are available only to a specialised flora. Cellulolytic bacteria and fungi are provided with an enzymatic complex that always comprises enzymes acting synergistically (Fig. 5.3). In aerobic microorganisms, of which *Trichoderma reesei* is a particularly well-known representative because of its use in the chemical industry, these enzymes are extracellular and released in the environment. In anaerobes (present in compost, muds and the rumen of ruminants) the cellulolytic enzymes are generally grouped on a common protein support affixed to the surface of the cells. Aerobic degradation of cellulose produces simple soluble glucides (cellobiose and glucose) which support a new sugar-consuming flora, different from the first because subjected to more severe competition. The end products of cellulose degradation inhibit the functioning of cellulases. By consuming them, the associated non-cellulolytic flora allows continuation of the activity of cellulolytic organisms. Under anaerobic conditions ethanol, organic acids, carbon dioxide and methane are produced. When these gases pass through less anoxic zones, the methane can be oxidised into carbon dioxide by bacteria close to chemolithotrophs (*Methanomonas*) and by nitrifying bacteria (*Nitrosomonas*).

Table 5.1 Main categories of organic substrates susceptible to decomposition by microbial activity

	Compounds not containing nitrogen	Nitrogenous compounds
Soluble compounds	simple sugars organic acids	amino acids
Slightly soluble or insoluble compounds easily degraded	fatty acids glucans (fungal walls) starch pectin	peptidoglucans (bacterial cell walls) chitin (animals and fungi) proteins
Insoluble compounds difficult to degrade	hemicelluloses (pentosans and hexosans) cellulose lignin suberin	microbial pigments

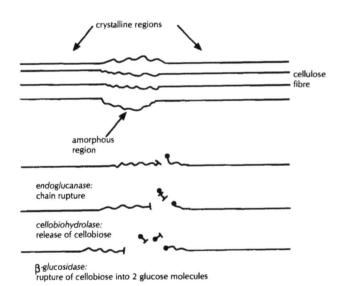

Fig. 5.3 Scheme of stages in enzymatic degradation of cellulose according to Montenecourt and Eveleigh (1979). A cellulolytic species generally holds several endoglucanases and several cellobiohydrolases of different structure, which act synergistically.

After cellulose, lignin is the most important component of plant tissues and is the most recalcitrant. Its degradation is possible only under aerobic conditions and apparently requires the presence of glucidic nourishment, as though an energy supplement were necessary. It is probably involved to some extent in the processes of cometabolisation. Unlike the polymerised compounds such as pectins, pentosans, hexosans or cellulose, lignin is not a polymer despite its high molecular weight. It is a heteroclitic congregation of analogous but non-identical compounds with phenolic nuclei bearing a lateral chain of three carbons (Fig. 5.4), to which sugars and amino acids

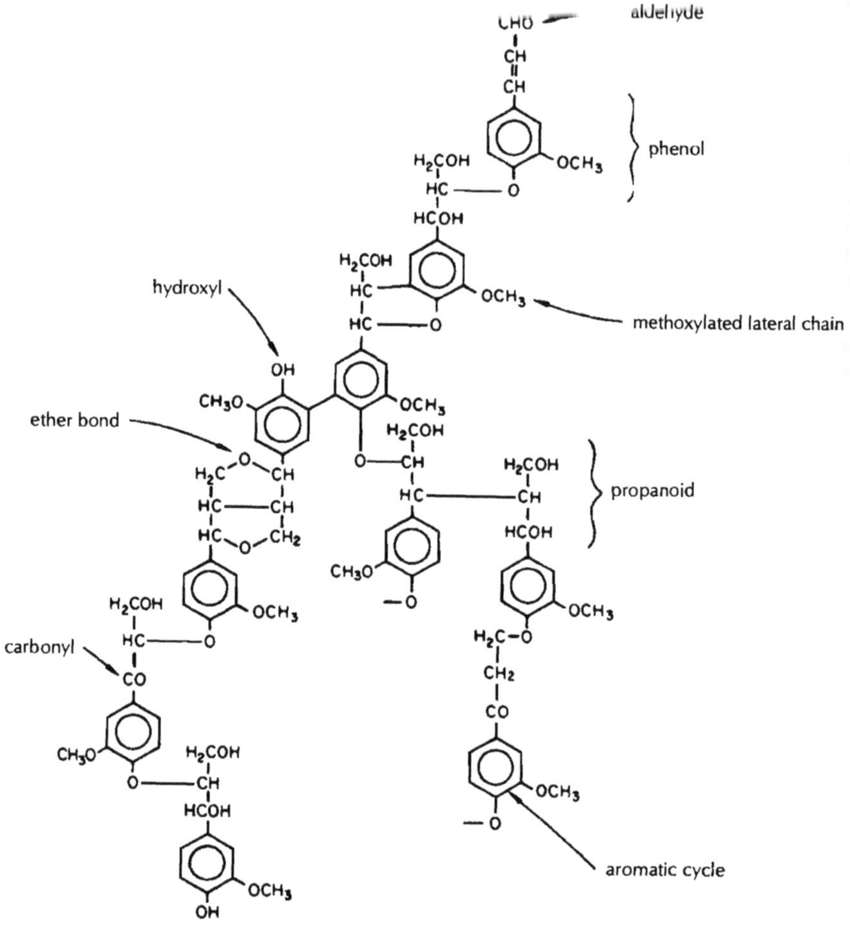

Fig. 5.4 Scheme of lignin, showing different examples of elementary molecules (phenylpropanoids) and the possible bonds (Paul and Clark, 1989).

may be attached. Thus there is not one lignin, but a quantum of different lignins depending not only on botanical species, but also on the time of their formation in a particular plant. Their decomposition is due to fungi which, in the presence of a feebly active glucidic microflora, consume the polyosides by scouring the lignin coating which protects the cellulose. The substrate transforms into a whitish fibrous mass consisting of a cellulose framework. Higher basidiomycetes as well as a few ascomycetes are responsible for this white rot. Imperfect fungi and ascomycetes, along with bacteria, are mainly involved in soft rots. In this process, the polysaccharides are rapidly utilised, whereas the lignin is subjected to partial alterations that release phenolic elements. When oxidised, they become brown and impart a dark colour to the disorganized mass that appears.

The principal ligninolytic enzymes in fungi responsible for white rot are the extracellular peroxidases. Besides lignin, they are capable of degrading numerous insoluble polycyclic aromatic hydrocarbons of industrial origin. Such fungi could therefore be used to depollute contaminated soils. One fungus very promising for this purpose is *Phanerochaete chrysosporium*.

Immobilisation of Carbon in the Biomass

Organic substrates serve two functions: one part is used for redox reactions and is finally released in the form of carbon dioxide; another part, more or less rearranged, is incorporated and utilised for the formation of cellular components. There is also a third part that serves neither of these two purposes and constitutes a useless waste for the bacteria, amoebae and fungi.

The efficiency of a microorganism is the ratio between the carbon it has successfully incorporated into its cytoplasm and the carbon initially contained in the substrate. This ratio depends, of course, on the quality of the organic matter under consideration: it will be much higher if the substrate is glucose rather than a lignocellulosic fragment. Efficiency also varies depending on the type of microorganisms: it is rather high in fungi (between 35 and 55%), low in aerobic bacteria (less than 10%) and really low in anaerobic bacteria (from 2 to 5%).

The fraction of the carbon substrate that has neither been mineralised nor assimilated by one species most often constitutes the food of choice for a second species. This second species in turn assimilates part of the leftover carbon, oxidises another part and discards the less useful fractions. A large variety of species thus follow, their number increasing depending on the complexity of the substrate (Fig. 5.5). The organic matter is thus gradually changed and finally completely disappears or, if it is very ligneous, a few non-assimilable aromatic residues remain which enter the humic compounds.

If the overall effect of all individual microbial activities is taken into account, it can be approximated that half of the carbon of an organic substrate is released in the form of carbon dioxide and the other half incorporated in the cells. But living matter is made up of elements other than carbon. For example, nitrogen is also indispensable in the ratio of 1 atom nitrogen per 10 atoms carbon. Thus consumption of a substrate containing 20 atoms carbon, of which 10 are oxidised and 10 incorporated, requires 1 atom of nitrogen. If the C/N ratio of the substrate is higher than 20, the microorganisms will take up nitrogen from the external medium, i.e., the soil solution, which is detrimental to the plants. This is what happens when an amendment rich in cellulose, such as straw, is buried in the soil (C/N ratio greater than 100). An addition of nitrogenous manure may then be necessary to avoid this competitive effect. This immobilisation can prove beneficial, however, in the absence of cultivation: from autumn to spring for example. Under these conditions leaching of soil nitrogen is prevented. This retention is only transitory since, after the substrate is consumed, the microbial

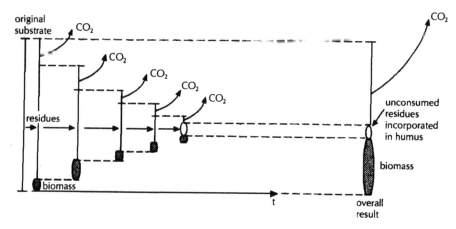

Fig. 5.5   Successive stages in the decomposition of a carbon substrate. The initial substrate is only partly mineralised and assimilated by the first microorganism. A residue remains that in turn is partially mineralised and assimilated by a second microorganism, leaving a residue utilisable by a third microorganism, and so on. A core of recalcitrant molecules remains and these can be incorporated in humic compounds. For an outside observer, the overall net result is that represented on the right side of the Figure. The interactions possible between microorganisms at each stage, have not been taken into account in this diagram.

populations decline and gradually release their stored nitrogen. If a green manure with a C/N ratio close to 15 is buried, there will be no immobilisation but rather a release of nitrogen readily utilisable by the plants.

The preceding C/N ratios are only an approximation and the intensity of the competition among plants and microorganisms depends on microbial activity, which again is dependent on the composition of the microflora and climatic conditions. On the other hand, the ease with which a substrate is decomposed does not depend solely on the C/N ratio, but also on the form in which the nitrogen, and especially the carbon, are present: the carbon of cellulose is not as easily hydrolysed as that of soluble sugars or pectins.

Similarly, there are also optimal C/S and C/P ratios. These are close to 200 and 300 respectively.

Formation of Humus

Humus which, bound to clays, largely determines the degree of structure and fertility of soils, is a product of the transformation of organic matter. It is a composite product, difficult to define chemically, and details of its formation are still not clear. Thus the information briefly presented below should be considered schematic (Fig. 5.6).

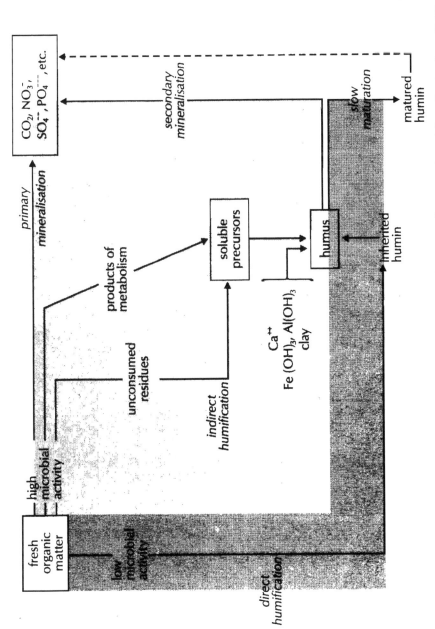

Fig. 5.6 Main paths for transformation of organic matter into humus and its mineralisation (adapted from Duchaufour, 1980).

Organic matter freshly introduced into the soil, as already seen, is partly converted into a mineral state (CO_2, NO_3^-, SO_4^{--}, PO_4^{---}, ...) by the process of **primary mineralisation**: another part is transformed into microbial cells while the rest, consisting mainly of tanins and polyphenolic residues of lignin combined with proteins, is not used. These soluble phenolic precursors, bound to products of microbial metabolism, gradually condense and form fulvic acids that precipitate in the presence of calcium. Hydroxides of iron and aluminium are of paramount importance as catalysts in the process of condensation and ensure a very stable fixation of the fulvic acids on clays. While effecting fragmentation of plant debris that has been rid of part of its aromatic components, which have an inhibitory effect, arthropods make the remainder more accessible to microflora and facilitate its decomposition, which accelerates during and after passage through the animal's alimentary canal. As will be seen later, animal predators have a stimulatory as well as regulatory effect on microbial populations (Fig. 5.7). Earthworms facilitate formation of clay-humus complexes by constant mixing of the clay and humic substances. During the course of these transformations, the humic compounds are enriched by the products of microbial activity: bacterial polysaccharides, fungal pigments and autolysates. Modifications and condensations follow and brown and grey humic acids form with increasing molecular weight, which precipitate at acidic pH. Due to incorporation of the derivatives of microbial activity, the humic substances are highly enriched with nitrogen compared to the initial materials: their C/N ratio is close to 10. Humus is equally rich in sulphur and phosphorus.

Besides this humification, which can be called **indirect** because it passes through intermediate stages of biological activity, **direct** humification can occur in regions where severe pedoclimatic conditions do not allow for very active microfloral growth. In this case, the humic compounds resulting from slow oxidation and physicochemical condensations of slightly biodegraded cellulosic and ligneous compounds have low molecular weights and are poor in nitrogen.

We must guard against assuming that humus is protected from biodegradation due to the fact that it is a complex product made up of high molecular weight constituents. On the contrary, it serves as a substrate for a microflora which can be very active under favourable environmental conditions. This biodegradation of humus constitutes **secondary mineralisation**. Humus therefore does not accumulate indefinitely: the thickness of the organic layer is a characteristic of the ecosystem. Under specified pedoclimatic conditions, there is an equilibrium between humification and loss due to secondary mineralisation. However, under certain conditions, humic acids can be subject to a slow maturation process, leading to formation of compounds with very high molecular weights that form extremely stable complexes with metallic ions. Using ^{14}C dating, it has been estimated that the age of certain constituents of this matured humin is more than 1000 years.

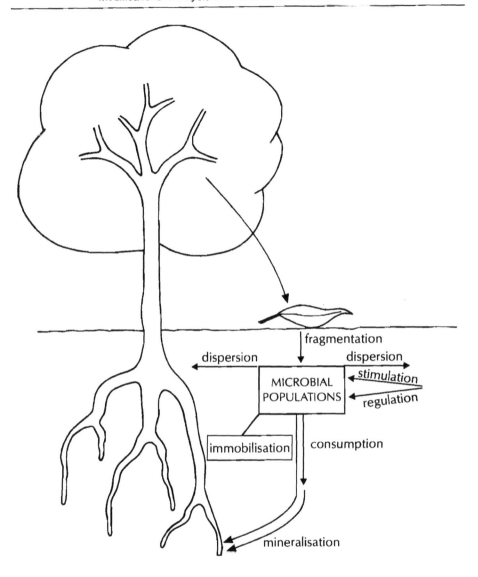

Fig. 5.7 Participation of soil microfauna in recycling organic matter (in green). Detritivorous microarthropods fragment the plant debris, and this significantly increases the contact surface with the microorganisms. While moving from place to place, nematodes and microarthropods contribute to the dispersal of bacteria and the fungal spores present on their teguments and in their excreta. While consuming the bacterial or fungal colonies, they stimulate their enzymatic activity while maintaining the population equilibria and, on the other hand, put back into circulation the mineral elements immobilised in the microbial biomass.

When a forest or grassland soil is cultivated, mineralisation of organic matter is accelerated. Earlier, this loss was compensated by the addition of manure to cultivated soils. But nowadays these improvements are not implemented and hence the humus content reduces slowly but steadily in many regions. In Lauragais (southern France), for example, this humus content has become lower than the critical level under which soil cohesion is insufficient to resist aeolian erosion and erosion due to rain. The mass of soil washed away during a single storm can be as high as 300 tons per hectare (Perny, pers. comm.).

Nitrogen Cycle

Like other essential elements involved in metabolic pathways, nitrogen is essentially assimilated by plants in mineral form. However, even though gaseous nitrogen constitutes almost 80% of the terrestrial atmosphere, it is not directly utilisable because of its inert nature, unlike the gaseous mineral form of carbon. Mineral nitrogen must be taken up from the soil solution by the roots, in the form of nitrates and ammonia salts. A small part of this soluble nitrogen is carried by rains. Meteoric nitrogen consists of nitrous and nitric acids formed during storms and represents 2 to 10 kg ha^{-1} y^{-1}, a quantity totally insufficient for the development of plant cover. (It should be added that in highly industrialised zones, similar quantities of ammoniacal nitrogen fall back to the Earth.) Most of the natural nitrogen added to the soil arises from activity of microorganisms, i.e., free-living or symbiotic prokaryotes. In the first case, the annual addition to the soil is in the range of 2 to 30 kg ha^{-1} and in the second, a few tens or even hundreds of kg ha^{-1}. However, the nitrogen thus incorporated into the microbial or plant biomass is not more accessible to other plants than atmospheric nitrogen. It has to be mineralised. The microbial decomposers first transform the organic nitrogen into ammonia, which in turn oxidises to nitrates more easily assimilable (Fig. 5.8). The resultant matter can then be utilised by non-symbiotic plants, then transmitted to other living links all along the food chain. This 'secondary' organic matter can itself be mineralised and recycled at each stage.

According to this scheme, a permanent flux of atmospheric nitrogen towards the soil is observed, where it is fixed and transformed into organic matter liable to indefinite recycling. In natural ecosystems, accumulation is avoided and equilibrium reestablished by the action of denitrifying organisms that ensure the return of excess nitric nitrogen back to the gaseous state. However, this is not true in the case of intensive agricultural systems where nitrogenous manures are supplied in addition to natural fixation. Only a part of the added nitrogenous manure is used by the plants. The rest is easily washed away by rain and irrigation water, escapes the denitrifying organisms (which, anyway, are not able to transform such high amounts) and pollute groundwater. Besides this diffuse pollution, local pollution can occur, which results from nitrification of pig slurry and household refuse.

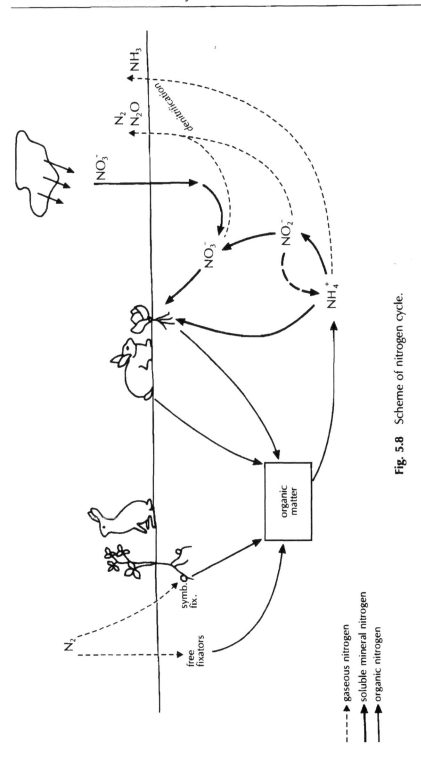

Fig. 5.8 Scheme of nitrogen cycle.

We shall therefore study the following successively: fixation of atmospheric nitrogen, ammonification and nitrification, constituting two stages of transformation of organic nitrogen into soluble assimilable nitrogen, and denitrification which ensures its return to the gaseous state.

Fixation of Atmospheric Nitrogen

Nitrogen-fixing microorganisms fulfil an ecologically irreplaceable function since, until the discovery of the process of industrial synthesis of ammonia, they were the only organisms capable of inducing entry of nitrogen into the biological cycles. Efforts exerted in the last few years to limit the addition of chemical fertilisers both for economic reasons and to preclude pollution of groundwater, continue to inspire research dedicated to this subject and numerous studies are underway.

Fixation of nitrogen is catalysed by an enzyme called **nitrogenase**. In *Klebsiella pneumoniae*, 20 different genes (*Nif* genes) determine the structure and function of nitrogenase. A remarkable conservation of the nucleotide sequences of the genes that code the structure of this enzyme has been demonstrated in all the nitrogen-fixing species studied to date (Eady et al., 1988). Nitrogenase consists of two proteins, one containing iron and the other iron and molybdenum. Another nitrogenase was discovered by Bishop and colleagues (1982) in mutants of *Azotobacter vinelandii* which were incapable of utilising nitrogen in usual Mo-containing culture media. This enzyme was later shown to exist in wild strains as well as in *A. chroococcum* and *Anabaena variabilis*, and its presence in *Clostridium pasteurianum* and other species appears likely. This 'alternative' nitrogenase also consists of two proteins (one containing iron and the other iron and vanadium) which are coded by *Vnf* genes very close to the *Nif* genes of the Mo-containing nitrogenase (Bishop and Joerger, 1990). A third nitrogenase, utilising only iron, was also discovered just recently.

Atmospheric nitrogen is reduced by nitrogenase in the presence of NADPH and ferredoxin that ensure the transfer of electrons (Fig. 5.9):

$$N_2 + 8H^+ + 8e^- \rightarrow 2NH_3 + H_2$$

This reaction consumes a huge quantity of energy (furnished by 16 molecules of ATP) and produces molecular hydrogen released in gaseous form. An excess of hydrogen competitively inhibits nitrogen fixation. For that reason, researchers are very interested in the existence of a hydrogenase in certain bacteria (*A. chroococcum, Bradyrhizobium japonicum* for example), coded by a *Hup* gene, that ensures recycling of hydrogen via the intermediary of respiratory metabolism.

All these nitrogenases are irreversibly denatured by oxygen. For this reason, aerobic fixator bacteria have systems of protection of the enzyme. The mechanisms can be very different: intensification of respiration (*Azotobacter*), synthesis of a protective protein (*Azotobacter*), polysaccharidic mucous membrane (*Beijerinckia*), confinement (nodules and vesicles of symbiotic organisms, heterocysts in some Cyanobacteria), etc.

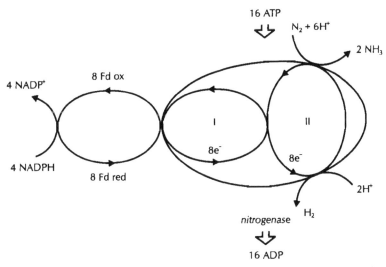

Fig. 5.9 Successive stages in the reduction of nitrogen into ammonia according to the overall reaction:

$$N_2 + 8H^+ + 8e^- \rightarrow 2NH_3 + H_2$$
$$16\ ATP \rightarrow 16\ ADP + 16\ Pi$$

NADPH: nicotinamide adenine dinucleotide phosphate; Fd: ferredoxin; I: protein – Fe of nitrogenase; II: protein – Fe + Mo (or Fe + V) of nitrogenase; ATP: adenosine triphosphate; ADP: adenosine diphosphate; Pi: mineral (inorganic) phosphorus

But a new enzymatic complex, allowing nitrogen fixation, was recently discovered in an aerobic chemolithotrophic thermophilic bacterium, *Streptomyces thermoautotrophicus*: not only is this new nitrogenase insensitive to oxygen, but the presence of this gas is indispensable for the nitrogen to be reduced to ammonia. In this system, nitrogen fixation is coupled to oxidation of carbon monoxide by a dehydrogenase. The electrons arising from oxidation of carbon monoxide are transferred to oxygen, thereby creating superoxide anions O_2^-. A superoxide oxidoreductase in turn oxidises the O_2^- ions and transfers the electrons to a dinitrogenase that reduces the nitrogen to ammonia. All these reactions, which can occur only in the presence of free oxygen, consume 4 times less ATP than classic nitrogenases (Ribbe et al., 1997).

The presence of large quantities of soluble mineral nitrogen (ammoniacal or nitric) inhibits the functioning of nitrogenase in a reversible manner: fixation is active only when the media are not very rich in soluble nitrogen. However, this is rarely the case where intensive agriculture is practised.

Nitrogen-fixing microorganisms are found either in a free state in the soil (or in water) or are closely associated with other organisms: fungi (lichens) or chlorophyllous plants.

Free-living Nitrogen Fixators

Many bacteria belonging to very different taxonomic groups are capable of fixing atmospheric nitrogen (Table 5.2). The most common and best known for a long time are organotrophs. Incapable of utilising cellulose, they require easily metabolisable carbon substrates, but many live in association with cellulolytic organisms and consume their by-products. These organotrophs are found in anaerobic or microaerophilic conditions as well as aerobic conditions. Nitrogen-fixing phototrophic bacteria are generally restricted to aquatic habitats. Some have an agronomically important role, in rice fields for example. Nitrogen fixators are also found among the methanogenic and halophilic archaea.

Table 5.2 Principal genera of free nitrogen-fixing bacteria. Not all of the species of the genera listed here are necessarily provided with nitrogenase

Nutritional classification	Respiratory metabolism	Principal genera	Special characteristics
Organotrophs	aerobic	*Azotobacter*	neutral and alkaline soils, temperate regions
		Beijerinckia *Derxia* }	soils of varying acidity, tropical regions
	microaerophilic	*Klebsiella* *Bacillus* *Azospirillum* *Burkholderia* *Rahnella* }	organisms generally associated with roots
	anaerobic	*Clostridium*	wide pH range
		Desulfovibrio *Desulfotomaculum* }	bacteria of sulphur cycle
		Methanococcus	archaea
Phototrophs	aerobic	*Gloeothece*	unicellular cyanobacteria
		Nostoc *Anabaena* *Aulosira* *Scytonema* }	filamentous cyanobacteria with specialised heterocysts
		Plectonema *Oscillatoria*	filamentous cyanobacteria without heterocysts
	anaerobic	*Rhodospirillum*	photoorganotrophic purple bacteria
		Chromatium *Thiocapsa* *Chlorobium* }	photolithotrophic green or brown bacteria (sulphur cycle)
Lithotrophs	aerobic	*Streptomyces*	actinomycetes of charcoal piles

The intensity of nitrogen fixation by free bacteria is very variable depending on the locality and climatic conditions. The annual supply generally represents a few kilograms per hectare but in certain cases can go up to 30 kg. In a watery environment, cyanobacteria can fix between 30 and 70 kg $ha^{-1} y^{-1}$. For this reason, they have an important role in the traditional methods of rice cultivation. The total quantity of nitrogen fixed by free bacteria each year represents nearly 50 million tons (Paul and Clark, 1989).

Azospirillum spp., which are closely associated with the roots of most Gramineae, have been the subject of intensive research for the last ten years. Numerous studies have shown that introduction of strains of *A. lipoferum* or *A. brasilense* into the rhizospheres of cultivated plants led to significant increase in grain yield as well as total dry matter. But the measurements obtained using ^{15}N tend to go against the hypothesis of accrued assimilation of nitrogen by plants so treated (Sarig et al., 1990). The stimulations observed are most likely due to the production of growth hormones by these bacteria (Tien et al., 1979).

SYMBIOTIC NITROGEN FIXATORS

Nitrogen fixation by symbiotic bacteria introduces 120 million tons of nitrogen into the biological cycles every year, which is more than twice the quantity supplied by free-living bacteria (Paul and Clark, 1989).

One of the oldest and major groups of nitrogen fixators is *Rhizobium*, comprising at present 6 genera showing either rapid growth (*Rhizobium, Sinorhizobium*) or slow growth (*Bradyrhizobium*) and associated with legumes. The discovery of a functional association between a species of *Bradyrhizobium* and a species of Ulmaceae, *Parasponia parviflora* (Akkermans et al., 1978) helped highlight the fact that the specificity of Rhizobiaceae is perhaps less strict than thought earlier. Rhizobia are bacteria often abundant near the roots of host or non-host plants, especially in soils where the pH is around 7. They can live perfectly in a saprophytic state but, in doing so, fix very little or no nitrogen. The various stages preceding establishment and starting up of the associative symbiosis have been very precisely studied in recent years. The process begins with the exchange of signals between the host plant and the bacteria (Fig. 5.10). The germinating seeds and the apical parts of the roots produce **flavonoids** that, even in nanomolar concentrations, selectively stimulate the rhizobia present in the surrounding soil. These signals, identified as lipochitooligosaccharides, have a chemotactic effect but, on the other hand, activate the transcription of a group of nodulation genes (*nod* genes: see box below). The products of these genes (Nod Rm-1, Nod Rm-2 etc.) trigger certain processes in the plant that lead to entry of the compatible bacteria. The most noticeable effect is deformation of the root hairs, which precedes infection (Fig. 5.11) but the products of the aforesaid genes also seem to play a role in fixation of the rhizobia on their sites of penetration. The polysaccharides superficially present on the rhizobia cells also probably play a role in limiting the defence reactions of the plant. Other proteins, yet to be identified, probably have a complementary action. The genetic information necessary for the establishment of symbiosis is

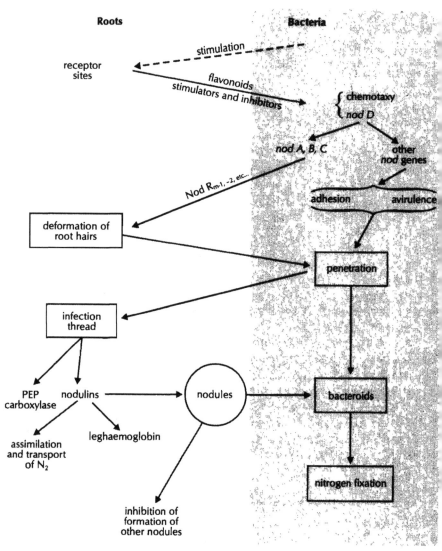

Fig. 5.10 Scheme of successive stages leading to formation of functional nodules (according to the syntheses of Rolfe and Gresshoff, 1988; Phillips, 1992). This representation is probably only a partial reflection of a much more complex reality. Indeed, other regulatory genes responding to isoflavones (the *nod* V and W genes) were recently demonstrated in *Bradyrhizobium japonicum*, whose action seems to combine that of *nod* D; *nol* repressor genes were also proved (Stacey et al., 1995).

contained in the genome of the host plant. A large number of genes contribute to the formation of the nodules (*enod* genes), as well as their functioning (*late nodulin* genes). In alfalfa or soybean, the bacteria attach to the root in a very localised area, at the limit of the elongation zone. They infect the absorbent root hairs after they have curled and assumed a characteristic

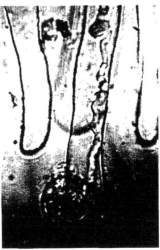

Fig. 5.11   First stages of infection of a legume by a compatible rhizobium. Left: deformation of root hairs caused by the presence of rhizobia (plate, G. Truchet, CNRS). Right: infection thread inside a root hair (plate, M. Obaton, INRA).

curved form (generally known as the 'shepherd's crook'). In the case of groundnut, the bacteria attach at the point of emergence of the lateral roots and in certain tropical leguminous plants such as *Sesbania*, infection takes place on the stems (the genus *Azorhizobium* was created for these bacteria). The bacteria penetrate via an invagination of the membrane of the infected cell. A rearrangement of the cell wall occurs, progressing from cell to cell. A kind of conduit is thereby constituted, called the **infection thread** in which the bacteria make their way. (Sometimes, in certain plants, e.g. groundnut, a differentiated infection thread is not formed.) The presence of the bacteria triggers the expression of genes whose products (nodulins) cause dedifferentiation of some cortical cells. These multiply and form an outgrowth, the **nodule**, which is connected to the vascular apparatus of the plant that supplies energy to the system. The infection thread traverses the cell walls and ramifies in the newly formed cells, which are tetraploid. The rhizobia are free inside these cells but remain separated from the cytoplasm by a membrane formed by the host. Their shape becomes modified and the enzymatic nitrogen fixator complex becomes activated: the bacteria become bacteroids (Fig. 5.12). Parallelly, under the action of a nodulin, the cells of the nodule produce a reddish-brown molecule similar to the haemoglobin, called the **leghaemoglobin**, which ensures transport of oxygen to the bacteroids while precluding inhibition of nitrogenase. The plants possess a complex system of regulation that is still not well understood, which limits the number of nodules formed and excludes the possibility of successive infections.

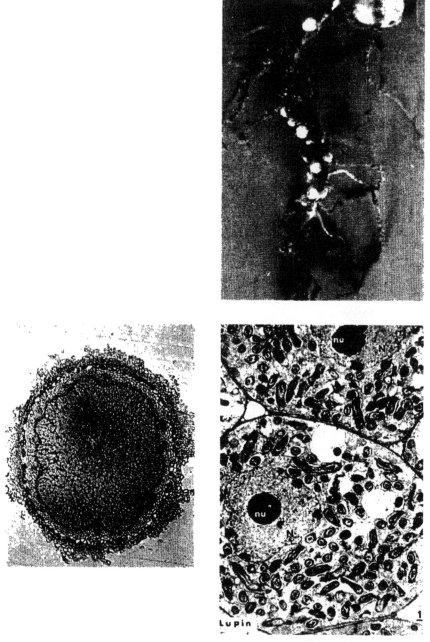

Fig. 5.12 Nodules, sites of symbiotic rhizobium-legume association. Top right: nodules formed on root of pea (plate, N. Amarger, photo library INRA). Below left: transverse section of a nodule of soybean (plate, J.J. Drevon, INRA). Below right: section of a nodule of lupin showing cells invaded by bacteroids; nu—nucleoli, N—nucleus (plate, M. Obaton, INRA).

Genes of Nodulation

These genes are carried by the bacterial chromosome or by large-size plasmids. They include:

- a *nod D* gene whose product specifically reacts with the flavonoids produced by the plant and regulates the functioning of other *nod* genes, as per the following scheme:

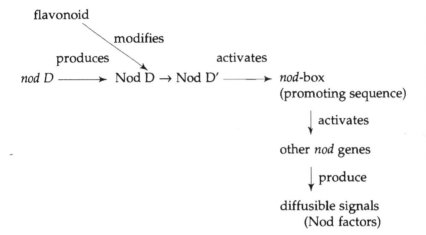

- *nod A, B* and *C* genes, often together in the same operon (and present in all the species studied), are hence called common genes. Activated by the *nod D* gene, they produce one or more messenger(s) responsible for deformation of the absorbent root hairs of the host. The chemical structure of these messengers which, like flavonoids, act at nanomolar concentrations, has been elucidated by Lerouge and colleagues (1990). They consist of four or five molecules of N-acetyl-D-glucosamine (basal element in chitin) to which various groups are attached.
- other *nod* genes, called specific genes. These determine the nature of the groups attached to the basal tetrameric or pentameric structure. Hence they are responsible for host specificity and recognition between bacteria and plant, a step prerequisite to infection. They are sometimes called *hsn* genes (host specific *nod* genes). Some *nod* genes also assure passage of the Nod factors across the bacterial membrane.
- other genes (activators, such as the *nod V* and *nod W* genes, or repressors such as *nol* genes) probably also contribute to regulation of expression of *nod* genes.

The Nod factors are hydrolysed in the vicinity of the roots by plant chitinases: a dynamic equilibrium is thus established between production and degradation.

The Nodulins

Nodulins are proteins expressed in the host plant in response to stimuli from symbiotic bacteria. Nodulins can be distinguished as precocious, preceding or accompanying infection, and tardy, appearing at the beginning of nitrogen fixation. They are products of genes involved in:

— deformation of root hairs
— formation of the infection thread
— morphogenesis of the nodule
— formation of the enveloping membrane
— energy supply for the bacteroids (PEP-carboxylase and sucrose synthase for example)
— transport of oxygen into the nodule (leghaemoglobins)
— assimilation of fixed nitrogen and its transport to organs of the plant.

The efficiency of nitrogen fixation greatly depends on the strains of *Rhizobium*. It also depends on the pair under consideration. In general, fixation is very high in tropical leguminous plants compared to temperate ones, and in leguminous forage crops compared to leguminous grain crops. Very often (in kidney-bean or soybean) harvesting the grains results in a greater export of nitrogen outside the field compared to the imports due to symbiotic nitrogen fixation. It should also be noted that not all leguminous plants are nitrogen fixators: while most Papilionoideae and Mimosoideae live in symbiosis with a Rhizobiacea, only 23% of Caesalpinoideae are capable of symbiosis.

Frankia are presently the subject of sustained interest despite being eclipsed for a long time by rhizobia and the handicap of difficulties in studying them (the method for culturing them has only been known since the last 15 years). These actinomycetes form symbiotic associations with trees and shrubs belonging to 8 different botanical families, distributed in all climatic conditions, and in general, capable of growing in poor, acidic, dry, hydromorphic or saline soils. Thus they constitute precious material for retrieval of waste or degraded land and for reforestation. As in *Rhizobium*, host infection occurs by various processes: passage between the epidermal cells in *Ceanothus* and *Elaeagnus*, penetration through absorbent root hairs in *Casuarina* and *Alnus* for example. The *Frankia* then infect the cells of the root cortex, but the nodules form from the modified lateral roots which retain their vascular system (Fig. 5.13). The process of recognition between the two partners of the association is still not well understood. The nodules are digitate, surmounted sometimes by roots that orient surface-wise (in *Casuarina* and *Myrica* for example). These nodules are perennials, and can become very voluminous and remain active for several years. However this activity is often subjected to seasonal variations, not only conceivably in hosts with deciduous leaves living in temperate countries, but even in species with persistent leaves. The mechanism of protection of nitrogenase of *Frankia* against oxygen are more varied than those of rhizobia: it can be included in vesicles (as in *Ceanothus americanus* for example), organs comparable to heterocysts of *Cyanobacteria*, or covered by thick and suberised walls of nodules (as in *Casuarina*); the transport of oxygen in that case is ensured by

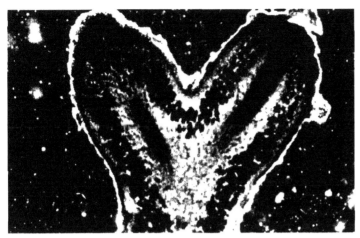

Fig. 5.13 Longitudinal section of a nodule of alder (*Alnus glutinosa*) induced by a strain of *Frankia*. The cells of the cortex, containing the hyphae and vesicles of the actinomycete appear dark in the photograph. A portion of the central steles is also visible (plate, P. Normand, CNRS).

a haemoglobin similar to that of leguminous plants. Intermediate situations also exist (Tjepkema et al., 1986). The *Hup* gene which helps in recycling hydrogen is more often found in symbiotic associations of *Frankia* than in symbiotic rhizobia. Nitrogen fixation, very variable, can reach 200 to 300 kg ha^{-1} y^{-1} in the case of some alders (*Alnus*) and beef-wood trees (*Casuarina*), which represent the most commonly studied species.

Cyanobacteria (*Nostoc* and *Anabaena*) also participate in symbiotic associations fixing nitrogen along with very varied partners: tropical Cycadales, ferns, liverworts, and even fungi (lichens). One of these associations is of particular interest, namely, that of the aquatic floating fern of genus *Azolla* with *Anabaena azollae*.

One bacterial colony is associated with the apical meristem of each stem of the fern. Growth of multicellular epidermal hairs that differentiate at the same time as the leaves in the vicinity of the colonies allows passage of bacteria into the ovoid cavities formed by retraction of the epidermis into the fleshy aerial part of the leaves (Peters and Meeks, 1989). The heterocysts in which nitrogen fixation takes place are formed on the as yet undifferentiated filaments of *A. azollae*. The bacteria apparently do not have a free form and are transmitted from one generation of fern to the next. When the sporocarps of *Azolla* are formed on the ventral surfaces of the leaves, they are colonised by a process analogous to the one just described. Contamination of the gametophytes and later of the new sporophyte after fertilisation, is thus assured. When soluble iron and phosphorus do not become limiting factors, nitrogen fixation can be very high. *Azolla* is widely used (for centuries) in rice cultivation in East Asia. When it is cultivated as a cover plant in association with rice, it can add 100 kg nitrogen per hectare. If used as green manure between two rice crops, the gain in nitrogen can reach even 330 kg per hectare (Paul and Clark, 1989) but in this case, it is necessary to maintain the fern in place for several months, then bury it—additional chores that could limit farmer interest.

Measurement of Biological Fixation of Nitrogen

Three different methods can be used for quantitative evaluation of microbiological fixation of nitrogen. One method uses mass spectroscopy for measurement of the ^{15}N isotope incorporated into the biomass after enrichment of the atmosphere with ^{15}N. This method is rather laborious to implement. In the second method, the property of nitrogenase is used to reduce trivalent substrates other than gaseous nitrogen: in an atmosphere of acetylene, the nitrogen-fixing bacteria produce ethylene that can be measured by chromatography in the gaseous phase. This technique, highly useful, allows in-situ measurements, but interpretation of the results is sometimes dicey. Further, the technique is not applicable in the case of bacteria such as *Streptomyces thermoautotrophicus* for which nitrogenase requires oxygen and does not reduce acetylene. A third method, more recent and more sensitive, involves measurement of hydrogen released during reduction of nitrogen. It is necessary to compare the total quantity of hydrogen produced by the activity of nitrogenase in the absence of nitrogen (under an atmosphere of oxygen and argon) with the quantity actually released in a normal atmosphere. This method is only utilisable in symbioses in which the *Hup* gene permitting recuperation of hydrogen does not intervene.

Principle of Measurement of Nitrogen Fixation by the Acetylene Method

The 'classic' nitrogenase catalyses not only reduction of molecular nitrogen and protons, but also a few other compounds such as acetylene, cyanides, azides and nitrous nitrogen.

In particular, when the atmosphere contains more than 10% acetylene, the enzyme is saturated and the other substrates are not utilised: ethylene forms to the exclusion of ammonia and hydrogen:

$$HC \equiv CH + 2H^+ + 2e^- \rightarrow H_2C = CH_2$$

The quantity of ethylene formed is proportional to the flux of electrons transferred by the nitrogenase and thus reflects its reducing activity.

In practice, a soil sample or soil clump together with the symbiotic plant is placed under a bell-jar made of a plastic film with an airtight valve. A known volume of acetylene is then injected into the bell-jar through the valve. After its introduction, the mixture of gases is siphoned out with a syringe at regular intervals to estimate the kinetics of fixation. The quantum of ethylene and acetylene in the air contained in the syringe is determined by gas-phase chromatography, by comparison with peaks in a control mixture of known composition. A mathematical formula gives the reducing activity from the slope of the straight line representing the ethylene/acetylene ratio as a function of time.

The presence of ethylene of microbiological origin in the soil must be taken into account but precision of the method is often limited by the difficulty encountered in establishing a good correlation between reduction of acetylene and reduction of nitrogen (coupled to production of hydrogen under natural conditions). Should the nitrogenase contain vanadium, the reducing activity can result in the appearance of a significant quantity of ethane H_3C CH_3. Lastly, the acetylene partially inhibits the activity of the two types of nitrogenases but not to the same degree. (Eady et al., 1988).

Another type of association was recently described. Microaerobic endophytic nitrogren-fixing bacteria of genera *Acetobacter* and *Herbaspirillum* have been isolated from the roots, rhizomes and leaves of sugarcane and a few other Gramineae (Cavalcante and Döbereiner, 1988). Species of *Azoarcus* have also been isolated from the roots of a plant (*Leptochloa fusca*) growing in the saline soils of Pakistan. *Azoarcus* also seem to be present in the roots of rice (Reinhold-Hurek and Hurek, 1998). *Acetobacter diazotrophicus* is the better known species among these endophytes. The presence of nitrates does not interfere with the fixation of nitrogen because this bacterium does not possess nitrate reductase: moreover, its nitrogenase is hardly susceptible to the presence of ammoniacal nitrogen.

Ammonification

A large number of bacteria and fungi are capable of transforming nitrogen from organic matter into ammoniacal nitrogen, under both aerobic and anaerobic conditions. Proteins represent the principal source of nitrogen, followed by amino sugars and purine and pyrimidine bases of nucleic acids. The enzymatic attack of these substrates is generally simple in vitro as they are supplied in a purified form. However, in nature, they are most often bound to complex molecules. Proteins in particular can be affixed to polyphenols that inhibit their degradation. Even the simple amino acids, when adsorbed on clay molecules, can resist the action of enzymes. Ammonification of organic matter is therefore relatively slow and even knowing the concentration of organic nitrogen in the soil, it is often difficult to calculate what fraction would be mineralised and how much nitrogen the plants would incorporate. It is generally thought that plants can benefit from 1 to 4% of the total organic nitrogen present in a soil at a given moment.

Estimation of the nitrogen available must indeed take into account the requirements of the microorganisms that preferentially use it in its reduced form for their own syntheses. It has been seen (page 125) that under certain conditions, microflora can enter into competition with plants and immobilise all the assimilable nitrogen for their own use. This immobilisation is nevertheless always temporary and the microorganisms usually play a very important regulatory role. Another part of the mineralised nitrogen is removed from the soil solution through adsorption on the clay-humus complex, or by trapping in the interior of the clay layers or by formation of a complex with humic substances (it is probably these captive NH_4^+ ions that are liberated under the effect of infra-red radiation and prolonged desiccation: see page 85). Lastly, one part can be lost in the form of gaseous ammonia if the pH is too high (Fig. 5.14). Loss of nitrogen due to release of ammonia can also occur in soils of neutral pH or even in slightly acidic soils, in the presence of a very rapidly decomposing organic substrate: the abundance of NH_4^+ ions can locally raise the pH by 1 or 2 units. Under these conditions, it is the NH_3 form that predominates and loss by volatilisation can be significant. Such a situation can occur when nitrogenous manure in the form of urea (readily hydrolysed) is added under conditions favourable to the active development of microflora: humid soil, high temperature and shallowly buried fertilizer.

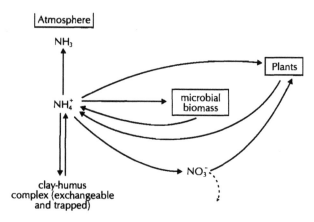

Fig. 5.14   Fate of ammonium ions in the soil.

Nitrification

Soil ammoniacal nitrogen, soluble or readily exchangeable, is more or less rapidly transformed into nitric nitrogen. This oxidation, whose biological origin was foreseen by Pasteur, is carried out in two stages by highly specialised chemoautotrophic bacteria. Even though apparently existing in all kinds of soil, these bacteria represent only a very small number of genera and species. Strictly aerobic, they use carbon dioxide or carbonates as their only source of carbon. The first group draws its energy from oxidation of ammonia into nitrite. The reaction does not occur directly but passes through at least one intermediary, hydroxylamine, whose incomplete oxidation can release nitric oxide (NO). The species *Nitrosomonas europaea* is by far the most widespread. The second group utilises oxidation of nitrites as a source of energy and transforms them directly into nitrates; the dominant species of this category is *Nitrobacter winogradskyi*.

These bacteria were first discovered by Winogradski in 1890. They are very difficult to obtain in pure culture and multiply slowly. Consequently, despite their great economic importance, they are still not well understood.

They proliferate in well-aerated soils and are mostly active during warm and humid periods, i.e., spring and autumn in temperate climates. Activity during spring is very good for crops as the nitric nitrogen produced is especially well assimilated by most plants. On the other hand, the renewed activity observed in autumn is often disastrous. It occurs during a period when the soil contains large quantities of harvest residue, viz., organic substrates, but no longer supports vegetation capable of utilising mineralised nitrogen. A large quantity of nitrates is therefore likely to be carried into soil depths by rain, always abundant in autumn. Nitric nitrogen, unlike ammoniacal nitrogen, is not retained by the absorbent complex. To avoid such loss, methods of retarding the action of nitrifying bacteria were sought. Many inhibitors were experimentally assayed; the most common was 2-chloro-6-trichloromethyl pyridine (see page 291). It should be remembered that burying a lignocellulose substrate helps in immobilising soluble nitrogen in the form of microbial biomass.

Besides nitrification effected by chemoautotrophic bacteria, nitrification caused by heterotrophic bacteria and certain fungi has also been demonstrated. This process, still little studied, appears negligible compared to autotrophic nitrification.

Some Types of Reactions in the Mineralisation of Nitrogen

• **Ammonification**

— *proteins* → peptides → amino acids:
$$NH_2-R-CONH-R'-COOH + H_2O \rightarrow NH_2-R-COOH + NH_2-R'-COOH$$

— amino acids → ammonia:
$$R-\underset{\underset{NH_2}{|}}{C}H-COOH \rightarrow R-\underset{\underset{NH}{||}}{C}-COOH \rightarrow R-\underset{\underset{O}{||}}{C}-COOH + NH_3$$

or even: $R-\underset{\underset{NH_2}{|}}{C}H-COOH \rightarrow R-CH_2-COOH + NH_3$

These two reactions can be coupled. Other deamination reactions are possible.

— *nucleic acids* → mononucleotides → P + sugar + nitrogenous base. Purine or pyrimidine base → urea → ammonium carbamate → ammonia:
$$NH_2-CO-NH_2 + H_2O \rightarrow NH_2-COO-NH_4 \rightarrow 2NH_3 + CO_2$$

• **Nitrification**

— *nitrosation*
ammonia → hydroxylamine → ? → nitrous acid:
$$NH_4OH \rightarrow NH_2OH \rightarrow ? \rightarrow HNO_2$$

— *nitratation*
$$(NO_2^- + H_2O) + 2 \text{ cytochrome } Fe^{3+} \rightarrow NO_3^- + 2 \text{ cyt. } Fe^{2+} + 2H^+$$
$$2H^+ + 2 \text{ cyt. } Fe^{2+} + 1/2O_2 \rightarrow 2 \text{ cyt. } Fe^{3+} + H_2O$$

Denitrification

When the oxygen content of the soil atmosphere becomes insufficient for it to fulfil the role of electron acceptor, the nitrates can be reduced to nitrites by a relatively large number of microorganisms. (This procedure involves reduction without the use of nitrogen, unlike the assimilative reduction effected by plants and microorganisms after the absorption of nitrates.) These nitrites can subsequently be converted to ammonia (Fig. 5.15). But if the redox potential continues to fall (from around 300 to 100 mV), certain bacteria reduce the nitrites into nitrous oxide and molecular nitrogen, which escape into the external atmosphere. These losses ordinarily represent 10 to 15% of the annual production of nitric nitrogen but can be much higher under favourable conditions, that is, if the addition of a readily decomposable organic matter is followed by a period of anoxia (after heavy rain) while the temperature is high. In a normally aerated soil, denitrification can continue in specific microsites: the interior of microaggregates and probably in the rhizosphere.

> A denitrification of purely chemical origin can also take place by the decomposition of nitrites at high pH (around 9, which can occur in limited places in the presence of excess ammonia-containing manure) or under very acidic conditions (pH lower than 5, which is rarer because in that case the nitrous acid-producing bacteria are poorly active). Unlike biological denitrification, chemical denitrification is possible in the presence of oxygen.

When nitrogen is a limiting factor, denitrification represents an overall loss of reserves. In industrialised countries, however, agricultural land is often characterised by the presence of excess rather than less nitrates.

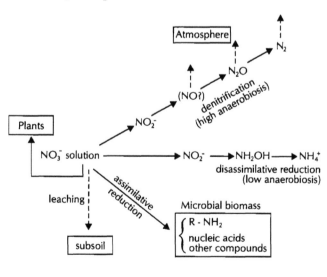

----- nitrogen losses

Fig. 5.15   Fate of nitrate ions in the soil.

Denitrification can therefore be considered a regulatory phenomenon that helps to preclude excessive pollution of the subsoil. This process could even be tried for stimulation in certain sensitive zones such as areas where pig slurry is spread. The chemolithotrophic bacterium *Thiobacillus denitrificans* is of specific interest for this role because under anaerobic conditions, it utilises nitrates to oxidise sulphur into sulphate:

$$5S + 6NO_3^- + 2H_2O \rightarrow 3N_2 + 5SO_4^{--} + 4H^+$$

In principle, this reaction only produces nitrogen whereas other bacteria such as *Corynebacterium nephredii* for example, produce nitrous oxide (N_2O) (Renner and Becker, 1970). This volatile compound can have harmful effects on the environment. It can go into the stratosphere, where it is dissociated by light and forms nitric oxide (NO), which can contribute to the destruction of the ozone layer according to the reaction: $NO + O_3 \rightarrow NO_2 + O_2$. At present, nitrous and nitric oxides of the atmosphere mainly arise from biological degradation of nitrogenous fertilisers. Therefore, it seems simpler and wiser to try and avoid excess nitrogen rather than try to resorb it a posteriori.

Phosphorus Cycle

Even though less abundant in the cells compared to the four elements C, H, O and N, phosphorus is no less basic for living organisms. It is a component of nucleotides of nucleic acids and energy transporters, present in phospholipids and in phosphate sugars and often found in plants in the form of inositol phosphate or phytin. Unlike carbon, nitrogen and (as we shall see) sulphur, phosphorus is found only in the soil and not in the atmosphere. Unfortunately, the major part of the soil phosphorus, mineral or organic, is insoluble. The concentration of soil solution is consequently low: on average, 0.03 mg L^{-1} (essentially in the form of $H_2PO_4^-$) and the requirements of plants are often difficult to satisfy (Hayman, 1975). Rapid depletion of the solution is avoided by gradual desorption of the ions retained by the adsorbent complex. But renewal of this stock requires solubilisation of the insoluble phosphorus. The phosphatases of the roots partly fulfil this function but microorganisms also play a not-so-negligible role. The phosphorus that goes into solution concomitantly originates from soil mineral reserves and from recycled organic phosphorus (Fig. 5.16).

Insoluble mineral phosphorus is constituted of phosphates (often fluorated) of calcium (neutral or alkaline soils), iron and aluminium (acidic soils). Numerous microorganisms are capable of solubilising these salts because of the organic acids (citric, oxalic, lactic, succinic, and especially carbonic acid produced by respiration) or chelates (2-ketogluconic acid) which they excrete. Under anaerobic conditions, phosphorus availability can increase due to a reaction of hydrogen sulphide (resulting from the biological reduction of sulphates) on iron phosphate. There is a precipitation of iron sulphide and release of phosphoric acid.

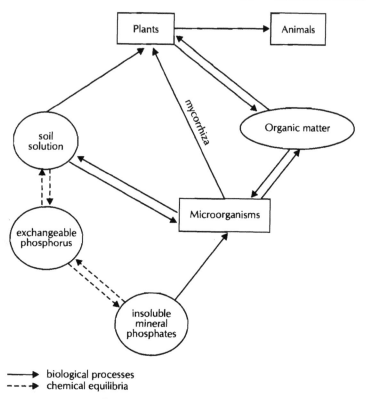

Fig. 5.16   The phosphorus cycle.

Mineralisation of organic phosphorus is brought about rather rapidly by hydrolysis of the phosphate bonds due to various phosphatases:

$$R—PO_4H_2 + H_2O \rightarrow R—OH + H_3PO_4$$

Depending on the pH of the environment, the phosphate ions are in the form of PO_4^{3-} (in alkaline soils), HPO_4^{2-} (in neutral soils) or $H_2PO_4^-$ (in acidic soils). The $H_2PO_4^-$ form is the one most easily utilised by plants.

Hydrolysis of phytins by phytases appears more difficult to effect so these compounds constitute an important fraction of organic soil phosphorus. Mineralisation is especially intense at pH close to neutrality.

Not all of the phosphate resulting from solubilisation is released into the surroundings. Part of it, large or small depending on the rate of multiplication of the bacterial population, is immobilised by them and temporarily unavailable to the plants. It can be considered that the quantity of phosphorus thus retained in the microbial biomass is equivalent to that present in the vegetation (Hayman, 1975). The C/P ratio of a substrate beyond which the microorganisms are in competition with the plants for phosphorus is in the range of 300. In practice, it is rarely surpassed except in cases of high cellulosic or very ligneous soil amendments.

Bacteria capable of mineralising organic phosphorus were the subject of great interest in the 1950s. Soviet researchers in particular were hopeful of improving the phosphate nutrition of plants using these organisms. A selected strain of *Bacillus megaterium* called 'phosphaticum' was used extensively during that period under the name 'phosphobacterine', which claimed augmentation of yields of the order of 10%. These trials, conducted later in other countries under controlled conditions, on the whole produced comparable results. But several other bacteria and a few fungi, not involved in mineralisation of phosphorus, were also shown to be capable of producing similar increments in yields (Brown, 1974). Now it is recognised that the positive effect of bacterisation is more likely attributable to: 1) production of growth substances which stimulate plant metabolism and, eventually, to 2) antagonistic action against parasitic microorganisms. Further, hormones of the gibberellin type were also demonstrated in the filtrates of cultures of *B. megaterium* and *Pseudomonas* spp. (Montuelle, 1966).

> The idea of using microorganisms to augment phosphate nutrition of plants has surfaced again in the last few years. Trials carried out with the fungus *Penicillium bilaii* have shown that it is possible to augment phosphorus assimilation and thereby improve yield in wheat, forage crops of family Cruciferae and various legumes (Kucey et al., 1989).

While bacteria are probably of secondary importance in the phosphate nutrition of plants, the fungi associated with the roots in **mycorrhizae** are irreplaceable auxiliaries (see page 278). Because of the low phosphorus content of the soil solution, the concentration of $H_2PO_4^-$ ions in the zone of absorbent root hairs diminishes very rapidly and the slowness of their diffusion does not allow for rapid enough renewal to assure necessary nutrients as and when required by the plants. The very dense and exhaustive network made up of mycelial filaments of the mycorrhizae allows for a significant increase in both the volume prospected and the absorption surface. Studies carried out with ^{32}P have shown that phosphorus can be scavenged from the root up to 7 or 8 cm by endomycorrhizae and up to 20 cm by ectomycorrhizae. These figures become significant when compared to the length of the absorbent root hairs, which is about one millimetre. The acid phosphatases of ectomycorrhizal fungi also contribute to mineralisation of phosphorus (Deransart et al., 1990). A synergistic effect has also been demonstrated between a bacterium (*Bacillus* sp.) and a symbiotic fungus (*Pisolithus tinctorius*) of *Pinus caribaea*: the association of the two microorganisms results in a significant stimulation of absorption of phosphorus from a phytate (Chakly and Berthelin, 1982).

Sulphur Cycle

Sulphur is one of the major elements necessary for living beings. It is found in three common amino acids (methionine, cysteine and cystine), in vitamins (thiamine, biotin) and in various other important organic compounds (glutathion, coenzyme A, ferredoxins, sulphates, glucosinolates of Cruciferae, . . .).

Traces of mineral sulphur are present in the atmosphere in the form of sulphur dioxide and plants can avail of it by foliate absorption. Their main source nonetheless is from sulphates, which represent only a small part of the sulphur reserves of the soil. Most of the sulphur is in the organic form and not directly assimilable (amino acids constitute a special case): hence evaluation of the total sulphur content of a soil gives no indication of its immediate availability.

Sulphur must first undergo mineralisation therefore. This conversion can be effected by a large number of bacteria and fungi when soil moisture and temperature are sufficiently high. Under aerobic conditions, the reduced sulphur of organic compounds is generally oxidised into sulphates. (Animal urine is an equally important source of sulphates.) Under anaerobic conditions, hydrolysis of amino acids produces only sulphides.

Only part of the sulphur mineralised from organic wastes is released in the soil. The rest is utilised by microorganisms for their own synthesis and therefore temporarily immobilised. As already seen in the case of nitrogen, the degree of immobilisation depends on the composition of the substrate. The microbial biomass contains around 1 atom sulphur per 100 to 150 atoms carbon. If the C/S ratio of the substrate is greater than 200, all the available sulphur will be utilised by the microorganisms; if this ratio is lower, mineral sulphur will be released. The N/S ratio can be similarly defined and is of the order of 10.

Atmospheric Sulphur Dioxide

Except in industrialised regions, the concentration of sulphur dioxide in the atmosphere does not exceed 3 $\mu g \ m^{-3}$. This gas arises from oxidation of biological releases of hydrogen sulphide; a small part also arises from volcanic activity. In the environs of large industries, atmospheric sulphur dioxide arises mainly from the combustion of coal or oil and treatment of ores. Its average concentration in such areas can be 100 times higher than the natural concentration. During the temperature inversion that occurred over London in December, 1962, during a period of smog that prevented escape of polluting gases, the level of sulphur dioxide reached 2.1 mg m^{-3} for 3 days, causing a number of deaths (Anderson, 1978). Since then the measures adopted in most eastern countries have helped restore the level of sulphur dioxide to more acceptable values, but nonetheless values still higher than that occurring naturally. Spontaneous oxidation of sulphur dioxide in the atmosphere gives rise to sulphuric acid which, along with nitric acid arising from oxides of nitrogen, results in the phenomenon of acid rains.

Atmospheric sulphur returning annually to the soil is of the order of 1 kg ha^{-1} in non-polluted zones but can exceed 100 kg ha^{-1} in some highly industrialised regions. Furthermore, most of the fertilisers used for intensive agriculture contain sulphates. Because of this dual supply (to which sometimes the sulphur of pesticides should also be added), sulphur deficiency is rare in developed countries, but not uncommon in less industrialised regions.

Lichens are very sensitive detectors of atmospheric pollution caused by sulphur. The north European species *Lobaria pulmonaria* disappears as soon as the concentration of sulphur dioxide exceeds 30 $\mu g \ m^{-3}$.

The reduced mineral forms of sulphur can be subject in the soil to biological oxidation in which several microbial communities intervene. This oxidation can be carried out by a large number of heterotrophic organisms that develop mainly in pH values close to neutrality. They convert the sulphur to a tri- and tetrathionate state and finally to sulphates. It is still not very clear as to which metabolic cycles are associated with these reactions, which seem to yield no energy. Oxidation of sulphur can also be effected by specialised aerobic chemolithotrophic bacteria that utilise the energy so produced to synthesise their organic compounds from atmospheric carbon dioxide. The most widespread of these specialised organisms is genus *Thiobacillus*. This genus, whose homogeneity is still debated, includes species with optimal pH values ranging from neutrality to very low values, and all of which are able to withstand acidity. Other specialised organisms, aerobics such as Beggiatoales, or anaerobics such as pigmented photolithotrophic bacteria and some archaea, are also capable of oxidising sulphides into sulphur and sulphates. They can have ecologically important functions but are confined to specific media: ponds, oceans and sulphurous sources.

Conversely, oxidised mineral forms of sulphur can be biologically reduced. To incorporate them in their amino acids, microorganisms (and plants) must reduce the sulphates absorbed by them. This involves an assimilative reduction whereby sulphur is reintroduced into organic matter. But a dissimilative reduction also takes place wherein the sulphates are not employed as material, but simply as electron acceptors under anaerobic conditions, at the end of a redox series (see page 88). Only a few genera of bacteria are capable of utilising sulphates in this manner. The best known are *Desulfovibrio* and *Desulfotomaculum*.

Like the carbon and nitrogen cycles, the sulphur cycle is a complex one with alternation between mineral and organic forms, oxidised and reduced forms, presence in soil and presence in the atmosphere.

ACTION ON XENOBIOTIC COMPOUNDS

The life of xenobiotic compounds in soil is extremely variable. Some highly polymerised plastic matter has exceptional longevity. Other organic synthetic products to the contrary, can be mineralised and recycled very rapidly. This mineralisation is both physicochemical and biological in origin.

When degradation occurs by abiotic pathways, it generally starts immediately and concentration of the product diminishes exponentially with time (Fig. 5.17).

Biodegradation

When the phenomenon is due to biological processes, ordinarily a period of time lapses—a few hours to a few days—during which apparently nothing

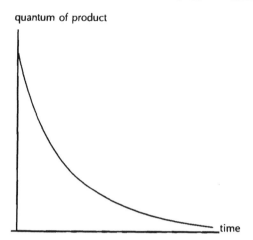

quantum of product

time

Fig. 5.17 Degradation of a product by physicochemical mechanisms. Concentration decreases exponentially: $C = C_0 e^{-Kt}$.

is produced: this is the **latency phase**. Thereafter the compound starts to disappear. Its rate of degradation increases until a maximal value is attained and then remains constant for some time. Lastly, after this linear phase during which the concentration of the product steadily decreases, degradation slows down little by little: the rate and concentration trend asymptotically to zero (Fig. 5.18).

Most of the agricultural pesticides presently used are susceptible to biodegration over a short or medium period of time as long as strong and rapid adsorption do not shelter them temporarily. A high level of clay augments their life period (but for the same reason, reduces their activity).

In very arid areas, such as Neguev in Israel for example, herbicides must be used, however, but with considerable caution. They can persist in the soil in phytotoxic doses since the microflora is not sufficiently active to degrade them between two crops and in the absence of irrigation.

It is easy to show that degradation of a compound is of biological origin. This can be done by sterilising a soil sample, then introducing the substrate and incubating the mixture under conditions close to those under which the phenomenon is observed in the field. If samples taken at regular intervals show no appreciable reduction in the concentration, this means that biological activity is necessary for disappearance of the compound.

Just how microorganisms metabolise substrates they have never previously encountered is a good question. In some cases the stereochemical structures of the xenobiotic compounds are close enough to those of natural compounds for the microbial enzymatic assets to suffice. Certain enzymes, inactive when confronted by a foreign substrate, can acquire a new specificity by mutation of their structural gene or regulatory genes. Examples of activation of previously unexpressed genes are also known, which is effected by

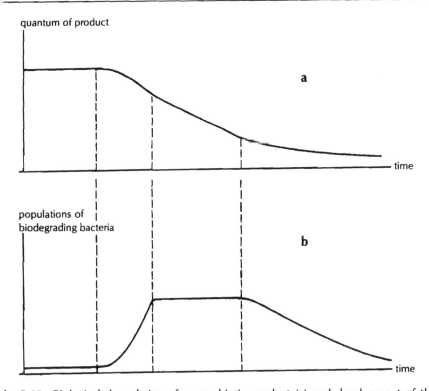

ig. 5.18 Biological degradation of a xenobiotic product (a) and development of the biodegrading microflora (b) during the corresponding period.

rearrangement on the chromosome or on the plasmids. There are also a few examples wherein unblocking of a catabolic process is due to acquisition by the microorganism of plasmidic genes that code for enzymes capable of acting over a more extended spectrum.

Agents responsible for biodegradation are found among bacteria and fungi, but only rarely among algae. A particular microbial population may be able to metabolise a xenobiotic product if by chance it is analogous to a natural molecule. After a period of induction, the necessary enzymes are synthesised. Most often only a few individuals are initially capable of participating in the biodegradation. Their proportion is initially very small since under ordinary circumstances they realise no selective advantage from their aptitude for metabolising a product that does not exist in natural ecosystems. However, in the presence of this new substrate, they begin to multiply and after a latent phase the microbial population exponentially attains a level at which its number stabilises: this is the stationary phase during which concentration of the product diminishes linearly as a function of time (Fig. 5.18). Thereafter, the population gradually decreases concomitant with disappearance of the substrate.

We have implicitly assumed throughout the preceding section that the foreign molecule is utilised as an ordinary nutritive substrate by microorganisms and assimilated either wholly or in part, in other words metabolised. It is known that quite a few bacteria and fungi are capable of utilising xenobiotic compounds, in particular pesticides, as the only source of carbon and sometimes of nitrogen. The genes responsible for this metabolisation are frequently carried by plasmids. Very often, an association of several different microbial species is necessary for complete action on these molecules, which is also the case with natural complex substrates (see page 122). However, in certain cases the disappearance of xenobiotic molecules in the ecosystem can also be due to another process called cometabolisation.

When degradation of an organic compound yields neither energy nor assimilable elements to a microbial population, this is termed **cometabolisation**. The phenomenon occurs when a microbial population developing on another substrate adequate for its growth, excretes enzymes into the medium, some of which, by chance, encounter receptor sites on the xenobiotic product and commence acting on it. This accidental biodegradation is very difficult to demonstrate because not at all profitable to the microorganisms responsible for it, i.e., it does not stimulate their growth. If such a substrate is supplied as the only source of carbon to bacteria in vitro they show no development whatsoever.

> The ability of microorganisms to degrade complex organic substrates is currently the subject of intense studies. These organisms can actually contribute very effectively to the depollution of industrial sites which have been contaminated by the accumulation of toxic products dumped either intentionally or accidentally: chlorated or non-chlorated hydrocarbons, creosote of obsolete gas factories, phenolic compounds etc. Bacteria, easily cultivated by fermentation, seem to be of particular interest and strains showing high performance have already been selected. But populations occurring naturally on polluted sites can also be utilised successfully. Several sites have been rehabilitated on a large scale in the last few years by employing these biological methods either in situ or in areas where treatment of effluents is carried out, or in bioreactors (Bourquin, 1990).

Enhanced Biodegradation

In the previous section, the various stages involved in the disappearance of a xenobiotic molecule introduced into the soil for the first time were described. If a second application is done shortly after the first, it can be seen that the latency period is much shorter. After a third application there may be no latency whatsoever. This is because the degrading flora, stimulated by the first application, is still present in the soil in a sufficiently high concentration for its effects to manifest immediately. Subsequent applications likewise augment the rate of metabolism of the product: this is called enhanced biodegradation.

Starting up of the Phenomenon of Enhanced Biodegradation

When a phytosanitary product is applied for the first time (T1), its concentration first remains markedly constant (latency time). It then starts to undergo biodegradation and gradually disappears. Below a certain concentration, termed the threshold of efficiency, the product has no more biological activity.

The time during which the concentration of the pesticide remains above the threshold of efficiency is termed remanence. Beyond this time, the molecule can still be detected in the soil but no longer fulfils its role.

If at time T2 a new treatment is carried out with the same active matter, the period of latency can be highly reduced: remanence is also shorter. With repeated treatments, the speed of biodegradation accelerates. As a result, for applications made at equal time intervals with the same dose of active matter, the period of biological activity of the pesticide (in grey) becomes shorter and shorter.

Implementation of this phenomenon depends on the type of soil and the climatic conditions and, of course, on the chemical nature of the pesticide. In certain cases, this is observed not only after repeated use of the same product, but also after a succession of different molecules, if they all belong to the same chemical family. Thus a treatment with carbofuran can lead to acceleration of biodegradation of aldicarb, bendiocarb, methiocarb and trimethacarb. All these insecticides are methylcarbamates.

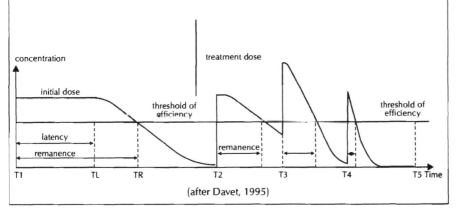

(after Davet, 1995)

The appearance of the phenomenon of enhanced biodegradation depends on the frequency of treatments and the survival capacity of the microbial populations implicated. Enhancement of degradation can be considerable. Thus, about 25 days are needed for the concentration of iprodione or vinclozolin, fungicides used against basal rot of lettuce, to diminish by half in a 'normal' soil. But this time-frame was lowered to less than 2 days in some market garden soils treated too frequently (Martin et al., 1990). Consequently, the protection ensured by the treatment became insignificant.

This phenomenon is, in a sense, reassuring as it shows that ecosystems are capable of rapidly purging at least part of the poisons they receive. But reaching such a crisis, this can be extremely troublesome, firstly because the efficiency of the phytosanitary protection is called into question since the range of active matter available for the treatment of soils is very small; and further because, as they do not clearly understand the reason for the setback, growers tend to augment the number of sprays and to increase doses, which only aggravates the problem and increases pollution. As a result, though biodegradation assists in removal of the synthetic products applied, it does not eliminate, for example, the chloride residues many of them contain.

Enhanced biodegradation does not threaten cometabolised compounds but theoretically concerns all products susceptible to metabolisation. In practice, it only affects a limited number of pesticides at the present time (Table 5.3) but some of these are difficult to replace.

Table 5.3   Main pesticides for which practical effectiveness can be compromised by enhanced biodegradation after a few treatments (a single treatment sometimes suffices to trigger the phenomenon)

Fumigants	Herbicides	Insecticides	Nematicides	Fungicides
1,3-dichloro-propene	chloridazone	aldicarb	aldicarb	benomyl
	chlorprophame	bendiocarb		
	chlortoluron	benfuracarb		carbendazime
	2,4-D	carbofuran	ethoprophos	
	dalapon	carbosulfan		dichloronitro
	difenamide	chlorfenvinphos		aniline
	EPTC	diazinon	oxamyl	iprodione
methyl	isoproturon(?)	dimethoate		
isothio-	linuron	furathiocarb		metalaxyl
cyanate	MCPA	isofenphos	fenamiphos(?)	
	metamitrone	mephosfolan		vinclozolin
	monolinuron	methiocarb		
	napropamide	phorate		
	propanil	trichloronate		
	propyzamide	trimethacarb		

For Further Information on the Effects of Microorganisms on the Environment

Burns R.G. and Davies J.A. 1986. The microbiology of soil structure. Biol. Agric. Hortic. 3: 95–113.

Cole J.A. and Ferguson S.J. (eds.). 1988. The Nitrogen and Sulphur Cycles. Cambridge Univ. Press, Cambridge, UK.

Coleman D.C., Reid C.P.P. and Cole C.V. 1983. Biological strategies of nutrient cycling in soil systems. Adv. Ecol. Res. 13: 1–55.

Gibson D.T. (ed.). 1984. Microbial Degradation of Organic Compounds. Marcel Dekker Inc., NY.

Lynch J.M. and Hobbie J.E. (eds.). 1988. Microorganisms in Action: Concepts and Applications in Microbial Ecology. Blackwell Sci. Publ., Oxford, UK.

Mansfield T.A., Goulding K.W.T. and Sheppard L. (eds.). 1998. Disturbance of the N Cycle. Proc. 3rd New Phytologist Symp. Cambridge Univ. Press, Cambridge, UK.

Pedrosa F.O., Hungria M., Yates M.G. and Newton W.E. (eds.). 2000. Nitrogen Fixation: from Molecules to Crop Productivity. Kluwer Acad. Publ., Dordrecht, The Netherlands.

Racke K.D. and Coats J.R. (eds.) 1990. Enhanced Biodegradation of Pesticides in the Environment. Amer. Chem. Soc., Washington, DC.

Schnug E. (ed.). 1998. Sulphur in Agroecosystems. Kluwer Acad. Publ., Dordrecht, The Netherlands.

Stevenson F.J. and Cole M.A. 1999. Cycles of Soil: Carbon, Nitrogen, Phosphorus, Sulphur, Micronutrients. John Wiley & Sons, NY, 2nd ed.

Wigley T.M.L. and Schimel D.S. (eds.). 2000. The Carbon Cycle. Cambridge Univ. Press, Cambridge, UK.

6

Interactions between Microorganisms

TYPES OF POSSIBLE RELATIONS

A group of individuals belonging to the same species *and* close to each other in space and time, constitute a **population**. All the populations of a defined ecosystem form a **community** (that can also be called a settlement). No matter at what level the ecosystem is considered, the populations are not merely juxtaposed; each one within a settlement interacts with the other populations. The more diversified a community, the more complex the multiple interactions, which simultaneously exercise pressures in varying directions. Overall, the effect of these divergent forces is to oppose all drastic changes and to cushion all the variations caused by modifications of environmental conditions: the functioning of a community is thus much more orderly than that of each of its components. The aptitude of communities to sustain stability is called **homeostasis**. This equilibrium, the result of dynamic processes, can be maintained only within certain limits, however. A very intense or very prolonged stress can destroy it and provoke either the disappearance of the settlement in its entirety, or uncontrolled proliferation of a particular population. Such phenomena can be observed, for example, following treatment with a pesticide.

In order to better understand the mechanisms of homeostasis, on which the stability of an ecosystem depends in part, it is important to know the reactions likely to exist among members of a community. Without this knowledge, we cannot elucidate the functioning of soils resistant to diseases nor envisage the possible utilisation of auxiliary microorganisms on a wide scale.

The overall effect of interactions manifested at a given moment among populations of the community can very schematically amount to a sum of vectors representing each, in a multidimensional space, an interaction between two populations. We shall now study these elementary bilateral actions.

Interactions between Two Populations

The effect that a population 'A' can exert on another population, 'B', can either be non-existent, favourable or unfavourable. Likewise, population B can have an unfavourable, favourable or negligible effect on population A. All these relations can be represented by the following matrix, proposed by Odum and Odum in 1953 in their treatise 'Fundamentals of Ecology' (Table 6.1):

Table 6.1 Matrix of possible relations between two populations, A and B (Odum and Odum, 1953)

Effect of population A on B	Effect of population B on A		
	Favourable (+)	Negligible (0)	Unfavourable (−)
Favourable (+)	+ +	+ 0	+ −
Negligible (0)	0 +	0 0	0 −
Unfavourable (−)	− +	− 0	− −

Apart from the case wherein the two populations are totally neutral vis-à-vis each other (because too far apart or because their ecological exigencies are totally different), all possible situations can be seen, ranging from that wherein each partner is benefited by the presence of the other, to that wherein each of the two is harmed by cohabitation.

Situations of Type + +

PROTOCOOPERATION AND MUTUALISM

These are situations wherein the two partners are capable of living independent of each other, but can mutually benefit from their proximity. Protocooperation implies relations more relaxed than mutualism. However, demarcation between the two states is not well defined and varies according to the investigator. The association of a methylotrophic *Pseudomonas* with a heterotrophic *Hyphomicrobium*, studied in vitro, is often cited as an example of protocooperation. Oxidation of methane by *Pseudomonas* produces small quantities of methanol. The methanol rapidly inhibits development of the *Pseudomonas*. But *Hyphomicrobium*, capable of utilising methanol and metabolising it as and when formed, detoxifies the medium and precludes inhibition (Wilkinson et al., 1974). Many examples of bacteria or fungi are known which, despite requiring vitamins for growth, are capable of developing on a minimal medium when associated. Thus *Proteus vulgaris* requires biotin but synthesises nicotinic acid; *Paenibacillus polymyxa*, requiring nicotinic acid, synthesises biotin (Yeoh et al., 1968). In a mixed culture these two bacteria can multiply with neither biotin nor nicotinic acid supplement as each supplies the vitamin required by the other.

SYMBIOSIS

In the case of symbiosis, the association between the two partners is very close and has an almost obligate character. The most well-known example of symbiosis in microorganisms is that of lichens, botanically considered true and definite species, but actually constituting a union of a fungus and a member of Chlorophyceae or cyanobacteria. There are also numerous examples of bacteria or algae included within protozoans. For instance, the free amoebae of genus *Acanthamoeba* are susceptible to organomercurial products but become resistant when sheltering bacteria such as *Aeromonas*, which possess enzymes that enable detoxification of these compounds. In exchange for the resistance to mercury that they provide, the bacteria benefit from being inside the amoebae in a stable and protected environment (Hagnère and Harf, 1992).

The hyphae and spores of the endomycorrhizal fungus *Gigaspora margarita* contain a very large number of organisms resembling bacteria. Analysis of their DNA showed that these organisms belong to genus *Burkholderia*, closely related to *Pseudomonas*. It was also demonstrated that these bacteria possess DNA sequences homologous to the *Nif* genes, suggesting that they can supply at least part of the nitrogen required by the fungus sheltering them (Gough et al., 2000).

Situations of Type + 0 or 0 +

COMMENSALISM

In this situation, population B is profited by the presence of population A, while population A is neither helped nor harmed by population B.

Numerous examples of commensalism exist. Many involve utilisation of secondary products of metabolism of one species by another. For example, *Saccharomyces cerevisiae* produces riboflavin, a vitamin necessary for the development of *Lactobacillus casei* (Megee et al., 1972). Fungi growing on litter or compost, such as *Chaetomium thermophile* or *Humicola insolens*, produce, from cellulose-containing substrates, organic acids and simple sugars that they do not utilise, but which other fungi such as *Thermomyces lanuginosus* (= *Humicola lanuginosa*) deprived of a celluloytic enzymatic complex, can consume (Hedger and Hudson, 1974).

In other cases, instead of profiting from the supply of a useful metabolite, the commensal profits from degradation of a toxic product or from modification of conditions in the environment that become more favourable.

Situations of Type – 0 or 0 –

AMENSALISM

Amensalism, the opposite of commensalism, is a situation wherein a population B is disturbed by a population A, with no particular advantage to population A. As a matter of fact though, it is often difficult to confirm that the disturbing microorganisms do not profit from the inhibition of their neighbours. Examples of true amensalism are therefore not numerous. Such cases nevertheless seem to occur during the decomposition of organic wastes rich in nitrogen: the ammonia formed by the action of proteolytic bacteria can diffuse and inhibit the growth of numerous fungi at a relatively significant distance from the substrate, well beyond the zone in which the fungi are likely to enter into competition with the bacteria. Strong acidification of the environment as a result of sulphur oxidation by *Thiobacillus thiooxydans* as well as lowering of the redox potential consequent to metabolisation of a carbon substrate by a very active microflora, can also be considered examples of amensalism.

It can be seen that amensalism is generally the result of non-specific mechanisms. Inhibition of one species by another, which produces antibiotics, often mentioned by Anglo-Saxon authors as a phenomenon of amensalism, mainly seems to highlight the situation (+ – or – +) described below. It is difficult to imagine that the ability to synthesise complex inhibitors such as antibiotics would continue to be transmitted from one generation to the next, were there no advantage to the organisms possessing this ability.

Situations of Type + – or – +

In this type of relation, population A grows at the cost of population B. There are numerous variants of this situation, which have been designated by terms whose meanings change according to the author. This perhaps simply demonstrates the fact that nature ignores classification and all kinds of intermediate models can be encountered among typical models on which the definitions are based. In some cases, the confrontation occurs in the presence of a substrate from which one of the partners is finally excluded. In other cases, one of the two protagonists is itself used as a substrate. Irrespective of whether exclusion or trophic relation is involved, these situations do not necessarily require intimate contact between the microorganisms (Table 6.2).

Table 6.2 Diversity in relations of type + – when one of the protagonists develops at the expense of the other

	Exclusion relations (antagonism)	Trophic relations (exploitation)
Action at some distance	antibiosis (antibiotics)	lysis (enzymes)
Action on contact	hyphal interference	predation, parasitism

ANTAGONISM

This term is often used in a very loose sense, notably in works dealing with biological pest control: it covers in that case almost all the relations of the + – and – – types. We shall use it in a much more restricted sense, to define a situation wherein one organism exercises an inhibitory effect on another that it tends to eliminate without consuming. Antagonism therefore represents a well-defined stage between amensalism (– 0 situation) and competition (situation wherein, as we shall see, the two partners are in difficulty). In this sense, the term is synonymous with *interference competition*, as defined by Lockwood (1981) and illustrated by Wicklow (1992).

Antagonism is essentially based on the secretion of soluble or volatile antibiotics that can readily be demonstrated in a large number of bacteria and fungi (Fig. 6.1). The antibiotics used in pharmacology are almost all synthesised by soil bacteria or fungi, and represent only a small fraction of the identified compounds. But measurement of them in situ is difficult and rarely has it been demonstrated in their natural environment that these compounds provide any ecological advantage to the organisms that produce them. Improvement of methods of analysis and utilisation of non-producing mutants obtained either spontaneously or by genetic manipulation, now render experimentation easier. It has thus been confirmed that the antagonism of *Trichoderma virens* against *Pythium ultimum* is due to the production of gliovirin (Howell and Stipanovic, 1983). The parasitic fungus *Gaeumannomyces graminis* var. *tritici* is inhibited by a phenazine produced by a fluorescent pseudomonad present on the roots of wheat. This inhibition has been demonstrated in the field and is sufficiently intense to afford protection to the wheat that harbours the bacterium (Thomashow et al., 1990). Another antibiotic, 2,4-diacetylphloroglucinol, secreted by another strain of *P. fluorescens*, likewise protects the roots of tobacco from the fungus *Chalara elegans* in certain soils (Keel et al., 1992). Antagonism and antibiosis can be considered synonymous in such examples.

Contact inhibition, also known as **hyphal interference**, likewise takes place in areas of antagonism but this does not seem to be caused by antibiotics. This phenomenon has been observed in some fungi capable of conquering substrates already colonised by other species. When the tips of the hyphae of the overpowered fungus come into contact with the mycelium of the invader fungus, growth of the former is abruptly arrested; within a few minutes its cytoplasm becomes granular and an increase in membrane permeability and degeneration of mitochondria are observed. This mechanism, first demonstrated in coprophilous or lignovorous basidiomycetes (Ikediugwu and Webster, 1970), is probably more widespread than thought earlier as it also exists in oomycetes such as *Pythium oligandrum* (Lutchmeah and Cooke, 1984).

LYSIS

Like antibiosis, lysis is manifested at some distance, but elimination of adversaries is accompanied by their **exploitation**. This phenomenon, peculiar to bacteria and fungi, can affect all groups of soil microorganisms. The cell wall or cuticle of the target organism is digested by extracellular enzymes (chitinases, cellulases and glucanases), sometimes accompanied by toxins targeted towards immobilising or killing the prey. The contents of the cells

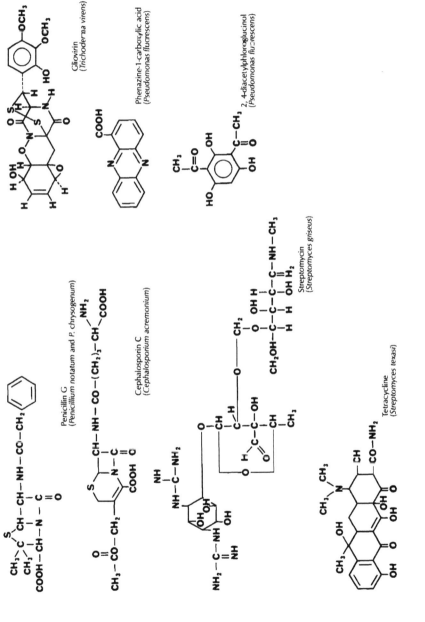

Fig. 6.1 A few examples of antibiotics of fungal or bacterial origin that are of medical or agronomic interest.

thus denuded, diffuse into the environment where other enzymes (notably proteases) assure their degradation. The products of digestion are absorbed by the fungi or bacteria responsible for the lysis.

Bacteria, and more particularly actinomycetes, are very important agents of lysis of fungi. As a general rule, species with pigmented cell walls impregnated with melanin are more resistant to lysis than hyaline species whose mycelium is quickly attacked. These susceptible fungi nevertheless very often possess organs of conservation (oospores, chlamydospores) whose thick walls ensure good resistance against the action of lytic enzymes. Several examples of lysis of bacterial colonies by fungi have been observed quite recently. The hyphae of these fungi are selectively attracted by certain bacteria (*Pseudomonas, Agrobacterium*). Once they have taken over a colony, they ramify profusely and the bacteria are rapidly lysed (Barron, 1992). This behaviour has so far been observed only among basidiomycetes, mainly the celluloytic ones. It could be a mechanism that ensures complementation in nitrogen.

The lysis just described is the result of an aggression by an exotic microorganism. It should not be confused with **autolysis**, a process of self-destruction discussed later.

PREDATION

Like lysis, parasitism (covered a little later) and predation also involve exploitation of one organism by another but the relations between exploiter and exploited are very close and necessitate intimate contact. Predation is characterised by **absorption** of the prey, in general smaller than the predator. It usually supposes an active search (with consumption of energy) for the prey by the predator.

Predation is a widespread behaviour among soil protozoans, especially naked amoebae (Fig. 6.2), nematodes and microarthropods. For example, an amoeba such as *Arachnula impatiens* can indiscriminately consume bacteria, algae, yeasts, filamentous fungi and even nematodes (Old and Chakraborty, 1986). But the diet is more often limited to one category of microorganisms: nematodes of genera *Pelodera* and *Acrobeloides* consume bacteria; nematodes of genera *Aphelenchus, Aphelenchoides* and *Ditylenchus*, collemboles (genera *Onychiurus* and *Folsomia*) and acarids (oribatids) feed on fungi. Many of these small predators manifest marked preferences: oribatids and collemboles prefer melanised fungi as prey and eschew less pigmented species. A nematode such as *Aphelenchus avenae* consumes the parasitic fungi *Rhizoctonia solani, Chalara elegans* and *Verticillium albo-atrum*, without touching oomycetes. Some amoebae have very specific, well-defined preferences; hence cultures of their preferred bacteria can be utilised to isolate them from the soil. Prey recognition is effected by chemical mediators: amoebae and nematodes orient themselves by chemotaxis, acarids and collemboles by olfactory signals. The composition of these signals can vary notably depending on the substrate on which the microorganisms consumed are grown, which elicits varying behaviour in the predators.

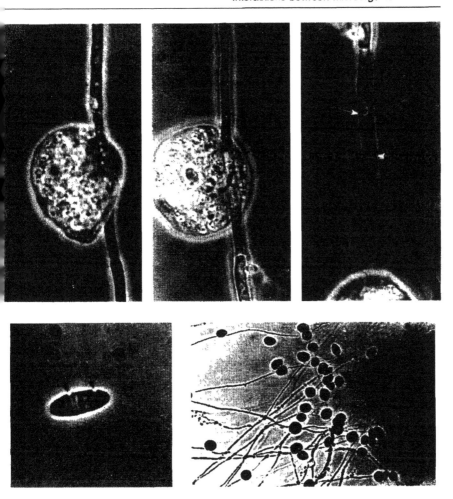

Fig. 6.2 Consumption of a fungus (*Fusarium oxysporum*) by mycophagous amoebae (plate, M. Pussard, INRA).

Above from left to right: Attack of a hypha of *F. oxysporum* by *Thecamoeba granifera* (for more details, see Pussard et al., 1979). Left: amoeba piercing the cell wall of the hypha. Middle: coagulated fungal cytoplasm is aspirated by the amoeba (arrow). Right: amoeba retires after consuming two adjacent parts of the hypha. Punctures made by it in the mycelial cell walls are clearly visible (arrows).

[lower Figures]

Bottom left: Conidium of *F. oxysporum* that has been emptied of its contents by *T. granifera*. An opening has been drilled in each of the chambers (arrows). Bottom right: A group of *T. granifera* in the process of consuming a thallus of *F. oxysporum* in a petri dish. The mycelium of the fungus is still intact at the left side of the photograph; right side, after passage of the amoebae, it is completely lysed.

Each individual ingests a considerable number of microbial propagules (this term is defined on page 252). Some of the propagules survive the trauma of passing through the digestive system and are excreted live at some distance from their place of capture. The predators thus contribute to the **dispersal of organisms** which per se are not very motile. For example, certain *Caloglyphus* (acarids) feed on a fungus, *Pythium myriotylum*, associated with shell rot of groundnut. They digest the mycelium but not the oospores of this fungus as the latter have a thick protective cell wall. While moving in the soil, the acarids disseminate the parasite, thereby contributing to contamination of new shells. Treatment with an acaricide helps lower the number of rotten shells significantly (Shew and Beute, 1979).

Predation exerts a significant **regulatory effect** on microbial communities. This phenomenon has been particularly studied in amoebae that maintain the population of bacteria at a level much lower than could be attained in their absence. However, even in the presence of significant quantities of predators, populations of rhizobia introduced into the soil are not eliminated but stabilise at a level characteristic of the soil and the bacterial strain (Crozat et al., 1982). Soil micropores definitely constitute a refuge that allows some of the bacteria to escape from the amoebae as the latter are too large to enter these pores. However, a threshold of predation can be shown to exist even in a liquid medium. According to Danso et al. (1975), this threshold is attained when the energy deployed by the amoebae to search for prey (the energy used increases with decrease in density of prey) is greater than the energy procured through their consumption. The effect of the protozoans on the fungal populations is more limited; however, maintenance of inoculum density of the fungal parasite *Gaeumannomyces graminis* var. *tritici* at a low level seems to be related, in certain Australian soils, to the action of mycophagous amoebae, unless the low density is an indirect effect resulting from stimulation of antagonistic bacteria by the amoebae (Pussard et al., 1994). By consuming the mycelium, mycophagous nematodes can also limit development of fungal parasites and under other circumstances, also depress mycorrhizal functioning. Preferential consumption of certain species can totally modify the structure of microbial communities. Thus dead leaves falling on the soil surface are colonised by a microflora essentially comprising highly pigmented fungi such as *Cladosporium cladosporioides* and *Epicoccum purpurescens.* In the absence of predators, these primary colonisers can survive on the leaves for several weeks. But when consumed by collemboles, they are rapidly eliminated and replaced by secondary colonisers constituted by less appetising hyaline species (*Trichoderma, Penicillium*) (Klironomos et al., 1992).

The regulation exercised by predators has an effect not only on the total strength, but also on the activity of the populations. For example, consumption of *Mortierella isabellina* by a collembole (*Onychiurus armatus*) provokes an increment in growth rate of the fungus and causes the mycelium to change its mode of growth from short, stubby and recumbent to aerial (Hedlund et al., 1991). This stimulation, observed in numerous predator-prey pairs, is probably exercised through the intermediary of chemical signals. Thus the culture filtrate of *Acanthamoeba castellanii* contains soluble and thermostable metabolites capable of enhancing the metabolic activity of its prey, *Pseudomonas putida*: respiration of the bacteria, production of ammoniacal nitrogen and synthesis of pyoverdine (a siderophore) increase very significantly under the effect of these metabolites (Levrat et al., 1992). Stimulation of the metabolic activity of the prey compensates the uptake caused by the predator to a threshold that varies depending on external conditions. A too high increment in predator population could nevertheless destroy this equilibrium.

Another important consequence of predation is acceleration of the **recycling** of mineral elements that accompanies it. This phenomenon, especially prominent in nitrogen and phosphorus, has several origins: the metabolic activity of the microorganisms is more intense in the presence of predators; predators participate in mineralisation by excreting ammoniacal nitrogen and phosphorous compounds; lastly, the waste produced by the microfauna serves as a substrate for new microbial populations which continue the work of mineralisation. In trials conducted in microcosms using earth containing pseudomonads, introduction of a mixture of protozoans resulted in 34% augmentation in mineralisation of organic soil nitrogen and an increment of 20% in nitrogen absorption by wheat grown in pots (Kuikman and Van Veen, 1989). An analogous phenomenon was observed when a nematode (*Mesodiplogaster lheritieri*) and a bacterium (*Pseudomonas paucimobilis*) were associated (Anderson et al., 1983).

A unique group of fungi that capture nematodes by means of adhesive mycelial rings similar to the snares of poachers can be linked to predators (Fig. 6.3). Once trapped, the victim is invaded by the fungus and completely digested. *Arthrobotrys* and *Dactylella* are the better known examples. The presence of free nematodes stimulates formation of the capturing organs. Induction of these traps results from the interaction between fungi and bacterial flora associated with the digestive tube of the nematodes (Table 6.3). Induction and predation are specific phenomena that depend on the microorganisms present. These fungal predators can also lead a saprophytic life. Many of them show strong cellulolytic or even ligninolytic activity, which prompts one to think that capture of nematodes (rich in proteins) constitutes a method of procuring nitrogen, which is very scarce in their substrates (Barron, 1992).

PARASITISM

The word parasitism evokes the idea of an organism, usually small, attaching to another organism, sometimes even inside it. Exploitation of the host supplies the parasite with all its resources or only part of them, essential or non-essential. The first case comprises strict parasites that are generally impossible to culture on artificial media. In the second case, the parasites are facultative, capable of multiplying on simple substrates, either enriched or not with vitamins and amino acids. As a general rule, the closer the trophic relations of the parasite with its host, the greater its specificity.

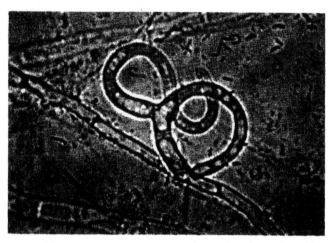

Fig. 6.3 Annular traps of a nematophagous fungus, *Arthrobotrys* (plate, J.C. Cayrol, photo library, INRA).

Table 6.3 Effect of presence of free nematodes and their associated bacterial flora on the percentage of larvae of *Meloidogyne* captured by the fungus *Arthrobotrys musiformis*. Larval trapping is extremely restricted in the absence of the free nematode *Diplogaster*. The addition of aseptic *Diplogaster* induces no stimulation. On the other hand, the quantity of larvae captured became significant after the addition of non-aseptic *Diplogaster*. Stimulation of formation of traps by the nematophagous fungus is apparently induced by the action of bacteria present in the digestive tract of *Diplogaster* (after Cayrol and Quilès, 1985).

Treatment	Period of observation		
	5 hours	24 hours	48 hours
Control: without *Diplogaster*	0	0.9	25.7
Diplogaster deprived of bacteria	0	6.9	34.1
Diplogaster with associated bacteria	51.6	75.2	94.3

Viruses represent a case of absolute parasitism. Bacteriophages have been well studied in medical and industrial bacteriology, but their incidence in soil bacterial populations is poorly known. Even though examples of viral parasites of fungi, protozoans and nematodes are known, little work has been achieved in assessing their importance.

Bdellovibrio illustrate the concept of parasitism very well. Discovered only about 40 years ago, *Bdellovibrio* are very small bacteria that traverse the cell wall of other bacteria, such as *Pseudomonas* or *Xanthomonas*, penetrate the cytoplasm and multiply there at the expense of the host. Several bacterial parasites of protozoans as well as bacterial parasites of nematodes (*Pasteuria penetrans*) are likewise known.

Numerous fungi are parasites of other soil fungi. Most are facultative parasites, capable of an active saprophytic life. *Rhizoctonia solani*, for example, can develop inside the relatively voluminous mycelium of certain zygomycetes. In turn, it may be parasitised by other fungi, such as *Trichoderma harzianum, T. hamatum* or *Penicillium vermiculatum*. Much attention has been paid during the last few years to parasites of phytopathogenic fungi and the successive stages of the process of parasitism intensely studied in the hope of selecting strains utilisable as agents of biological control (see box below and page 314). Similar interest has also been shown in fungal parasites of the larvae or eggs of nematodes. In most cases the species concerned sporulate heavily and possess adhesive conidia which affix on the surface of the host. They then produce a germinative tube that penetrates the interior of the animal, ramifies and gradually digests the host. This group has representatives among genera *Acrostalagmus, Verticillium* and *Paecilomyces*. *Nematoctonus* are lignivorous and nematophagous fungi whose perfect stage, termed *Hohenbuehelia*, is similar to *Pleurotus*. Here again, the nematodes probably constitute the nitrogen source necessary for assimilation of the lignocellulosic substrate (Barron and Dierkes, 1977).

Mycoparasitism in Trichoderma

Mycoparasitism is a complex phenomenon that progresses in several stages. The mechanisms have been particularly well studied in interactions between *Trichoderma* (fungi that might possibly be used as agents of biological control) and the pathogenic fungi *Rhizoctonia solani* and *Sclerotium rolfsii* (Chet, 1987).

1. Stimulation: the *Trichoderma* on perceiving its host, directs its hyphae towards it by chemotropism. The nature of the stimulus is not known.

2. Recognition: This is manifested by adhesion of the mycoparasite to the cell walls of its host. Attachment is achieved by bonding of an agglutinin of the pathogenic fungus to certain sugars present in the walls of *Trichoderma*. It can be easily shown that the cell walls of *R. solani* contain a lectin by placing its hyphae in a suspension of human erythrocytes of blood group O. The erythrocytes agglutinate instantaneously on the mycelium. These cells are characterised by the presence of L-fucose, a sugar that plays a role in the composition of the cell walls of *T. harzianum* and *T. hamatum*. The hyphae of *R. solani* previously placed in a solution of L-fucose so as to saturate the lectin, agglomerate neither the erythrocytes nor the mycelium of *Trichoderma*.

   The cell walls of *S. rolfsii* contain a lectin of a different nature. This lectin is a glycoprotein similar to concanavalin A. It does not agglutinate any type of erythrocyte but can

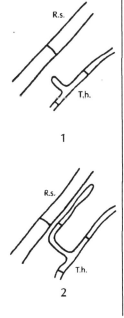

agglutinate certain bacteria such as *Escherichia coli* of type B as well as strains of *Trichoderma* capable of attacking *S. rolfsii* (but not the non-parasitic strains). It recognises D-glucose and D-mannose, sugars also present in the cell walls of *Trichoderma*.

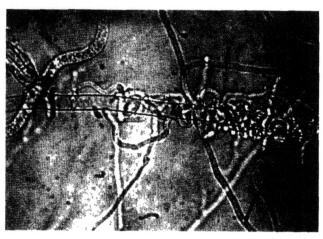

3

3. Rolling up (photograph above): A *Trichoderma* sp. rolls itself around the hyphae of its host and forms falcate structures that ensnare it. Inbar and Chet (1992) have shown that rolling requires prior sugar-lectin recognition. They fixed concanavalin A or purified agglutinin of *S. rolfsii* on nylon fibres by covalent bonds. The hyphae of *Trichoderma* attached to these fibres and encircled them as though they were the mycelium of their host. On the other hand, untreated fibres provoked no reaction (plate, P. Camporota, INRA).

4. Penetration: The extremities of the falcate formations secrete chitinases and β-1,3 glucanases that dissolve the cell wall of the host, enabling them to penetrate the hyphae.

5. Lysis: Other extracellular enzymes are then secreted (protease, lipase) and the cellular contents of the host rapidly lysed.

Trichoderma also produce a vast range of antibiotics of differing chemical composition. A wide variability in the nature and quantity of the antibiotics synthesised is observed depending on the isolates.

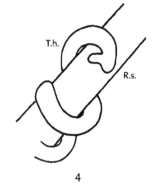

4

Situation of Type – –

COMPETITION

Organisms are in competition when an element indispensable to their development is present in insufficient quantity in the environment. The organism most adept at rapid utilisation of the limiting element or at making it inaccessible to others, takes it away from competitors. This preferential utilisation sometimes necessitates resorting to special mechanisms not needed in situations of plenty: we shall see an example with siderophores. Such a strategy nevertheless implies expenditure of supplementary energy.

Even though competition is a familiar notion and a phenomenon implicitly considered frequent, it is not easy to demonstrate. Among all the elements present in the environment, it is necessary to identify which (or which ones) is (are) limiting. This supposes precise knowledge of the requirements of the microorganisms; it is also necessary to quantify development of the populations present and to determine the quantity of the limiting factor. Competition can be considered a universal phenomenon: in a situation of acute scarcity it can occur among various species of the same community as well as among individuals of the same population. It is even more intense when the exigencies of the organisms are similar. In other words, competition occurs essentially within one **ecological niche**.

The term ecological niche should not be confused with habitat, which describes the place where a species lives. Ecological niche refers to the function exercised by a species within the settlement. It can only be represented in a space with n dimensions, each dimension being related to a particular characteristic of the species (nutrition, conditions of development, reproduction, susceptibility to stress etc.). Many species can share the same habitat (micropores of the soil for example) but have very different ecological niches.

Among microorganisms, competition is much greater for nutritive elements than living space. However, it is also likely that competition for space occurs when parasites or symbionts are jostling for sites of penetration on the surface of a plant organ: competition between strains of *Rhizobium* for nodulation of roots of a legume is one example, and of paramount agronomic importance. As a matter of fact, successful introduction of a strain selected for its nitrogen-fixing qualities, depends primarily on its competitive power vis-à-vis wild strains of *Rhizobium* already present in the soil. It would avail nothing to introduce a high-performance strain if it were incapable of colonising the roots of its host in the presence of indigenous microflora.

The qualities that enable an organism to succeed in a competitive situation can be highly varied. If conquest of a new substrate, still unoccupied, is the objective, the successful organism is that one most sensitive to the stimuli emitted by the substrate and which reaches it most rapidly. Occupancy of terrain is synonymous here with obtaining nourishment. For instance, siphomycetes that show very rapid growth are generally the first invaders of leaves and plant debris fallen on the soil. In an environment already colonised, the balance is in favour of organisms capable of eliminating their competitors by various means, antibiotics for example: demarcation between antagonism and competition is not always clear. Victory may also go to a species capable of detoxifying the antibiotics of its adversaries but without itself being inhibitory. Yet another strategy is possible: it involves collection and confiscation of all traces of the vital element present in limited quantity in the environment. This is what happens in the case of iron. Competition for ferric iron, a cofactor for numerous enzymes, is very high because this ion is extremely insoluble at pH values close to neutrality. Microorganisms respond to the Fe^{3+} deficit by synthesising and excreting chelating compounds, the **siderophores**. Siderophores are powerful agents of sequestration of Fe^{3+}. However, their capability for complex formation with iron varies markedly. The affinity for iron shown by siderophores of fungi (for example fusarinine of *Fusarium oxysporum*) is generally much lower than that of siderophores of bacteria (for example pseudobactin of *Pseudomonas*): this signifies that fungi are quite often less competitive for iron than bacteria. The variability of siderophores is also based on their stereochemical constitution, which is recognised by specific membrane receptors. After capturing the ferric ions, the chelates bind to these receptors on the surface of the cells, from where the ions are transported into the cytoplasm. The more competitive strains are those that possess a siderophore with very great affinity for iron and a structure that other strains cannot recognise and which, furthermore, have a system of transport capable of utilising the siderophores of the other strains (Leong, 1986).

Complex Situations

Equilibria among microbial populations in the soil are managed by the elementary mechanisms just described but the real situation is far more complex than these schematic representations. Environmental factors can considerably modify the degree of interaction between microorganisms. Thus predation by protozoans is active only in soil sufficiently moist. At low temperatures, a fungus can become dominant whereas it is dominated by a more competitive species at higher temperatures. On the other hand, several different types of interactions can occur simultaneously between two populations. We have already studied the *Paenibacillus polymyxa–Proteus vulgaris* association as an example of mutualism. However, *P. vulgaris* excretes a protein into the medium whose inhibitory effect retards development of *P. polymyxa*. Therefore, both mutualism and amensalism occur (Yeoh et al., 1968). In the relation between *Acanthamoeba castellanii* and *Pseudomonas putida*, described earlier, stimulation of the bacteria (commensalism) and predation coexist.

Introduction of a third partner into a system also modifies the nature of the relations between the first two. Thus, when the two bacteria *Escherichia coli* and *Azotobacter vinelandii* are cultivated together in a chemostat where glucose is a limiting factor, *E. coli* rapidly supplants *A. vinelandii*, which does not assimilate glucose to the same extent (Fig. 6.4). If *Tetrahymena pyriformis*, a predatory protozoan, is added to the chemostat, the populations of the two bacteria, consumed simultaneously in proportion to their number, evolve parallelly, as shown by damped down oscillations (Fig. 6.5). *A. vinelandii* is not eliminated (Jost et al., 1973).

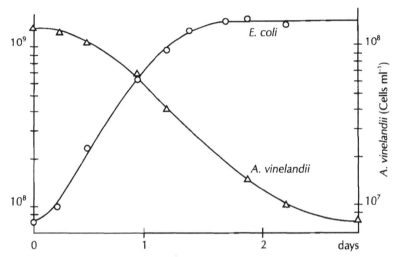

Fig. 6.4 Development of populations of *Azotobacter vinelandii* and *Escherichia coli* when they are in competition in continuous culture (after Jost et al., 1973).

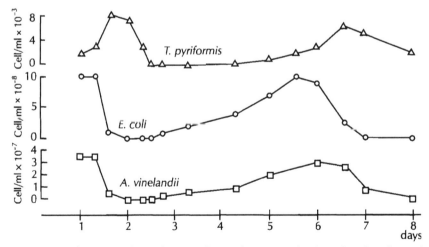

Fig. 6.5 Development of populations of *Azotobacter vinelandii* and *Escherichia coli* in continuous culture, in the presence of a predatory protozoan, *Tetrahymena pyriformis* (after Jost et al., 1973).

Another example of these complex relations is given by control mechanisms of the 'quorum sensing' type. In such systems, the expression of some of the genes depends on the size of the bacterial population.

This system was first demonstrated in a symbiotic marine bacterium, *Vibrio fischeri*, which forms luminous organelles in the fish or the squid that shelters it. This bacterium possesses a *lux I* gene that assures the synthesis of an N-acyl-homoserine lactone (AHL) which plays the role of a signal. When the bacteria are free and dispersed in water, the signal is diluted and nothing happens. When they are united in large quantities in the confined space of an organelle, the AHL accumulates in sufficient quantities to interact with the product of a *luxR* gene: the *lux* operon, which codes for luciferase, is then activated and light is emitted.

This type of regulation by the AHL was subsequently demonstrated in a large number of telluric bacterial species: activation of a gene or a sequence of genes does not occur unless population densities are sufficiently high, since such densities are necessary for the signal to accumulate at an adequate level. But these signals are not very specific and a bacterial population can respond not only to its own AHL, but also to those arising from non-isogenic populations in the vicinity. In other words, a gene subjected to a regulation of the quorum sensing type can be expressed in a population of low density if the signal is emitted by other, more abundant bacteria present in the same ecological niche. Inversely, exotic populations can render a signal inactive by intercepting or degrading it (Pierson et al., 1998). AHLs are the most commonly studied signals but other diffusible ones of diverse chemical nature are also known.

The structure of microbial communities is a result therefore of a **dynamic equilibrium** assured by multiple connections that are often difficult to demonstrate.

FUNGISTATIC EFFECT OF SOILS

The preceding pages lead us to believe that soil is a permanent bed of intense activity. In reality, it should be understood that most of the time this activity is quite weak, in spite of the multiplicity of species that abound there. Numerous studies conducted on fungi in natural soils have helped in establishing that germination of their propagules and growth were much more limited than previously thought, even though the conditions of temperature, pH and water content appeared favourable. Dobbs and Hinson coined the term **fungistasis** (or mycostasis) in 1953 to designate this phenomenon and numerous studies have followed since.

Fungistasis is encountered in all types of soil, at all latitudes, and applies to most fungi; it is therefore not specific. To demonstrate the latter, spores were placed on water agar discs or on moistened porous membranes placed on the soil surface, and their germination studied. Spores can also be introduced directly into the soil or buried in a nylon net. The inhibitory effect is annulled by sterilisation of the soil and restored when the sterilised soil is reinoculated with a fraction of the original soil: this is the result of biological activity. Experiments have also shown that by and large the smaller the spores, the more susceptible to fungistasis; likewise, the susceptibility of mycelial filaments is inversely proportional to their diameter.

Explanation of Fungistatic Effect by Action of Inhibitors

Since inhibition had been observed, it was logical to search for inhibitors in the soil.

Realising that almost all the antibiotics used in human and veterinary medicine are produced by microorganisms living in the soil, antibiotics were the first to draw scientific attention. However, even though they are numerous, producers of antibiotics represent only a small fraction of the microflora and the compounds they synthesise are rapidly adsorbed on the colloids or denatured by the soil enzymes. On the other hand, their action is specific: it never encompasses all the species present. Lastly, it is possible to re-establish fungistasis in a sterilised soil by resowing exclusively with species that do not produce antibiotics. Antibiotics may therefore be implicated in local phenomena, but they cannot explain generalised inhibition.

Unidentified soluble inhibitory products of varying molecular weights have been demonstrated in soil percolates harvested from lysimetric tanks (Vaartaja, 1977). Basidiomycetes and oomycetes are susceptible while some species of *Penicillium* and *Gliocladium*, on the contrary, can be stimulated. Many aromatic compounds arising from the decomposition of lignin (such as vanillin, eugenol, cinnamic aldehyde) have an inhibitory action and incorporation of sawdust in the soil effectively increases the fungistasis (Lingappa and Lockwood, 1962). These products very likely contribute to fungistasis in humus-rich soils. Their role is doubtful in soils poor in organic matter.

Demonstration of volatile compounds capable of exerting an effect at a greater distance from their source of release than soluble molecules, revived interest in research on inhibitors. A small air current containing gases emitted by a soil sample placed in a closed chamber, then passed over spores in a neighbouring compartment, sufficed to preclude their germination (Romine and Baker, 1973). Ammonia released during decomposition of proteins and urea inhibits germination and development of numerous fungi (see page 94). This effect is higher in alkaline soils. However, inhibition does not affect all species (Schippers et al., 1982). Ethylene is also quite frequently found in soils. It is produced by anaerobic bacteria in a reducing environment and in the presence of profuse organic substrates. Fungi and bacteria also produce ethylene under aerobic conditions from simple substrates such as methionine. For some time inhibitory properties were attributed to this gas. Subsequently it was acknowledged that ethylene is not fungistatic but may have an indirect action: in soils incubated in the presence of ethylene, volatile compounds such as allyl alcohol form, which are themselves strong inhibitors (Archer, 1976). Volatile antibiotics may also be present in the atmosphere of certain soils, acidic or alkaline, at concentrations sufficient to exercise a fungistatic effect (Liebman and Epstein, 1992). Their chemical nature has not yet been determined.

Explanation of Fungistatic Effect by Phenomena of Competition

The inhibition caused by volatile or soluble compounds is too closely related to specific characteristics of certain soils for a significant part of the fungistatic effect universally observed to be attributed to it. For that reason, Lockwood proposed another explanation in 1964, supported by numerous subsequent studies, briefly summarised here (a detailed synthesis was published by Lockwood in 1977). Two types of spores must be considered: small spores, with hardly any reserves, which do not germinate in distilled water and need a supply of external energy, glucose, for example; and voluminous spores containing nutritive reserves and capable of germinating in pure distilled water.

Spores Lacking Reserves

The conidia of *Trichoderma viride*, for example, germinate if placed on an agar disc containing water and 1 to 2 g L^{-1} glucose and incubated in a petri dish or on sterilised soil (Fig. 6.6). If the same glucose-agar disc is placed in contact with a non-sterile soil sample, germination is very poor. No germination takes place if the disc is incubated for 1 or 2 days on non-sterile soil before placement of conidia. Nor can an antibiotic be detected in such discs. Furthermore, they contain almost no more glucose. This shows that microbial activity of the soil creates a nutrient sink that rapidly deprives the glucose-agar disc of its nutritive elements as a result of diffusion. The diffusion gradient created prevents the conidia from finding the energy supply required for germination. The same effect can be realised by incubating the agar discs on activated animal charcoal or by prolonged washing with distilled water before sowing the conidia.

Spores with Nutritive Reserves

Larger conidia, such as those of *Cochliobolus victoriae* for example, germinate when placed on a filter membrane floated on distilled water or applied on sterile moistened sand (Fig. 6.7). But the rate of germination is very low if the membranes are incubated on a non-sterile soil. The rate is likewise very low if the membranes are placed on sterile sand washed by a current of distilled water. If washing is halted forthwith, germination again becomes possible. But if washing continues for several days, germination occurs only after a supply of glucose. This shows that the nutritive reserves of the spores, for the most part soluble, are rapidly lost due to the nutrient sink, making germination problematic. By culturing *C. victoriae* on a substrate containing radioactive carbon, it is possible to obtain conidia marked with ^{14}C. The exudates from these spores are used within a few minutes by the soil microflora and mineralised into $^{14}CO_2$. Respiration of the spores is another source of loss adding to the leakage represented by the exudates. For reasons still not clear, it is always more intense in a non-sterile soil (Mondal and Hyakumachi, 1998). The spores are thus progressively reduced to the

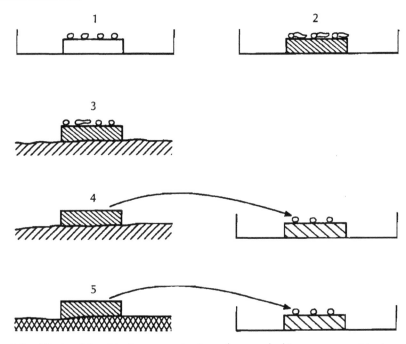

Fig. 6.6 Effects of fungistasis on germination of spores lacking reserve nutrients.
1. Placed on distilled water agar discs, the spores do not germinate even under aseptic conditions.
2. Spores germinate under aseptic conditions if placed on a disc of glucose water agar.
3. If the nutritive agar disc containing spores is placed on the surface of a non-sterile soil, germination is very poor.
4. Nutritive agar disc maintained for 2 days on the surface of a non-sterile soil before being sown. It was then placed in a petri dish and the spores disposed on it; it can be seen that no germination occurred.
5. The same situation can be reproduced by maintaining the agar disc for 2 days in contact with sterile active charcoal.

state of spores without reserves. They can no longer germinate except in the presence of an external source of energy. Such would be the case in our scheme, were washing of sand done with a nutritive solution instead of distilled water.

These experiments lead us to infer that soil microflora is in a permanent state of scarcity. The fungistatic effect is due to the intense nutrient sink caused by **chronic energy deficiency**. The soil thus always functions well below its capacity. Under these conditions, the introduction of the simplest substrate into such a medium should result in a violent spurt of activity, all the more intense the greater the biomass. This is well illustrated by the rapid and intense release of carbon dioxide that follows the addition of glucose, reflecting the intensity of the nutrient sink.

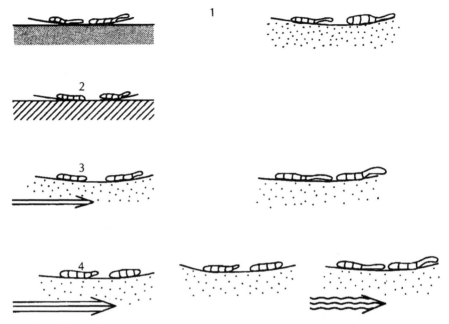

Fig. 6.7   Effects of fungistasis on germination of spores containing reserve nutrients.
1. When the conidia are placed on a porous membrane, they germinate in the presence of sterile distilled water; they also germinate if the membrane is placed on sterile moistened sand.
2. Germination is poor if the membrane holding the spores is placed on the soil surface.
3. Germination is likewise poor if the membrane is placed on sterile sand washed with a current of distilled water. However, if the washing is terminated after a few hours, germination resumes.
4. If washing is intense or if it lasts for several days, germination is very poor and does not resume once it ceases. However, if a current of glucose water is passed through the sand, germination occurs normally.

Autolysis

The absence of nutritive resources not only causes a blockage of germination and mycelial growth, but also causes lysis of the previously formed mycelium. It involves autolysis, caused by the action of endogenous enzymes that are generally stocked in specialised vesicles, the lysosomes. This can be readily demonstrated by placing a young thallus of *Fusarium solani* for example, on a filter membrane that is itself placed on sterile sand. If the sand is washed slowly with a nutritive solution, the thallus grows. If the nutritive solution is replaced by distilled water, lysis of the mycelial filaments occurs soon, in the absence of any contamination. Autolysis stops and growth resumes once the nutritive supply is resumed (Ko and Lockwood, 1970).

Autolysis is therefore a response to nutritional stress. It can ultimately result in the death and disappearance of the colony. However, most commonly it is the first stage in a process that leads to new distribution of the cytoplasmic material. Nitrogen, carbon and phosphorus of the lysed parts are thus transferred towards the tips of the hyphae which continue to grow and in which these elements are immediately reassimilated. This phenomenon of translocation helps in the exploration of the environment by the fungi and allows them to search for new substrates least costly as possible (Paustian and Schnürer, 1987). This often results also in the formation of structures of conservation such as sclerotia or chlamydospores, which is made possible by recycling of the cytoplasmic material. Thus the macro-conidia of *Fusarium* (fragile organs of dissemination) spread on a filter membrane and subjected to nutritional stress, undergo autolysis rapidly and chlamydospores (organs of conservation with thick cell walls) are formed. Mineral salts are nevertheless necessary, but the soil solution normally contains them (Hsu and Lockwood, 1973).

Ecological Consequences of Fungistasis

Susceptibility of spores to fungistasis appear to be an adaptation to the nutrient sink in the soil. Should the spores germinate spontaneously, their growth could not continue in the absence of nutrients, the germinative tubes would undergo lysis and the stock of fungal propagules in the soil would rapidly decrease to complete disappearance. The exogenous nutritive stimulus, which alone allows germination, would guarantee the spore that an adequate substrate (plant debris for a saprophyte, host plant for a parasite or a symbiont) is present in the environment. It is also imperative that after germination, the fungus be capable of surmounting competition and colonising the substrate. Anyway, it would have dispensed its energy with some chance of success if it germinated but was unsuccessful in colonising. In case of failure, the autolysis induced by the impossibility of continuing exploitation of the substrate must allow it to eventually constitute a few new organs of conservation. There are a few strains of *Cochliobolus sativus* in which the conidia are non-susceptible to fungistasis and germinate spontaneously. Well adapted to aerial life (transmitting the infection from leaf to leaf), they disappear rapidly if incorporated into the soil. On the other hand, conidia of the forms of *C. sativus* susceptible to fungistasis can survive for many months under the same conditions (Chinn and Tinline, 1964).

Conclusion and Generalisation

The hypothesis of generalised competition, though accepted by a large majority of researchers, does not exclude the possibility of chemical inhibition coming into play under specific circumstances and in specific sites. It should be remembered that the fungistatic effect is universally present but can be temporarily suspended by the addition of an energy-giving substrate.

Besides the numerous studies carried out on fungi, a few have also been done on other soil microorganisms. Ko and Chow (1977) showed that bacteria, isolated from a soil, then reintroduced, are incapable of multiplying in this soil unless previously sterilised or a solution of glucose and peptone added. Thus there is also a bacteriostatic effect of soils. More recently, ciliatostasis has also been proposed to explain the phenomena that provoke encystment and inhibit decystment and development of protozoans. It can therefore be said that a general **microbiostasis** occurs in the soil. The intensity of this effect varies depending on the species subjected to it, the types of soil and environmental conditions.

In a way, growth is therefore an accidental phenomenon between two periods of inactivity. Its main purpose is the accumulation of reserves necessary for the constitution of new forms of conservation before the state of dormancy returns. We shall see that in this ocean of poverty, the roots of the plants represent islands near which the nutritive status of the soil is profoundly modified.

RESISTANT SOILS

A few soil fungi and bacteria can cause serious plant diseases with grave economic consequences. However, under certain conditions, the pathogenicity of these microorganisms does not manifest: there are soils in which, despite the presence of a pathogenic agent, the susceptible plants remain unharmed even under climatic conditions favourable for expression of the disease. These soils are said to be resistant to disease development or, more simply, resistant (i.e., suppressive soils). On the other hand, soils in which the disease manifests are termed susceptible (i.e., conducive or permissive soils).

In order to demonstrate the resistance of a soil, the pathogenic agent has to be introduced and a susceptible variety of host plants cultivated under conditions of temperature and water content most propitious for manifestation of the disease. Serially increased quantities of the pathogenic inoculum can help determine the dose at which resistance is overcome (Fig. 6.8). For a given level of inoculum (calculated from preliminary trials), the progress of symptoms as a function of time can also be ascertained. Based on these data, statistical models enable a comparison of multiple soils and their classification according to degree of resistance (Corman et al., 1986).

Soil resistance may be **constitutive** or **acquired**. In the first case, the biological and physicochemical environment is such that the infective agent, if introduced, is not able to survive, or if already present, cannot express its pathogenicity. In the second case, the soil initially allows the disease to develop. Then after some years, because of certain techniques employed or, more commonly, because of repeated cultivation of the host plant in the same field, the soil becomes progressively suppressive.

We have already encountered (see page 99) soils wherein resistance to *Ralstonia solanacearum*, the agent for bacterial wilt, is related to the presence of certain types of clays. We shall now present two examples in which microorganisms are involved: constitutive resistance in one case (fusariosis) and acquired resistance in the other (take-all disease). Soils resistant to other

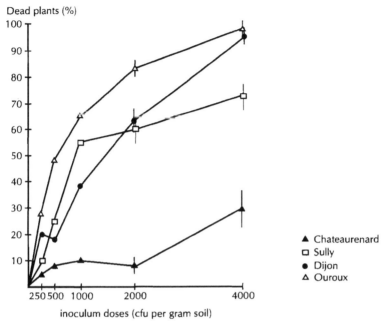

Fig. 6.8 Receptivity of 4 soils to vascular fusariosis of flax: percentage of plants dead 6 weeks after inoculation of soils with increasing doses of *F. oxysporum* f. sp. *lini*. Resistance of the Châteaurenard soil begins to surrender only at inoculum doses greater than 2000 cfu per gram soil, whereas in the other soils, severity of the disease increases rapidly with concentration of the pathogenic agent. However, it may be observed that starting from 1000 cfu g^{-1} the progression slows down in the Sully soil: it may consequently be assumed that Sully soil is less susceptible than the soil of Ouroux (Alabouvette, 1986).

diseases caused by fungi or nematodes are also known (Cook and Baker, 1983). But the mechanisms of conferral of this property are still, by and large, not known.

Soils Resistant to *Fusarium oxysporum*

Soils resistant to diseases provoked by diverse special forms of *Fusarium oxysporum* are definitely the better known models. They have been the subject of intense research over the last twenty years. So, it is impossible to summarise in a few lines all the works devoted to them, notably by the team of R. Baker at Fort Collins (Colorado, USA), that of J. Louvet, followed by C. Alabouvette at Dijon (France), and that of B. Schippers at Baarn (The Netherlands). The results obtained at Dijon from a market garden soil of Châteaurenard have already given rise to several reviews (Alabouvette, 1986; Lemanceau and Alabouvette, 1993; Alabouvette et al., 2001).

The main points are as follows:
— *Resistance is specific*: the Châteaurenard soil precludes development of all the diseases caused by *F. oxysporum*, irrespective of the host studied.

But this soil is not resistant to diseases caused by other fungal parasites, not even those caused by other species of *Fusarium* (*F. solani* or *F. roseum*).

— *Resistance is of biological origin*: this totally disappears if the soil is sterilised by autoclaving, fumigation or exposure to gamma rays.

— *Resistance is transmissible*: a small proportion of the soil from Châteaurenard suffices to impart resistance to a susceptible soil if it has been sterilised before mixing, and this confirms the biological nature of resistance.

— *Pathogenic agent survives in resistant soil*: the evolution of a population of pathogenic *F. oxysporum* incorporated in a resistant soil, over a period of time, is comparable to that observed in a susceptible soil. The fungus is still present in each of these soils even one year after introduction.

— *Fungistasis is higher in resistant soil*: a very small proportion of chlamydospores of *F. oxysporum* (less than 10%) is capable of germination in the presence of resistant soil. In the presence of a susceptible soil, this proportion is much higher (it can even reach 35%). An addition of glucose expedites germination; if a sterilised soil is used, inhibition totally disappears. Furthermore, it is necessary to add 5 to 10 times more glucose to the resistant soil than to the susceptible, in order to achieve a similar level of germination, or a comparable development of the populations. All this shows that the resistant soil has a much higher fungistatic power than the susceptible soil.

— *Competition for iron is also evident*: if EDDHA, a ligand with a very high affinity for iron, or pseudobactin, a siderophore produced by fluorescent pseudomonads, is added to a susceptible soil, the severity of fusariosis is reduced. On the other hand, the addition of iron chelated by EDTA, readily assimilable by *F. oxysporum*, reduces the resistance of Châteaurenard soil (Fig. 6.9). In an agar medium poor in Fe^{3+} or in soil, germination and/or elongation of the germinative tubes of the chlamydospores of *F. oxysporum* is very poor in the presence of certain strains of *Pseudomonas*, their purified pseudobactin, or EDDHA. The Fe^{3+} ion is thus a limiting factor for *F. oxysporum* and the fluorescent pseudomonads are comparatively more competitive.

— *Analysis of microbial flora responsible for resistance*: the rapid respiration method of Anderson and Domsch (1978) was used to compare the Châteaurenard soil with a susceptible soil. The initial respiratory response to the addition of glucose was 3 to 4 times more in the resistant soil, which shows that it holds a higher microbial biomass (Fig. 6.10). The respiratory activity peaked in 12 hours in the resistant soil, then decreased very rapidly, falling below its initial value after 24 hours. In the susceptible soil, the release of carbon dioxide was gradual, the maximum hardly remarkable, but the respiratory activity remained at a level higher than the initial level for several days. Fungistasis was thus much more rapidly re-established in the resistant soil than in the susceptible in which competition for carbon was less intense, and the chlamydospores of *F. oxysporum* had time to germinate.

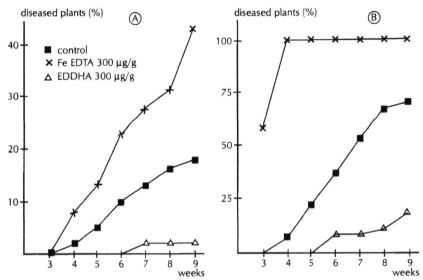

Fig. 6.9 Receptivity to vascular fusariosis of flax of a resistant soil (A) and a susceptible soil (B) modified with EDDHA ligand or Fe EDTA chelate. Graphs depict the evolution of the percentage of diseased plants after infestation by *F. oxysporum* f. sp. *lini* as a function of time. Resistant soil became infected at a dose of 10,000 cfu per gram soil and the susceptible soil, at a dose of 5,000 cfu per gram (Lemanceau et al., 1988).

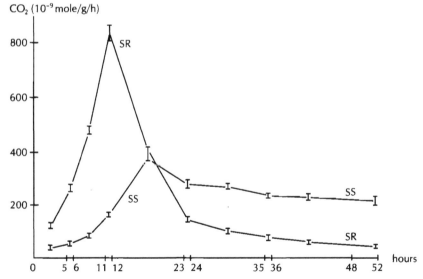

Fig. 6.10 Kinetics of release of carbon dioxide by the resistant soil of Châteaurenard (SR) and by a susceptible soil (SS) after the addition of glucose at the rate of 1 mg per g soil (Alabouvette, 1986).

Gradual sterilisation of the Châteaurenard soil using a mixture of air and vapour at different temperatures revealed that resistance is essentially related to the presence of fungi: it disappears after treatment at 55°C, a temperature at which almost all the fungi are eliminated, whereas most of the bacteria survive. Mycological analysis demonstrated the abundance of saprophytic forms of *F. oxysporum* and *F. solani*, which represent 25 to 40% of the total fungal flora, whereas the normal average is close to 10%. It is possible to restitute almost all the protective (suppressive) properties in a sterilised soil sample of Châteaurenard by introducing a mixture of saprophytic *F. oxysporum* and *F. solani*. One single strain of *F. oxysporum* has been shown to be effective if properly selected: not all isolates have the same qualities. Analysis of bacterial flora gave a rather unexpected result, considering all that has been said in the preceding section: the resistant soil is relatively poor in fluorescent pseudomonads.

— *Competition for the receptor sites*: though it has not been actually demonstrated, it would seem that a root has a determined number of sites available for colonisation. Colonisation of a host plant by a strain, or a mix of strains of *F. oxysporum*, does not exceed a certain limit, regardless of the quantity of inoculum added. Non-pathogenic *F. oxysporum* are capable of colonising not only the surface of the roots, but also the tissues of the root cortex (without ever penetrating the stele). They thus interpose themselves between the plant and the parasite (Olivain and Alabouvette, 1997).

— *Interpretation*: The protective role exercised by the Châteaurenard soil thus appears extremely complex. The determining factor seems to be competition for energy, which is very stringent because of significant global microbial biomass. The pathogenic *F. oxysporum* are all the greater disadvantaged as they represent only a very small fraction of the total *Fusarium*, whereas fusarial flora is abundant: competition is indeed still stiffer among individuals belonging to the same ecological niche. Under these conditions, the low affinity of siderophores of *F. oxysporum* to iron constitutes an additional handicap: the lack of iron (necessary for the cytochromes of the respiratory chain) could result in utilisation of other oxidation pathways of the hydrates of carbon that are less efficient than the normal pathway (Lemanceau et al., 1993). Competition for iron would thus make competition for carbon more arduous, especially for pathogenic forms, which seem to be more susceptible than saprophytes. In addition to competition for energy resources, as we have seen, there is competition for the receptor sites on the surface or inside the roots. But this is not all: since the antagonistic microflora is capable of colonising the cortical tissue of the roots, it induces defence reactions in the host plant. The non-pathogenic strain Fo 47 of *F. oxysporum* for example, provokes accumulation of proteins in tomato, which correspond to pathogenesis-related proteins implicated in the establishment of a systemic acquired resistance (Duijff et al., 1998).

Soils Resistant to *Gaeumannomyces graminis* var. *tritici*

The take-all disease caused by *Gaeumannomyces graminis* var. *tritici* is known in the major wheat-producing areas and very severe crop damage may result. However, this disease does not progress as per the general rule, according to which symptoms gradually worsen if a susceptible plant is sown consecutively for many years in contaminated soil. Contrarily, in the case of take-all, a lull in the disease has been observed at the end of a few years, in which attacks stabilise to such an extent that their impact on yield can be considered negligible. This phenomenon, known as 'take-all decline' has been the subject of many studies. Nevertheless, its 'modus operandi' still remains unclear; none of the proferred explanations is wholly satisfactory (Hornby, 1979; Cook and Weller, 1987; Simon and Sivasithamparam, 1989). The main points established thus far are these:

— *Resistance is related to monoculture*: if monoculture of wheat is practised in a field in which the parasite is present, a strong increase in symptoms is seen during the first few years, then a rather rapid reduction, and finally stabilisation of attacks at a fairly low level. Four or five successive crops are generally necessary to attain this state, but sometimes more. Monoculture is definitely indispensable for development of resistance: its effect is suppressed by interruption (Fig. 6.11).

— *Resistance is specific*: resistance is induced by cultivation of a susceptible plant (barley or more commonly wheat) and is specific to take-all disease.

— *Resistance is of biological origin*: it is suppressed by disinfecting with methyl bromide or treatment at 60°C for 30 min, which eliminates the fungi and non-sporulating bacteria. It is transmissible to a susceptible soil previously sterilised: an addition of 1 to 10% resistant soil suffices.

— *Parasite populations are modified qualitatively and quantitatively*: it appears that monoculture of the host plant initially favours the less pigmented forms of the fungus (*G. graminis* var. *tritici*) adapted to parasitism, to the detriment of highly melanised, poorly aggressive saprophytic forms. The fungus does not possess specialised structures of conservation and as hyaline hyphae are less persistent than pigmented ones, the chance of survival of the fungus between two wheat crops will, thereby diminish in spite of increase in proportion of tissue colonised in the roots.

— *Antagonists seem to be essentially associated with roots of wheat*: for a given density of inoculum of *G. graminis* var. *tritici*, it has been noted that on average there are as many primary lesions on the roots of plants grown in a resistant soil as on those grown in a susceptible soil. But in the resistant soil these lesions do not develop and secondary lesions are not observed, which leads to the thought that antagonism is exercised only on (or in) the roots of wheat.

— *Analysis of microflora associated with wheat*: the total number of microorganisms associated with the roots does not vary significantly with the appearance of resistance and it is difficult to demonstrate

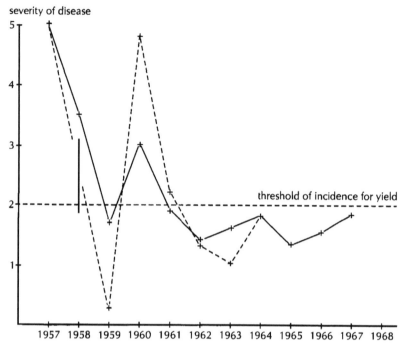

Fig. 6.11   Evolution of the severity of take-all as a function of time in a wheat monoculture. The severity of the disease is estimated by observation of plants sampled on 2 linear metres and after taking into account the state of the roots and the proportion of shrivelled ears. It is calculated on a scale of 0 (no attacks) to 6 (all plants attacked and yield zero). Readings were obtained at the agronomy station of Quimper, starting in 1957, after the first cultivation of wheat in 1956:

(—)   uninterrupted growth of wheat up to 1967;

(- -)   wheat in 1956 and 1957, interrupted by potato cultivation in 1958, followed by uninterrupted cropping of wheat up to 1964 (according to data of Lemaire and Coppenet, 1968).

notable changes in microbial communities. However, in the great cereal belt of north-west America, a relative increase in populations of fluorescent pseudomonads is observed. These populations contain a larger proportion of strains capable of inhibiting *G. graminis* var. *tritici* in vitro.

Possible methods of action of pseudomonas: these pseudomonads produce siderophores and competition for iron could be one of the mechanisms responsible for resistance. The resistance of a susceptible soil can, as a matter of fact, be augmented by adding a purified pseudobactin to it. Inversely, a resistant soil can be made susceptible by adding iron in the form of Fe-EDTA. Also, several strains of *Pseudomonas* isolated from the roots of wheat produce phenazines (Thomashow et al., 1990) or phloroglucinols (Harrison et al., 1993), strong inhibitors of *G. graminis* var. *tritici* in vitro. Production of antibiotics has been demonstrated in situ in resistant soils; on the other hand, antibiotic production has not been detected in the rhizosphere of wheat cultivated in susceptible soil (Raaijmakers et al., 1999).

— *Interpretation*: it can be supposed that the wounds caused on the root cortex by the fungal parasite are a source of exudates that stimulate development of secondary invaders, notably comprising fluorescent pseudomonads. Growth of the bacterial populations augments production of antibiotics based on a cell density-dependent gene regulatory mechanism (see page 174): above a certain threshold, the combination of a diffusible N-acyl-homoserine lactone with a transcription activator leads to biosynthesis of phenazine (Pierson and Pierson, 1996). After the death of the roots, these bacteria remain in the plant debris. Here they contribute to the regression of populations of *G. graminis* var. *tritici* which, as indicated above, probably become less competitive after their passage into the plants. A new host-parasite-antagonist equilibrium is established at the end of a few years provided that selection pressures stay similar throughout; that is, if the monoculture of wheat (or barley) is continued. Maintenance of the equilibrium requires the presence of a minimum number of lesions even if their size remains small. It is also interesting to note that a monoculture of wheat in land where *G. graminis* var. *tritici* is not present, does not make the soil resistant to the parasite (Gerlagh, 1968). It is mandatory that the host plant, its parasite, and the fluorescent pseudomonads all be present concomitantly for the take-all decline to establish.

This explanation of the lull in take-all disease, based on works conducted in north-west America, does not take into account facts observed in other soils and under other climates. In England, for example, pseudomonads represent only 2% of the bacteria of the rhizosphere of wheat in resistant soils and their role seems more problematic (Hornby, 1979).

For Further Information on Interactions between Microorganisms

Alabouvette C., Hoeper H., Lemanceau P. and Steinberg C. 1996. Soil suppressiveness to diseases induced by soil-borne plant pathogens. In: G. Stotzky and J.M. Bollag (eds.). Soil Biochemistry, vol. 9, pp. 371-413. Marcel Dekker Inc., NY.

Anderson J.M., Rayner A.D.M. and Walton D.W.H. (eds.). 1984. Invertebrate-microbial interactions. Cambridge Univ. Press, Cambridge, UK.

Bazin M.J., Saunders P.T. and Prosser J.I. 1976. Models of microbial interactions in the soil. CRC Crit. Rev. Microbiol. 4: 463-498.

Bull A.T. and Slater J.H. (eds.). 1982. Microbial Interactions and Communities, vol. 1. Acad. Press, London.

Carroll G.C. and Wicklow D.T. (eds.) 1992. The Fungal Community: Its Organization and Role in the Ecosystem. Marcel Dekker Inc., NY.

Hutchinson, S.A. 1973. Biological activities of volatile fungal metabolites. Annu. Rev. Phytopathol. 11: 223-246.

Khan M.W. (ed.). 1993. Nematode Interactions. Chapman and Hall, London.

Lockwood J.L. 1977. Fungistasis in soils. Biol. Rev. 52: 1-43.

Pierson L.S. III, Wood D.W. and Pierson E.A. 1998. Homoserine lactone-mediated gene regulation in plant-associated bacteria. Annu. Rev. Phytopathol. 36: 207-225.

Vančura V. and Kunč F. (eds.).1988. Soil Microbial Associations. Control of Structures and Functions. 'Developments in Agricultural and Managed-Forest Ecology', vol. 17. Elsevier, Amsterdam.

7

Interactions between Microorganisms and Plants

MICROORGANISMS OF SOIL AND OF ORGANIC SUBSTRATES

Immobilisation of Mineral Elements

Fresh organic matter is rapidly colonised. Part of it is assimilated by microorganisms, another part mineralised, while a more recalcitrant fraction participates in the formation of humic acids. The proportion of mineral elements released and immediately available for plants depends on the nature of the substrate. If the contents of nitrogen, sulphur and phosphorus are small in the residues compared to their carbon composition, all these elements will be immobilised in the microbial biomass. The addition of mineral salts would be necessary in this case to preclude competition between the plants and the microorganisms (see page 125). However, this immobilisation is only temporary and the mineral elements that could be leached are conserved in the fertile soil horizons. Renewal of the biomass ensures progressive minera-lisation of these elements that gradually become available for the plants.

Toxicity Phenomena

General Phytotoxicity

Decomposition of large quantities of organic matter accumulating on the soil surface could result in temporary anaerobiosis, especially if high humidity favours microbial proliferation when diffusion of oxygen is slower. Under these conditions, phytotoxic organic acids can be formed. These acids could be the reason for poor wheat gemination observed sometimes in autumn, when the current crop is sown directly on the unburied straw of the previous crop. However, according to Cook and Veseth (1991), these incidents are due not to phytotoxicity but to fungal parasites (*Pythium* spp., *Gaeumannomyces graminis*), which pass directly from the straw of the preceding crop to the seedlings of the new crop because of the no-tillage technique. It is also possible that the phytotoxicity resulting from anaerobic decomposition of straw renders the young seedlings more susceptible to attack by fungal parasites. Penn and Lynch (1982) found that germination of winter cereals was very poor in parts of fields weeded with glyphosate consequent to intense invasion by couch grass (*Agropyron repens*). The decomposing residues contained no herbicide but possessed high concentrations of acetic acid—up to 42 mM g^{-1}, an acid toxic at just 5 mM concentration. *Fusarium culmorum* was also present and soaking wheat roots for 1 hour in acetic acid at 5 mM concentration greatly augmented plant susceptibility to this minor parasite.

Toussoun and Patrick (1963) showed that application of aqueous extracts of decomposing plants to kidney-beans increased exudation of sugars and amino acids from the roots and the hypocotyls, resulting in enhanced colonisation of the tissues by various pathogenic fungi. Vaughan et al. (1983) later demonstrated the presence in soil solutions of aromatic acids, which are toxic at low concentrations (less than 1 mM), arising from the biodegradation of lignin. This could partially explain why the residues of older plants are generally more phytotoxic than those of younger ones, with slightly lignified tissues. In vitro, aqueous extracts of decomposing plants caused lesions on the roots of various plants, in the absence of any parasite (Patrick et al., 1963). The apical meristems were particularly susceptible. In all cases, toxicity of plant wastes was of short duration, ceasing 1 to 3 weeks after deposition on the soil or burial, and disappearing as fast as oxidation of the compounds formed took place. Rapid disappearance of the toxic effect and the fact that toxicity is observed only with the addition of massive quantities of fresh organic matter, resulting in temporary anaerobiosis, support arguments in favour of a biochemical rather than parasitic origin of toxicity. But the mere identification of toxic compounds in decomposing waste does not constitute sufficient evidence for concluding that they have a biological effect. It is also necessary to show that these compounds are present in the soil in concentrations sufficient to induce manifestation of toxicity. For example, patulin produced by *Penicillium urticae*, a fungus that develops on straw, was claimed responsible for defective cereal germination (Norstadt and McCalla, 1969) but its concentration in the soil was much lower than that necessary for toxicity. Similarly, corky root of lettuce, mainly observed in the United States, was attributed by Amin and Sequeira (1966) to the formation of toxic compounds from the residue of the preceding lettuce crop. But it seems more likely that the bacterium *Rhizomonas suberifaciens*, identified later (van Bruggen et al., 1988), was responsible for the symptoms observed by Amin and Sequeira.

The weak diffusion of phytotoxic organic acids limits their effect, should such arise, to the immediate vicinity of the residues. This is not so however, if the metabolites resulting from biodegradation are gases. We have already seen (see page 94) that excessive ammonification can have harmful consequences. Ethylene, produced by a large number of bacteria and fungi, is biologically oxidised when the redox potential of the soil is sufficiently high, but accumulates under anaerobic conditions. The effects on plants of ethylene produced by microorganisms are difficult to distinguish from the effects of ethylene of plant origin on the one hand, and from the effects of anoxia, with which ethylene production is associated, on the other (Jackson, 1985). High concentrations could inhibit root growth.

Specific Phytotoxicity: Replantation Problems

It is often difficult to grow a new plantation of fruit trees in a plot previously occupied by an orchard comprising trees of the same species: the trees do not grow well, produce little fruit and sometimes die. These replantation problems, especially encountered with peach and apple trees, are usually attributed to **soil fatigue (sickness)**, a convenient term for masking gross ignorance by purportedly scientific precision.

> According to the Koch principle, a disease is caused by a pathogenic agent obtainable in pure culture from the diseased organs. Inoculation of healthy subjects with the pathogenic agent originating from a pure culture, must reproduce all the symptoms of the disease. Lastly, it should be possible to re-isolate the pathogenic agent from the experimentally infected tissues in order to clearly confirm that the organism is responsible for expression of the symptoms.
>
> But the problems of replantation and soil sickness are the result of multiple causes, which may occur at different times, be cumulative in effects, and the general outcome a single non-specific symptom, namely laboured growth. Koch's principle is therefore not applicable here. Bouhot (1979a) proposed a test that takes into account the diverse aspects of soil sickness and enables an initial diagnosis. As a first step, samples showing soil 'fatigue' are subjected to general, non-specific treatments, then compared with untreated controls to ascertain whether the problem is essentially one of toxicity, or parasites, or nutritional imbalance or poor physicochemical constitution of the soil (the samples are treated with active charcoal, a biocide, a nutritive solution and organic matter respectively). In the second step, the hypothesis arrived at based on the first step is improved, using specific treatments. For example, the hypothesis that parasites are the cause can be further diversified into nematological, bacterial or fungal origin, and so on. Several hypotheses can be formulated: if they are correct, the combination of corresponding treatments would effectively eradicate the symptoms.

In peach trees, soil sickness seems to be caused by the phenomenon of phytotoxicity. Application of old roots or their aqueous extracts to a virgin soil severely disturbed the growth of a young plantation of peach trees. The bark of the roots of peach contains a heteroside, prunasine, whose content gradually increases with age. Bacterial hydrolysis of this heteroside liberates two phytotoxic compounds—benzaldehyde and hydrocyanic acid: 1 g root of medium size contains nearly 13 mg prunasine, yielding 1.15 mg HCN

(Gur and Cohen, 1989). The bacteria responsible for biodegradation of prunasine (and another very similar compound, amygdalin) are particularly abundant in the vicinity of young peach roots, but less active in the rest of the soil: this explains why toxicity persists even if a delay of 2 or 3 years is observed before planting a new orchard and also explains why other categories of fruit trees, associated with a different microflora, are not affected.

> These problems of replantation could also be related to the presence of a parasitic nematode, *Pratylenchus penetrans*. While puncturing the living tissue of the roots, it allows the heterosides to come into contact with a cellular emulsine that causes their hydrolysis, liberating the two toxic compounds (Patrick et al., 1964).

Replantation problems also occur in apple orchards. The numerous explanations proposed are based either on phytotoxicity or parasitism. As in the case of peach, it was thought that phytotoxicity could be due to the biodegradation of a compound present in the old roots that remain in the soil. This compound, phloridzine, liberates phloretine, phloroglucinol, p-hydroxycinnamic and p-hydroxybenzoic acids that are toxic to young trees (Patrick et al., 1964). But the concentrations measured in the soil are lower than the activity threshold. It has also been proposed that a microflora that becomes dominant in old plantations causes the formation of toxic compounds such as patulin (Čatska et al., 1982), or that various bacteria (*Pseudomonas* spp., *Bacillus subtilis*) and fungi (*Penicillium janthinellum, Constantinella terrestris, Pythium irregulare, Cylindrocarpon lucidum*), which are slightly pathogenic when considered individually, progressively accumulate and become injurious to young tissues due to the addition of their effects (Utkhede and Li, 1989; Braun, 1991). A group of fungi more conventionally recognised as pathogens (*Cylindrocarpon destructans, Rhizoctonia solani, Phytophthora cactorum* and *Pythium* spp.) seems to be responsible for the apple replant disease in the state of Washington (USA) (Mazzola, 1998).

The decline of old asparagus plantations could be due to modifications of the microflora caused by the accumulation of phenolic compounds liberated by the plant in the soil. The soil of an asparagus plantation contained 1.5 to 2.4 times more phenols compared to a similar soil without asparagus. Among these phenols caffeic acid and, more importantly, methylenedioxycinnamic acid, had a significant inhibitory effect on colonisation of the roots by the symbiotic fungus *Glomus fasciculatum* and on the functioning of root mycorrhizae (Pedersen et al., 1991).

Crop Residues, Source of Inoculum

The symbionts and parasites, protected by their host, face a critical situation after its death, namely confrontation with microbiostasis and a struggle for conquest of new substrates. The tissues colonised while the plant was living play an essential role in the survival of these organisms, acting as a bridge between two successive hosts. Even after the plant's death these tissues contain living cells and protect the microorganisms they shelter from other invaders for a certain period of time. Tissue resistance disappears only gradually and the wholly saprophytic species progressively commence takeover. Several strategies are possible, studied mainly in the case of parasitic fungi but partial extrapolation to bacteria has been possible (Cooke and Rayner, 1984).

— The fungus, incapable of resisting the flood of saprophytes, remains in a dormant state in the form of more or less developed organs of conservation: chlamydospores, oospores, melanised pseudoparenchymas, sclerotia or microsclerotia. Such a situation occurs in *Pyrenochaeta lycopersici*, for example, which is responsible for the corky root disease of tomato.

— The fungus is sufficiently competitive to take advantage of being the first occupant and to achieve colonisation of the substrate despite the arrival of saprophytes. This situation has been particularly well studied by Garrett (1970) and the Cambridge school for fungal parasites of cereals. In this case, competitive exploitation of straw by the microorganisms is closely related to their ability to efficiently hydrolyse cellulose, ensuring a slow but regular metabolisation of resources by synthesising a minimum of enzymes. Nitrogen can become a limiting factor for the synthesis of enzymatic proteins and the assimilation of a substrate with a very high C/N ratio, such as straw. This is the reason why fungi such as *Fusarium culmorum* or *Gaeumannomyces graminis* var. *tritici* survive better in stubble on soils rich in nitrogen than in soils poor in it. On the other hand, nitrogen reduces the survival of another parasite of cereals, *Cochliobolus sativus*. Garrett (1966) resolved this apparent contradiction by showing that the ratio (cellulolysis adequacy index) between cellulolytic activity and general metabolism varies depending on the species (see box below). A low ratio signifies that the supply of soluble sugars does not cover the potential requirements of metabolism, in which case an addition of nitrogen stimulates cellulolysis and thereby provides a better response to the organism's demand and improves its chances of survival. This is what is observed in the first two fungi. A high ratio indicates that the cellulolytic activity releases more hydrates of carbon than the organism can consume. The addition of nitrogen in this case would accelerate degradation of the straw, which already tends to be high, leading to premature depletion of the substrate, followed by the disappearance of the fungus, as has been observed in the case of *C. sativus*.

— If the colonised tissues are aqueous, poor in lignin and cellulose, and therefore rapidly decomposable, they cannot ensure long term protection to pathogenic agents. Hence, in order to survive, the pathogenic agents need to rapidly produce efficient structures of conservation (sclerotia, for example, in *Sclerotinia minor*, responsible for collar rot in lettuce), or be sufficiently competitive (e.g. *Rhizoctonia*) to jostle with saprophytic microorganisms for the organic substrates available. There is thus an alternation between parasitic and saprophytic life.

— A fourth type of strategy is encountered in lignivorous fungi. Provided with a significant quantity of reserve nutrition in the stumps and killed roots, they can produce rhizomorphs capable of prospecting the soil over great distances in search of new hosts. Well studied in the case of *Armillaria mellea*, the rhizomorphs are threads of one or more millimetres in diameter, formed from a very large number of hyphae arranged parallelly and advancing in a synchronised manner. The external hyphae are impregnated with melanin in such a way that the rhizomorph has no interaction with the ambient surroundings. Nourishment of the growing tip is ensured by the transfer of nutritive elements from the basal extremity. The rhizomorphs thus extend in the soil, sometimes for many metres, escaping fungistasis.

Cellulolysis Adequacy Index

Even after the death of their host, parasitic or symbiotic fungi are relatively sheltered in the tissues they have colonised. Yet it is necessary that they utilise the accumulated reserves properly. According to Garrett (1966, 1970), survival of the fungi in the straw of cereals (or, broadly speaking, in the residue of a crop rich in cellulose) depends on the proper balance between their cellulolytic activity and their overall metabolism. Cellulolytic activity determines the rate at which the hydrates of carbon from straw are solubilised and made available to the fungus. The overall metabolic rate regulates the rate at which the sugars are consumed for growth and respiration.

The rate of cellulolysis can be estimated by measuring the loss in weight of the filter paper discs on which the fungi have been maintained for 7 weeks under controlled incubation conditions. On the other hand, the linear rate of growth can be considered representative of the overall metabolic rate. The 'cellulolysis adequacy index' (C.A.I.) is the ratio between rate of cellulolysis and rate of growth. It can be seen from the Table below that the fungi which show improved survival in straw after the addition of nitrogen have a low C.A.I. On the other hand, *Curvularia ramosa*, which is indifferent to the addition of nitrogen, and *Cochliobolus sativus*, whose survival is shortened by nitrogen, have high C.A.I. values.

Pathogenic agent	Effect of nitrogen on survival in saprophytic phase	Loss in weight of filter paper-A (% of initial weight)	Linear growth-B (mm/24 h)	C.A.I (A/B)
Gaeumannomyces graminis var. *tritici*	Strongly positive	2.8	7.3	0.38
Fusarium culmorum	Strongly positive	7.5	11.0	0.68
Cercosporella herpotrichoides	Mildly positive	0.9	1.3	0.69
Curvularia ramosa	Insignificant	6.6	5.3	1.25
Cochliobolus sativus	Strongly negative	7.5	3.6	2.10
(after Garrett, 1970)				

MICROORGANISMS OF THE RHIZOSPHERE

The Rhizosphere

Rhizosphere Effect

Contrary to the quasi-inactivity imposed by microbiostasis in the rest of the soil, the surroundings of the roots are areas of intense microbial life. This soil zone, engirdling the roots and directly or indirectly subjected to their influence, is termed the **rhizosphere**, a term introduced in 1904 by the German biologist Hiltner. The rhizosphere can be approximately defined as the soil that remains attached to the roots when the plant is carefully uprooted, then gently shaken.

If the microorganisms present in 1 g soil of the rhizosphere thus defined are counted (indicated by R) and compared to the microorganisms present in 1 g of the same soil taken some distance from the roots (indicated by S), it can be stated that overall the R/S ratio will always be much greater than 1. The ratio R/S expresses the **rhizosphere effect**. This effect is particularly distinct with respect to bacteria (in their case, on average the R/S can be of the order of 10^3). But the effect is also evidenced in protozoans, fungi and nematodes.

Energy Flux, Foundation for the Rhizosphere Effect

The considerable increase in microbial populations in the rhizosphere can only be explained by the abundant presence of energised substrates in that region. Numerous studies, carried out mainly on annual Gramineae, but also on a few other annual or perennial plants, have shown that the roots are actually the source of carbon and nitrogen flow.

A significant portion of the compounds photosynthesised in the aerial parts of a plant is transported to the roots through the intermediary of phloem. The importance of these transfers varies depending on the plants, their stage of development and external conditions. Probably half of the carbon fixed in the chlorophyllous tissues is transferred towards the roots. Some is stored as a reserve or incorporated in new tissues. Another part is consumed by cell respiration and more or less rapidly transformed into carbon dioxide ejected into the soil atmosphere. A third portion can be considered lost in the external environment. This loss of substance, which may take very different paths, has been termed **rhizodeposition.** The quantity of carbon introduced into the soil, both in the form of carbon dioxide and by rhizodeposition, can often far exceed the quantity of carbon retained in the root tissues, especially in annual plants (Fig. 7.1). It can reach 40% of the net quantity of carbon fixed by the aerial parts (Table 7.1). Consequently, an estimation of the quantity of organic matter brought to the soil by a crop will be underestimated if only root biomass is taken into account.

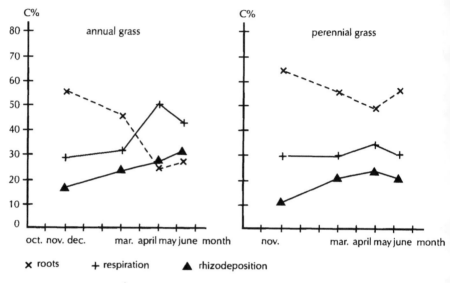

Fig. 7.1 Distribution (in percentage) of carbon transferred in the roots of an annual Gramineae (*Bromus madritensis*) and a perennial Gramineae (*B. erectus*) during the course of one year. The plants were sown in autumn and divided into four lots. At the beginning of tillering (December), during tillering (March), during stem growth (April) and at the end of flowering (May), one of these lots was exposed for 2 weeks to an atmosphere of $^{14}CO_2$, then harvested and analysed (after Warembourg et al., 1990).

Rhizodeposition

The trophic supply from the roots to the rhizosphere results from the following three processes:

— *Exudation*: This term is often overused as a synonym for rhizodeposition. As a matter of fact, it denotes the passive diffusion from cells of low molecular weight soluble compounds, for example sugars or amino acids.

— *Secretion and excretion*: Secretion is an active process that consumes energy. It generally involves compounds of high molecular weight such as enzymes or mucilage. The cells can also actively excrete protons or organic acids in order to maintain their ionic equilibrium. These processes essentially take place in the elongation zone but the border cells, which remain metabolically active for several days, also modify the environment in front of the root tip.

Table 7.1   Transfer of carbon from the aerial parts of plants towards the roots and from the roots to the soil (adapted from Whipps, 1990)

Plant	Age (days)	C transferred towards roots (as % of net C fixed) (a)	C transferred from roots to soil (as % of C transferred towards root)		C lost by roots (as % of net C fixed) (d)
			respiration (b)	rhizodeposition (c)	
Wheat	21	59	39	29	40
Barley	21	51	37	30	34
Maize	28	29	33	28	18
Tomato	14	43	20	70	39
Pea	28	44	53	29	36

Measurements were done under controlled conditions (14-18°C, light for 16 h) in non-sterile soil. A constant specific activity of atmospheric $^{14}CO_2$ was maintained while ensuring that the soil and roots remained isolated from that atmosphere. Thus the radioactivity measured in the soil and the roots could only arise from $^{14}CO_2$ assimilated by the aerial parts. a: net C fixed is equal to total fixation minus quantity released in the form of CO_2 by the aerial parts, which cannot be measured by available techniques; b: calculation taking into account respiration of roots and microorganisms that mineralise the organic C supplied by the roots; c: the C returned to the soil in inert form or in the form of biomass; d: $d = \dfrac{(b+c) \times a}{100}$.

Lysis: Cap cells and absorbent hairs are very short-lived while border cells and the external root cortex degenerate less rapidly. These desquamous cells sometimes represent a very important supply of organic matter, which remains localised in the immediate proximity of the root since most of the cellular carbon components are insoluble.

Measurement of Carbon Flux in the Rhizosphere

Study of plants maintained aseptically in mineral solutions or on sterilised substrate has gradually given way to experiments in which the plants are cultivated in a non sterile soil. It has been observed that the physical environment and the presence of microflora notably modify the volume and nature of the rhizodeposition. However, under non-aseptic conditions, it is sometimes difficult to ascertain whether the compounds identified have come directly from the roots or are metabolites arising from microbial activity. Moreover, simple molecules are consumed very rapidly (thereby eluding analysis) and transformed into microbial biomass, then mineralised. Therefore it is imperative that these increases in microbial biomass and microbial respiration be taken into account while considering the overall picture. To determine the carbon flux, the plants are placed in an atmosphere containing a known proportion of $^{14}CO_2$ and then, using a scintillation counter, the ^{14}C in the different compartments or, more simply, the $^{14}CO_2$ that appears in the soil atmosphere, is measured. The nature of the results obtained depends on the exposure time. A short exposure reveals the passage of ^{14}C into the roots, then into the rhizosphere and its incorporation in the biomass. The arrival of hydrates of ^{14}C in the roots results in a primary emission, extremely early, which corresponds to cellular respiration. After about 12 hours a second emission of $^{14}CO_2$ is observed, which arises from mineralisation of the exudates and root secretions effected by the microorganisms (see Figure below). A few days later a third emission of $^{14}CO_2$ could be noted, due to mineralisation of the cellular and microbial lysates. However, in order to obtain precise information about the distribution of carbon in the various compartments, an exposure of short duration does not suffice. For this it is necessary to maintain the plants for many weeks in an atmosphere wherein the specific activity of $^{14}CO_2$ is maintained constant while ensuring that no direct contact takes place between the soil and that atmosphere. This method was used to obtain the results presented in Table 7.1 and Figure 7.1.

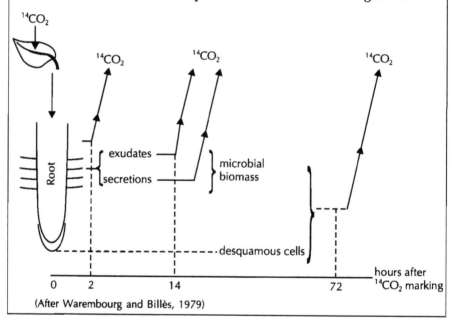

(After Warembourg and Billès, 1979)

All the phenomena described above occur in living roots. It should not be forgotten that the root system is perpetually renewing itself and that most of the fine roots die completely in just a few weeks. This is another important source of organic matter for the soil but the process is totally different from that of rhizodeposition and does not give rise to a rhizosphere effect.

Due to these diverse mechanisms, a very large number of soluble, insoluble or gaseous molecules, the inventory of which is still far from complete, become available in the immediate vicinity of the roots. The most common compounds are amino acids, simple or polymerised glucides, a large variety of organic acids, enzymes, vitamins, sterols, nucleic derivatives and proteins arising from lysis of the cells. More specific compounds have also been identified in certain rhizospheres: flavonoids, thiophenes, benzofurans, organic sulphides, heterosides etc.

The nature and abundance of rhizodeposition depend largely on the genome of the plant and its stage of development. Rhizodeposition is also affected by all the events capable of modifying plant functioning: modification of light intensity, temperature, availability of water (Fig. 7.2), redox potential, deficiency of minerals, phytotoxicity and reduction of leaf surface (by hail, insects or diseases). Some environmental factors have an impact on photosynthesis, others on metabolism, and yet others on cellular permeability. No general rule can be formulated because experimental results vary markedly depending on the species studied. Thus, when the temperature of the soil is around 37°C, exudation increases in the roots of kidney-beans and cotton but, contrarily, decreases in the roots of pea (Schroth et al., 1966).

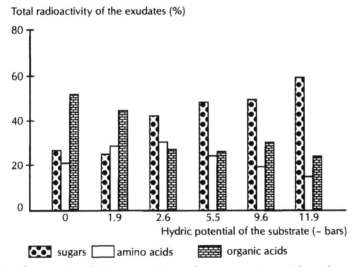

Fig. 7.2 Changes in relative proportions of sugars, amino acids and organic acids released in the external medium by the roots of *Pinus ponderosa* when subjected to continuous hydric stress for 6 days. Pine plants, one year old, were placed in an atmosphere containing $^{14}CO_2$ for 4 days before exposure to stress induced by the addition of polyethylene-glycol to the nutritive solution (Reid, 1974).

Modulation of Rhizosphere Effect

We took the precaution of stating earlier that, *'give or take a little'*, the R/S ratio is always much higher than 1'. We did so because not all species within a large biological group are stimulated to the same extent. Thus, among bacteria, the *Bacillus* and Gram-positives in general are not influenced or are sometimes inhibited while contrarily, Gram-negative non-sporulating organisms respond very strongly. Among fungi, genera *Fusarium*, *Aspergillus*, *Penicillium* and *Gliocladium* are generally very well represented in the

On the other hand, the abundance and composition of the microflora vary considerably depending on the family to which the plants belong. In a given soil, leguminous plants will by and large have a very high R/S ratio, indicating that their rhizospheric flora is very abundant, whereas crucifers will have a low R/S ratio, signalling a less marked effect. A significant variability also exists among species of the same botanical family and likewise, eventually, among diverse varieties of the same species. There is good reason to believe that the genetic characteristics of the plant largely determine the composition of the microflora associated with it, based on the quality and quantity of rhizodeposition (see box below).

Effect of Plant Genome on Composition of Rhizospheric Microflora

The composition of the rhizospheric microflora is controlled by the genome of the plant. This was demonstrated by Neal et al. (1973) for spring wheat. They replaced a pair of chromosomes of the Cadet cultivar with a homologous pair from the Rescue cultivar (the 20 other pairs of chromosomes of Cadet were not modified). When the substitution involved the 5B pair, the microflora of Cadet doubled and was qualitatively identical to that of Rescue. When the substitution was carried out on the 5D pair, no change was observed with reference to the normal Cadet cultivar.

These results are illustrated by the two Figures below. The histogram shows the general increment (with respect to a bare soil) in bacterial populations due to the rhizosphere effect, as well as modulation of this effect depending on the genotype of the wheat lines. The graph illustrates the relative importance of a few functional groups of rhizospheric bacteria (the axes are graded in logarithms).

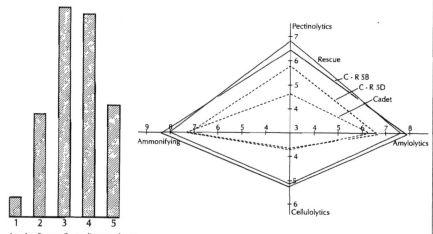

1 and ordinary soil, at a distance of roots
2 and ---- rhizosphere of Cadet
3 and ____ rhizosphere of Rescue
4 and ____ rhizosphere of Cadet containing the 5B pair of chromosomes of Rescue (C - R 5B)
5 and ---- rhizosphere of Cadet containing the 5D pair of chromosomes of Rescue (C - R 5D)

It can be seen that the rhizosphere of Rescue and that of the Cadet line possessing the 5B chromosomes of Rescue (C-R5B) harbour many more pectinolytic and cellulolytic bacteria than Cadet. Deacon and Lewis (1982) showed that senescence of the cortical cells was of paramount importance in these two lines compared to Cadet. They concluded that the abundance of these bacterial populations is probably related to the abundance of a substrate constituted of dead cortical cells.

It is also interesting to note that Rescue and C-R5B were susceptible to root rot caused by *Cochliobolus sativus* whereas Cadet and the Cadet line containing the 5D pair of Rescue chromosomes (C-R5D) proved resistant.

Since rhizodeposition is by and large a reflection of the photosynthetic activity of the plant, all factors that modify this activity exert an influence on the energy flux in the rhizosphere: light, temperature, water content of the soil, mineral nutrition and all kinds of stress. Herbicides, antibiotics or soluble fertilisers sprayed on the leaves can be partially transferred to the roots and appear in the rhizosphere or modify the nature of the exudates.

Lastly, the rhizosphere effect does not manifest with the same intensity throughout the life of the plant. In annual Gramineae such as wheat, it increases during the period of vegetative growth, then decreases during the reproductive and maturation phases (Rivière, 1960). The effect disappears when the plant becomes senescent (Fig. 7.3).

pH of Rhizosphere

The pH of the rhizosphere differs from that of the adjacent soil. This difference is more or less noticeable depending on the buffering power of the soil and the type of plant (differences seem more pronounced among dicotyledons). The extremity of the roots is definitely more acidic than the rest of the soil and a pH gradient is observable along the root, measurable with microelectrodes or, in vitro, with coloured indicators. The amplitude of variation can reach or even exceed 2 units of pH. This modification is essentially due to the lengthwise growth of the root and the absorption of ions from the soil solution. Elongation of the cells situated behind the cap is preceded by an intense excretion of protons and the greater the elongation, the greater the flux. Part of these protons is reabsorbed by tissues that have already differentiated. This results in an electric current that circulates towards the inside of the root in the growing zones (meristems, zones of elongation and formation of absorbent hairs) and towards the outside in older or mature zones (Pilet et al., 1983). The field created by this current is in the range of 1 to 100 mV per cm.

Absorption of mineral ions is an active process: it is coupled to hydrolysis of ATP by a membrane ATPase that, while transferring the protons from the cytoplasm towards the outside of the cell, creates a pH and electric potential gradient (Fig. 7.4). The cations penetrate and follow the gradient. The anions accompany the return of the protons towards the cytoplasm or are exchanged

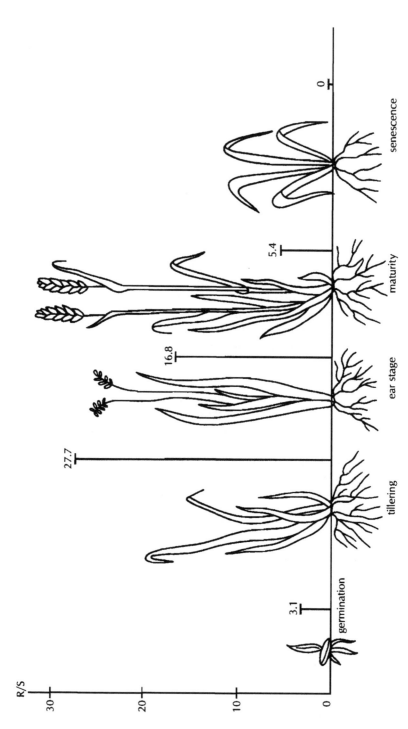

Fig. 7.3 Variation in intensity of rhizosphere effect (in grey) on wheat as a function of the developmental stage. In this experiment, the rhizospheric effect was measured for populations of bacteria and was expressed by the R/S ratio, where R indicates the number of bacteria present in 1 g rhizospheric soil and S the corresponding number for 1 g soil at a distance from the roots (after Rivière, 1960).

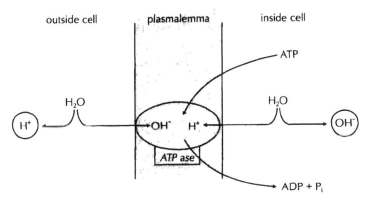

outside cell   plasmalemma   inside cell

Fig. 7.4 Diagram of functioning of a membrane ATPase. Hydrolysis of ATP (adenosine triphosphate) results in production of protons H$^+$ outside the cell. ADP (adenosine diphosphate) and inorganic phosphorus P$_i$ are formed (after Callot et al., 1982)

for hydroxyl ions OH$^-$. (For certain ions such as K$^+$, Ca^{++} and Cl$^-$, specific ionic canals have also been demonstrated; their functioning differs). As a result of all these ionic exchanges, the concentration of H$^+$ ions outside the roots is not constant. The pH of the rhizosphere at the surface of absorption (root hairs and young tissues) can vary considerably depending on the nature of the ions taken up by the roots. This effect can be clearly shown by modifying the nitrogenous nutrition of the plant: if it is supplied with ammoniacal salts, the pH diminishes in the rhizosphere; if the fertiliser added is constituted of nitrates, the pH increases (Fig. 7.5). In leguminous plants, supplied with nitrogen by their symbiotic bacteria, consumption of NO$_3^-$ is very low or zero: these plants therefore absorb many more cations than anions and hence their rhizosphere, rich in H$^+$ ions, is always acidic (Hauter and Mengel, 1988).

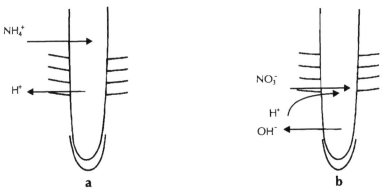

Fig. 7.5 Diagram illustrating modifications in the pH of the rhizosphere resulting from supply of a nitrogenous fertiliser. a: acidification in the presence of ammoniacal nitrogen; b: alkalisation in the presence of nitric nitrogen.

In addition to these two general mechanisms, there are some more specific reactions. When a deficiency of iron or phosphorus is experienced, some plants excrete protons or organic acids (caffeic acid, citric acid) to improve the solubility of these elements. Acidification is limited to the apical zone in some cases (sunflower, for example) or, on the contrary, may extend the entire length of the root (cucumber, groundnut: Marschner et al., 1982).

Lastly, it should not be forgotten that the emission of carbon dioxide due to respiration of the roots and their associated microflora, constitutes a supplementary factor contributing to acidification of the environment.

Other Characteristics of the Rhizosphere

The redox potential of the rhizosphere differs from that of the rest of the soil. It is generally lower because of the notably high oxygen consumption due to respiration of the roots and microflora. The environment can become particularly reducing in compressed or engorged soils where diffusion of gases is not easy. But in aquatic plants provided with an aerenchyma, such as rice, the rhizosphere can, on the contrary, be a less reducing zone than the hydromorphic soil in which the roots advance.

Absorption of water necessary to compensate evapotranspiration creates a permanent flux from the soil solution towards the roots: the matrix potential of the soil therefore decreases by and large closer to the roots. However, in the dry superficial soil layers, releasing of water drawn from the deeper, wetter layers has been observed. This phenomenon is termed **hydraulic lift**. The plant sorts out the dissolved substances, resulting in a local qualitative change in composition of the liquid phase and a modification of osmotic pressure. Some ions can accumulate. An increment in concentration of Ca^{++} ions can, under certain conditions, induce precipitation of a sheath of calcium carbonate that engirdles the roots, a phenomenon which in turn may have two important consequences: dramatic rise in local pH and eventual entombment of the roots whose cells, impregnated with limestone, become fossilised (Callot et al., 1982).

Roots develop in the area of least resistance and preferentially invade pre-existent pores: these may be galleries made by animals or microfissures caused by alternation of wetness and desiccation of clays. As their diameter increases, the roots exert pressure on the walls of the channels, which results in some compaction and reorientation of the clayey soil phase (Callot et al., 1982). Compaction is especially significant at the apex of the root: it can locally increase soil density to an extent comparable to the compaction caused by the passage of a tractor (Young, 1998). All these mechanisms together constitute a system of drainage in which the degree of moisture can differ significantly from that in the rest of the soil.

The physicochemical characteristics of root environment are therefore inexorably conditioned by the soil, the substrate in which roots develop. Hence we should approach with great caution the conclusions drawn from studies on the rhizosphere carried out with plants cultured in nutritive solutions wherein the environment differs markedly from that under natural conditions.

Consequences of Rhizosphere Effect on Microorganisms

In a soil subjected to rigorous constraints of microbiostasis, the area near the roots, which is nourished by the continuous flux of rhizodeposition, constitutes a haven of abundance. The organic sources of nitrogen, carbon, phosphorus available in the rhizosphere significantly stimulate the development of populations: the rate of multiplication of bacteria, considered in toto, is twice higher in the rhizosphere of *Pinus radiata* than in ordinary soil (Bowen and Rovira, 1976). However, competition for soluble minerals (especially iron) is intense and the richness of the rhizosphere is not uniformly shared among all the soil microorganisms. As shall be seen a little later in the section devoted to rhizospheric affinity, these minerals are only accessible to a small number of privileged organisms adapted to these specific surroundings. Consequently, the rhizosphere is an area in which population densities are extremely high but species diversity lower than in the rest of the soil. This reduction in diversity of species is accompanied by a lowering of intraspecific variability. The rhizosphere effect and the relatively stable conditions that exist at the surface of the roots exert a selection pressure that is always oriented in the same direction. On the contrary, populations in the soil are subjected to permanent variations in the environment; they cannot adapt except through sufficient genetic diversity. Hence it can be expected that the heterogeneity of isolates of the same species diminishes on passing from the soil to the rhizosphere, and from the rhizosphere to the root. Such a reduction in diversity was shown by Mavingui et al. (1992) in strains of *Paenibacillus polymyxa* associated with the rhizosphere of wheat, by using acidification criteria of the carbonated substrate as well as serological reactions or DNA restriction fragment length polymorphism analysis. Similarly, the diversity of fluorescent pseudomonads isolated from roots of flax and tomato was less than that of isolates of the soil (Latour et al., 1996). Results concerning populations of *Fusarium oxysporum* are likewise similar: the structure of populations associated with the roots of wheat and tomato differed from that of populations isolated from the surrounding soil. Each of these plants favoured a particular genetic type, representing up to 57.5% and 40% of the isolates respectively, whereas the most abundant group in the soil did not exceed 25% of the isolates. However, the fusarian flora associated with flax and melon did not differ from that of the soil (Edel et al., 1997).

Microorganisms adapted to the rhizosphere, however, have a high metabolic activity and multiply very rapidly: the cells of pseudomonads, for example, divided every 5 hours on average in the rhizosphere of *Pinus radiata* versus division only every 77 hours in non-rhizospheric soil (Bowen and Rovira, 1976). The remarkably high density of organisms in an active growing stage, which characterises the rhizosphere, is an eminently favourable factor for genetic exchanges among individuals, whereas the probability of such exchanges occurring in the rest of the soil is much less likely. The transfer of plasmids by conjugation between different strains of rhizobia, bradyrhizobia or pseudomonads was thus obtained experimentally in non-sterile soil in the presence of plant roots. The transfer of the symbiotic plasmid of *Rhizobium leguminosarum* to other soil bacteria has also been observed: the transconjugants became capable of producing nodules in white clover even though not previously associated with the roots (Rao et al. 1994). Transfer of the K84 plasmid from the auxiliary bacterium *Agrobacterium radiobacter* which produces agrocine (see page 315), to the bacterial pathogen *A. tumefaciens*, which seems to occur under natural conditions in Greece (Panagopoulos et al., 1979), can only be explained by a phenomenon of conjugation between the two bacteria.

The root exudates of some plants can influence the composition and activity of the microflora of the roots of other plant species growing nearby. Thus the exudates of *Ambrosia psilostachya, Aristida oligantha, Bromus japonicus, Digitaria sanguinalis, Euphorbia supina* and *Helianthus annuus*, pioneer plants in the recolonisation of abandoned plantations in Oklahoma state (USA) inhibited *Rhizobium* and significantly reduced the number of nodules formed on various leguminous plants. As a result, the nitrogen content of the land left fallow remained particularly low (Rice, 1968). A fungitoxic factor present in the rhizosphere of heather (*Calluna vulgaris*) prevented mycorrhization of the roots of nearby trees: heather benefited from their poor growth (Robinson, 1972).

Gene regulation by the 'quorum sensing' mechanism seems widespread among rhizospheric bacteria. As already mentioned, expression of a particular gene may be triggered by AHL signals produced by non-isogenic as well as isogenic bacteria. Because of the great number of active colonies coexisting on the roots, such cross-communication phenomena are not uncommon and can modify the outcome of competitive interactions in the rhizosphere.

Rhizospheric Microflora

Subdivisions of the Rhizosphere

At the beginning of this chapter we defined the rhizosphere as that part of the soil surrounding the roots and subject to their influence. It is now clear that this definition requires greater precision. The extent of the rhizosphere is determined both by the quantity and nature of the rhizodeposition as well as the intensity of the response of the organisms selected as indicators: a rhizosphere effect can be mentioned insofar as a significant modification in the status of the soil microorganisms can be highlighted. If microorganisms sensitive to specific volatile messengers such as the fungus *Sclerotium cepivorum* are taken into account, the rhizosphere effect can be demonstrated at distances greater than one centimetre from the emitting roots (but will be zero in the presence of a plant incapable of synthesising the messenger). If, on the contrary, the stimulatory effect is due to soluble products, as is most often the case, it manifests only at a distance of about 1 to 3 mm from the root, depending on the diffusion capacity of the compound and the sensitivity of the organisms.

When the root is carefully washed to remove the rhizospheric soil, mycelial hyphae and bacterial colonies strongly adhering to its surface are revealed (Fig. 7.6) and its floristic composition may differ from that of the rhizosphere. Thus the surface of the root constitutes a specific habitat known as the **rhizoplane**.

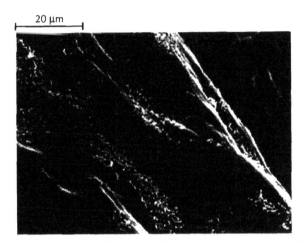

Fig. 7.6 Colonisation of the surface of a wheat root by free nitrogen-fixing bacterium *Rahnella aquatilis* (plate, Achouak and Villemin, CNRS).

Colonisation of the rhizoplane starts very precociously but is never complete. On the contrary, even in a soil very rich in microorganisms the microbial cover never formed a continuous sheath around the root and represented less than 10% of the surface of young roots (Rovira et al., 1974) and no more than 37% in 3-month-old roots of *Pinus radiata* (Bowen and Rovira, 1976). Once in place, the colonies usually extend little. The rate of root lengthening is generally much higher than the rate of growth of the microorganisms: for example, the advance of various fungi along the roots of broad-bean was at the most 3 mm day^{-1} while the roots lengthened at the rate of 9 mm day^{-1} (Taylor and Parkinson, 1961). In cultivated cereals, the rate of root elongation is of the order of a few cm day^{-1}. Bacterial colonies always remain very small and their ability to disperse along the root length seems to be quite limited even when they have flagella and the water content of the soil is high. When *Pseudomonas fluorescens* was deposited on the roots of pea in a soil maintained at a matrix potential of –6 kPa, one week later no colonies were found beyond 3 cm below the point of inoculation; concomitant root elongation ranged from 8 to 23 cm depending on experimental conditions (Bowers and Parke, 1993).

The microflora of the rhizoplane is therefore constituted of many independent communities. But these epiphytic communities are not distributed uniformly on the root surface. They are preferentially implanted at the junction of the epidermal cells and at the base of the absorbent hairs, i.e., regions where exudation is most intense and where protective mucilage is thickest. The sites of emergence of lateral roots and lesions caused by mechanical abrasion and animal or microbial parasites are also zones of intense exudation and hence readily colonised.

Nevertheless, even after disinfecting to deliberately eliminate the flora of the rhizoplane, it is possible to obtain colonies of bacteria and fungi from root fragments, which are not necessarily parasites but often simple saprophytes. These organisms live at the expense of the senescent cells. Senescence of the cortex, which occurs even in roots maintained under axenic conditions, is a natural phenomenon. It commences with the death of absorbent hairs, continues with that of the epidermal cells, followed by cortical cells (Fig. 7.7). The rapidity of senescence of the cortex is a genetic characteristic of the plant species: it is very high in cereals and much lower in dicotyledons. It varies within a species depending on the genome of the cultivar (Deacon and Lewis, 1982). Once introduced into the cortical parenchyma, fungi and bacteria can penetrate into meatuses between the living cells, utilising their exudates without provoking lysis. The term *endorhizosphere* is sometimes used to designate this part of the rhizosphere that extends inside the plant tissue. The term expresses the very important idea that it is possible for the microorganisms to advance progressively and imperceptibly from the soil to the interior of the root. The epidermis merely constitutes a provisional frontier between the plant and the external medium. The rhizoplane, often considered an interface between the soil and the root, plays this role only in the youngest roots. The real limit of the continuum soil-root is actually the endodermis.

> The term endorhizosphere evokes the sense of gradual passage from the soil to the inside of the root. It is, however, strongly criticised by Kloepper et al. (1992) for being:
> — incorrect: 'endo-rhizosphere' literally designates 'a zone inside the rhizosphere'; or the rhizosphere is a region of the soil and not of the root;
> — badly defined: depending on the authors, the endorhizosphere ends at the endodermis or includes the central cylinder and pertains either to all the microorganisms present in the root or only to microorganisms that are neither parasitic nor symbiotic;
> — and therefore useless: words that designate the precise anatomical region where the microorganisms are found should suffice.
> It should be noted, however, that the endodermis clearly demarcates a frontier crossed neither by rhizobia, endomycorrhizal fungi nor saprophytes (except accidentally). The term endorhizosphere was probably badly coined and is now seldom used. Regrettably, no other word exists for designating this particular space where the soil and the plant interpenetrate.

Generalisation of the Notion 'Rhizosphere'

Any surface delimiting a frontier between a living organ (or organism) and the world external to it constitutes a biotope populated by a particular microflora. Thus there is a phyllosphere on the surface of leaves and a spermosphere enveloping germinating seeds. Similarly, fungal hyphae modify their immediate environment and in so doing, create a hyphosphere.

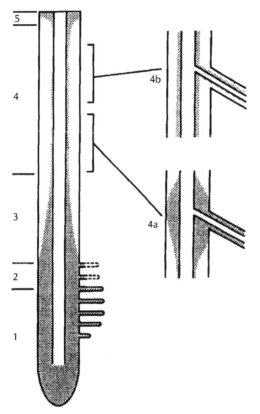

Fig. 7.7   Senescence of the cortex. This diagram, in which the scale is arbitrary, represents the distribution of dead cortical cells in the seminal roots of wheat. Grey zones correspond to living cells that have nuclei (Henry and Deacon, 1981).

1. Apex, differentiation and elongation zones and living absorbent hairs.
2. Zone in which absorbent hairs and non-pilose cells of the epidermis lose their nuclei.
3. Zone of gradual death of cortical parenchyma (from the 2nd to the 5th cellular layer).
4. Zone in which the 2nd to 5th cellular beds have lost their nuclei but the 6th layer of cortical cells, which is in contact with the endodermis, retains them. The nuclei continue to exist in several cellular beds around the points from which lateral roots emerge in the young parts of the root (4a) but are absent in older regions (4b).
5. Zone, situated just below the seed, where the cortex continues to live.

The same concept can be applied to animal microbial ecology. The human skin, for example, is covered with bacteria (on average 5×10^8 per cm^2). In most saprophytes, they constitute a stable characteristic microflora, comprising almost exclusively Gram-positive bacteria (Ducel, 1987). The surface and superficial layers of the epidermis are populated by aerobic bacteria (*Propionibacterium*). Normally, microorganisms are not found in the dermis. Thus the different regions of the skin can be considered analogous to the rhizoplane, cortex and central cylinder of plant roots. This microflora, which produces bacteriocins and fatty acid bactericides, maintains the microbial equilibrium of the skin just like the bacteria of the rhizosphere which ensure relative protection of the root. The flora of the intestinal mucosa is another example of an ecosystem situated at the frontier of two extremely different media.

Even the endodermis does not constitute a impenetrable barrier. Thus, several species of fungi were found in the central cylinder of the roots of tomato after the stele had been dissected from the cortex: some (*Fusarium oxysporum, Verticillium dahliae, Cephalosporium* sp.) are normal hosts of the vascular system; others (*Aspergillus, Penicillium*), saprophytes, and still others (*Colletotrichum coccodes*) colonisers of the cortex. Rupture of the endodermis during emergence of secondary roots probably constitutes the main route of access to the central cylinder. The extent of colonisation of the vessels, minimal when the roots were healthy, increased dramatically when they were attacked by a cortical parasite such as *Pyrenochaeta lycopersici*, probably capable of opening supplementary breaches (Davet, 1973). A rigorous demonstration of the possible spread of common microorganisms of the rhizosphere into the plant via its vessels was recently carried out using a strain of *Pseudomonas aureofaciens* genetically modified in such a manner that it could be readily identified without disturbing its activity. Even though the *P. aureofaciens* strain is not a parasite, bacteria were detected in the stems and even in the leaves of a large variety of plants just a few days after its deposition in the vicinity of their roots (Kluepfel and Tonkyn, 1992).

Even a perfectly healthy root is therefore never an axenic one!

Consequences of Biological Activity in the Rhizosphere on the Plant

RECYCLING OF MINERAL ELEMENTS

Rhizospheric activity almost always results in better assimilation of mineral elements by the plants. Experimental results have led to the following hypothesis (Bottner and Billès, 1987). Energy supplied by rhizodeposition enables increase in the microbial biomass, rapidly followed by proliferation of predators. Stimulation and rapid renewal of populations under the effect of predation, is accompanied by intense enzymatic activity which facilitates lysis of the desquamous cells of the roots and biodegradation (partly by cometabolisation) of soil organic matter present in the vicinity. The C/N ratio of the predatory microfauna, which releases large quantities of carbon in the form of CO_2, is slightly elevated and almost equal to that of its prey (close to 10). As a result, this microfauna absorbs too much nitrogen and jettisons the excess in the form of ammonia and urea, both of which are assimilable by the plants and less likely to be leached than nitrates (Fig. 7.8). Intensification of mineralisation of organic matter in the rhizosphere pertains not only to nitrogen, but all other mineral elements as well.

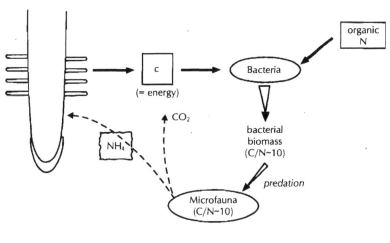

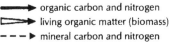
organic carbon and nitrogen
living organic matter (biomass)
mineral carbon and nitrogen

Fig. 7.8 Acceleration of mineral cycles due to predation. Rhizodeposition supplies the carbon substrates for the rhizospheric bacteria. Some of these bacteria fix atmospheric nitrogen but most derive nitrogen from the organic matter of the surrounding soil (non-utilisable by the plants). Bacterial populations are subjected to severe predation by amoebae, as well as nematodes and microarthropods. The C/N ratio of the microfauna is approximately equal to that of the bacteria: from 8 to 12. As the microfauna respires intensely, it loses carbon in the form of CO_2. In order to maintain its C/N ratio value, it must therefore excrete nitrogen in the form of NH_4, which can be recovered by the plants (after Clarholm, 1985).

LOWERING OF REDOX POTENTIAL

Respiration of roots and microbial activity consume significant quantities of oxygen, provoking a reduction in redox potential. This phenomenon can favour absorption by the plants of certain cations. When the environment tends to become anaerobic, iron and manganese are reduced by some bacteria and in this form are much more soluble than in the oxidised state (see page 88). Dissolution of crystals, always impure, simultaneously release oligoelements which may also be present.

In a soil of normal structure, gaseous exchanges, albeit slow, generally allow diffusion of sufficient oxygen so that the average redox potential in the root vicinity does not reach negative values. But heavy rain or excessive irrigation, causing drastic engorgement of the soil, could temporarily render it highly reductive. In saline soil, rich in sulphates (not uncommon in certain low plains, in Tunisia for example), this situation could lead to sudden withering of a crop as a result of proliferation in the rhizosphere, of sulphato-reducing bacteria that produce toxic sulphydric acid (Dommergues et al., 1969).

SYNTHESIS OF GROWTH REGULATORS

Most of the growth regulators known in higher plants can be synthesised by bacteria or fungi: as a matter of fact, gibberellin was extracted from culture filtrates of the fungal parasite *Gibberella fujikuroi* even before it was shown to exist in plants. Indoleacetic acid, formed from tryptophan is produced like other similar compounds, by a large number of rhizospheric microorganisms. These auxins increase cell permeability, which could explain why much larger quantities of exudates are always found in the presence of rhizospheric flora ·than in sterile conditions: rhizodeposition can even double in the presence of microflora (Heulin et al., 1987). But this increment could also be the indirect result of stimulation of plant functions under the effect of auxins. Many bacteria and fungi can also synthesise cytokinins from adenine. Lastly, ethylene is a common volatile product in the soil atmosphere. Its precursor is commonly methionine but a trial carried out on a large number of compounds showed that most of the amino acids as well as other organic acids present in root exudates, stimulate ethylene production without necessarily being direct precursors (Arshad and Frankenberger, 1990).

PRODUCTION OF CHELATING AGENTS

Siderophores (see page 172) are the most well known of the chelating agents synthesised by microorganisms. Besides iron, manganese and zinc can also be more readily assimilated in the presence of certain bacteria (Barber and Lee, 1974). Collection of poorly soluble mineral ions can occur through the intermediary of organic acids such as oxalic acid or ketogluconic acid. Many rhizospheric bacteria are thus capable, in vitro, of solubilising mineral phosphates of the soil. Some, such as *Bacillus megaterium*, are under experimentation for improving assimilation of phosphorus by plants, but as yet have not yielded convincing results.

PRODUCTION OF TOXIC COMPOUNDS

In addition to antibiotics that can prove phytotoxic under certain conditions, rhizospheric microflora sometimes synthesises metabolites that behave like true toxins. Some have been well identified and their mechanism of action is known. Such is the case of rhizobitoxin, emitted by certain strains of *Bradyrhizobium japonicum*, responsible for chlorosis in soybean. It is a specific amino acid which inhibits formation of homocysteine, a precursor of methionine (Owens et al., 1968). Other toxins are known mainly by the effects they produce. For example, a pseudomonad of the rhizosphere of wheat produces a soluble metabolite which retards development of the root system (Bolton and Elliott, 1989). The simple presence of *Pythium myriotylum* in the rhizosphere of tomato suffices to induce necrosis of the roots even before penetration of the tissues. This very stable toxin is active even after the death of the fungus (Csinos and Hendrix, 1978) so that protection of the plants using biological antagonists would not be effective. There also exist bacteria and fungi capable of producing cyanides and volatile hydrocyanic acid from glycine, serine, methionine or threonine sometimes present in root exudates.

The few examples presented above represent toxic products directly elaborated by microflora. Toxins resulting indirectly from the action of microbial enzymes on organic plant substrates should also be noted; bacteria that hydrolyse prunasine in the rhizosphere of peach trees provide an example (see page 191).

DETOXIFICATION

Roots of some plants excrete phenolic compounds that are toxic for other plant species: aerial parts, washed by rain, can also add toxins to the soil. This happens in walnut and many Ericaceae of the east coast of the United States, such as *Arctostaphylos glandulosa* for example (Chou and Muller, 1972). This phenomenon, termed **allelopathy**, is similar to microbial antibiosis and produces the same result, namely exclusion of competitive species. These toxins are biodegradable, however, and some plants, though susceptible, succeed in coexisting because bacteria capable of metabolising the toxins are present in their rhizosphere. Sometimes detoxifying microorganisms are found in the vicinity of the producing plant. Thus coumarine, extremely inhibitory in nature and excreted by *Anthoxanthum odoratum*, is detoxified by the pseudomonads present in its own rhizosphere (Rivière and Chaussat, 1966).

Useful and Deleterious Organisms

The intense metabolic activity occurring in the rhizosphere is not without consequences for plant development. As already mentioned, rhizospheric microflora increases the quantity of root exudates. It is very difficult to determine whether the flux of organic compounds dispersed in the external medium represents a dead loss or a productive investment for the plant.

It is certain that at least part of the rhizospheric microflora exerts a **deleterious effect** on plant growth. Suslow and Schroth (1982), for example, isolated a large number of bacteria from the rhizoplane of sugar-beet. When these bacteria were inoculated under controlled conditions, growth retardation and a reduction of up to 48% in the weight that could be attained with reference to controls were recorded on one-month-old plants; furthermore, the root system was poorly developed and less ramifying, the absorbent hairs fewer and tissue senescence accelerated. These effects prompted coinage of the term 'deleterious bacteria'.

Few studies have been carried out on these bacteria. They are difficult to demonstrate because, unlike parasites, they are not intimately associated with the tissues. Moreover, their effects depend largely on environmental conditions, are often cumulative and hence not easy to reproduce. For these reasons, their mode of action has not yet been clearly elucidated. A direct toxic action could be demonstrated in a few cases. In some pseudomonads of the rhizosphere of winter wheat, the toxic action is associated with production of an antibiotic (Fredrikson and Elliott, 1985). The tobacco disease known as frenching, which results in plant stunting and blanching of the limbs while the veins of the leaves remain green, seems to be caused by *Bacillus cereus*, a saprophytic bacterium of the rhizosphere (Steinberg, 1951). This bacterium excretes a toxin which, absorbed by the plant, interferes with synthesis of amino acids by competing with isoleucine. Fungi have shown similar behaviour. Thus, 1-amino-2-nitrocyclopentane-1-carboxylic acid produced by *Aspergillus wentii* could be responsible for the slow growth and discoloration and deformation of the leaves sometimes observed in chrysanthemums in Florida (Woltz, 1978). Microorganisms per se hardly pathogenic, upon localising in the cortical tissue are also capable of excreting metabolites that damage the living cells and cause extensive necrosis. This was observed during development of *Cylindrocarpon destructans* in the senescent cortical layers of strawberry roots (Wilhelm, 1959) and a similar situation seems to arise for several bacteria of the rhizoplane or the cortical tissues.

Quite likely, competition also occurs in the rhizosphere between plants and microorganisms for certain mineral elements (aside from phosphorus and nitrogen, which are quite rapidly recycled). Thus, under experimental conditions, the pseudobactin extracted from fluorescent pseudomonads locks up iron and impedes its assimilation by young plants of pea and maize, provoking a generalised chlorosis (Becker et al., 1985).

Deleterious bacteria also seem to favour colonisation of the roots by parasitic fungi, either by facilitating their entry into the tissue by their necrogenic action or by a direct stimulatory effect.

On the other hand, other rhizospheric bacteria can exert an overall **stimulatory effect** on plant growth. Some strains of fluorescent pseudomonads quintupled growth of potatoes cultivated in a greenhouse in just 15 days. In the field, the gain in yield attributable to inoculation with such strains reached 17% (Kloepper et al., 1980). These bacteria, even though they represent only a small proportion of the rhizospheric flora, seem to be present on the roots of most plants, and their effects are often more significant than would be expected from their number. They are commonly designated by the abbreviation PGPR (Plant Growth-Promoting Rhizobacteria) and have aroused great interest over the last twenty years.

It has been remarked in numerous cases that the stimulatory effect of PGPR manifests only when the inoculated plants are cultivated on a non-sterile substrate. One may therefore deduce that the role of these bacteria is to counteract the influence of deleterious organisms or parasites normally present in the soil. Numerous studies, mainly concerning *Pseudomonas fluorescens* and *P. putida*, concur in this hypothesis. Detailed genetic analyses have shown that their protective action is mainly exercised via siderophores and antibiotics. We have already studied examples concerning specific pathogenic agents, viz. *Fusarium oxysporum* and *Gaeumannomyces graminis* var. *tritici* (see pages 181 and 185). The same mechanisms function with respect to the general microflora: some pseudomonads significantly reduce populations of fungi (*Aspergillus, Penicillium, Fusarium*) and especially Gram-positive bacteria present in the rhizoplane (Kloepper and Schroth, 1981). More recently, it has been shown that strains of fluorescent pseudomonads, while growing on the surface of roots, can induce a systemic resistance in plants against vascular diseases as well as against fungal, bacterial and viral pathogens of the aerial parts. This resistance differs from the 'induced systemic resistance' that occurs after localised parasitic aggression: it is coupled neither with a hypersensitive reaction nor the appearance of pathogenesis-related proteins. The mechanisms responsible for 'induced systemic resistance' due to PGPR are still not well understood. Some siderophores (pseudobactin, pyochelin, salicylic acid) seem to be involved as this induced resistance manifests mainly when the Fe^{3+} content is low. But in certain cases, the purified membrane lipo-polysaccharides of *P. fluorescens* are as effective as living bacteria (van Loon et al., 1998).

Protective action is not limited to bacteria alone. Diverse rhizospheric fungi can also defend the roots against eventual invaders either by competition or by antagonism. Protection is sometimes due to simple enzymatic reactions. For example, the fungus *Talaromyces flavus*, which protects egg-plant from verticilliosis, secretes large quantities of glucose oxidase. In the presence of glucose of the root exudates, this enzyme produces hydrogen peroxide which weakens or even destroys the microsclerotia of *Verticillium* (Fravel and Roberts, 1991). The non-pathogenic strain of *Fusarium oxysporum* Fo 47, which protects tomato from vascular fusariosis, is also capable of inducing systemic resistance. In this case certain pathogenesis-related proteins associated with classic 'systemic acquired resistance' were detected (Duijff et al., 1998).

Demonstration of strains of bacteria and fungi capable of promoting plant growth not only in the soil, but also under wholly axenic conditions, has resulted in the proposal of other modes of action besides protection against a hostile microbial environment. Mineral assimilation, especially that of phosphorus, is augmented in the presence of certain bacteria (Lifshitz et al., 1987). It is now known that this phenomenon results from increase in the absorption surface of roots—more numerous, longer and with enhanced absorbent root hairs. This stimulation is definitely hormonal in nature, but it is still not precisely known whether PGPR induce an increased production of growth substances by the plant or if they synthesise hormones that act directly (Abbass and Okon, 1993). The rhizosphere also harbours bacteria capable of fixing atmospheric nitrogen. These bacteria have mainly been studied in Gramineae, in which their presence is associated with low nitrogenase activity.

Cyanide-producing Bacteria: Useful or Deleterious?

The following example illustrates how the role of rhizospheric bacteria is sometimes ambiguous and that interpretation of the various phenomena involved is tricky.

Schippers et al. (1987) demonstrated that bacteria, notably pseudo-monads, occurring in the rhizosphere of potatoes produce hydrocyanic acid from the root exudates. This production requires the presence of Fe^{3+} ions. Absorbed by the root cells, the cyanides can prevent mitochondria from producing ATP, the energy source for the principal metabolic processes. The mitochondria also have an alternative pathway which is not susceptible to cyanides, but the efficiency is much less. Under these conditions, the plant is undernourished and the tubercles few. Some strains of *Pseudomonas*, applied to the tubercles at the time of plantation, enable recovery of normal production. These strains, as well as their purified pseudobactin, prevent production of cyanide by the deleterious bacteria under in vitro conditions. It is probable that this inhibition is due to sequestration of Fe^{3+} by the siderophores of auxiliary bacteria.

On the other hand, the Swiss team of G. Défago isolated a *Pseudomonas fluorescens* (strain CHAO) from a soil resistant to root rot of tobacco caused by *Chalara elegans* (=*Thielaviopsis basicola*). This strain, which develops in the rhizosphere of tobacco, produces hydrocyanic acid and precludes development of the disease under controlled conditions. The presence of Fe^{3+} improves protection, which excludes intervention of siderophores. Mutants incapable of producing cyanide can only ensure limited protection, but they recover effectiveness if the gene coding for production of HCN is reintroduced into their genome. The CHAO strain and complemented strains induce abundant development of absorbent hairs, i.e., have a direct effect on the plant (Voisard et al., 1989).

Production of HCN in the rhizosphere thus entrains deleterious or beneficial effects depending on the bacterium-host plant pair involved as well as the prevailing environmental conditions.

At this stage in our report, it appears possible to readily distinguish three categories of microorganisms in the rhizosphere: those which are indifferent, those which are deleterious and those which stimulate growth. The situation is in fact much more complex and numerous observations have led to the idea that demarcation between deleterious and auxiliary bacteria is quite vague. The same metabolites can in actuality lead to diametrically opposed results. Since 1970, Hussain and Vančura have maintained that the culture filtrate of a fluorescent, indoleacetic acid-producing pseudomonad can be toxic or stimulatory, depending on the concentration. Siderophores are one of the weapons of PGPR against undesirable microorganisms. But we have seen (see page 214) that pseudobactin can provoke ferric chlorosis under in vitro conditions, and this at the same concentration (10 μM) as that which has proven antagonistic to deleterious bacteria (Becker et al., 1985). Antibiotics are another commonly used weapon. The strain CHAO of *P.*

fluorescens produces pyoluteorine and 2-4-diacetyl-phloroglucinol, both of which are effective against *Pythium ultimum*. Maurhofer et al. (1992) made a new strain using this bacterium, which produces more antibiotics and inhibits *P. ultimum* more effectively than the wild strain. But despite the protection it ensures against the fungal parasite, this recombinant strain induces a strong reduction in plant growth This is explained by increased production of antibiotics which renders it phytotoxic. By the way, it is remarkable that deleterious bacteria and PGPR belong to the same genera and sometimes even the same species: *Pseudomonas fluorescens*, *P. putida*, *Bacillus subtilis*, *Klebsiella*, *Enterobacter* and *Arthrobacter* are classified in turn in one or the other category. Likewise, among fungi, *Fusarium oxysporum*, depending on the strain, can parasitise the cortex or the vascular system, be an agent of biological control (Alabouvette, 1986) or a growth stimulator (Davet, unpubl.). pH, soil water content and extent of the absorbent complex are essential regulatory factors but temperature seems to be of paramount importance. Thus the same strain of *Bacillus subtilis* can significantly stimulate seed germination at 20°C and inhibit it at 35°C (Schiller et al., 1977). *Phialophora radicicola* var. *radicicola*, a harmless coloniser of the rhizoplane, actively prevents penetration of *Gaeumannomyces graminis* var. *tritici* in temperate regions, but becomes an aggressive parasite in warm climates (Sivasithamparam, pers. comm.).

The foregoing shows that knowledge of the mechanisms controlling functionality does not suffice; an understanding of the ecological exigencies of the microorganisms of the rhizosphere is absolutely imperative for realisation of their maximum potential.

Rhizospheric Affinity

The environmental conditions in the rhizosphere exert a selective effect on soil populations. Only a part of the microflora and the telluric microfauna (which varies depending on the plants and their physiological state) is represented in this environment. Several successive stages have to be overcome for an organism to establish and maintain itself in the rhizosphere.

Microorganisms subsist in a dormant state in the soil until they receive a **signal** that triggers the process of chemotaxis or active growth, preferably directed towards the root. This signal may be barely specific, constituted of sugars and amino acids for example that diffuse in the soil solution; in this case a rapid response would result in successful occupancy of the site since competition is stiff. On the other hand, the signal may be highly specific; often volatile, it can trigger a reaction of the organism targeted at a relatively great distance from the root but not representing a handicap since competition in this instance is negligible. This type of message is generally the result of a long coevolution between the plant and its rhizospheric hosts, frequently resulting in relations more intimate than simple cohabitation. We shall return to these relations in the next chapter.

In order to survive in the rhizosphere, microorganisms that have responded to the stimulus must be capable of withstanding the particular conditions occurring there: acidic pH, large-scale uptake of mineral ions by the roots, toxicity of some exudates, products of secretion or cellular lysates (which might also include toxicity due to hydrogen peroxide and superoxide ion produced by the peroxidases of the root surface). Little information is available as yet on the mechanisms employed by the microorganisms for combating these hurdles, especially with respect to saprophytic organisms. It is known, however, that parasites and symbionts do overcome unfavourable conditions by installing themselves inside the host tissue. Microorganisms residing outside the host are not able to survive unless 'rhizosphere competent'. According to Ahmad and Baker (1987), rhizosphere competence, contracted here to **rhizocompetence**, represents the aptitude of a microorganism to grow, function and maintain itself in the rhizosphere as and when the root lengthens.

A rhizocompetent organism must be able to withstand the physicochemical environment of the root as well as the strong competition that reigns there. This ability can be in the form of production of antibiotics (pseudomonads, bacteria abundant in the rhizosphere, synthesise a large variety of these), resistance to foreign antibiotics as well as optimal utilisation of available molecules, comprising the simplest ones. Indeed, sugars and organic acids, even though often facilely utilised in vitro, do not have an equivalent value for all microorganisms in a competitive situation. The differential effect of these simple molecules, constituting the main components of the root exudates, led to numerous studies in the 1960s and 70s, whose results are, quite wrongly, somewhat forgotten nowadays. For example, populations of bacteria enumerated in a soil differed depending on whether the soil was percolated with a solution of saccharose and amino acids arising from exudates of cultivated tomato (*Lycopersicon esculentum*) or from a wild tomato plant (*L. hirsutum*) (Fig. 7.9). Gram-negative bacteria (including pseudomonads) were strongly stimulated by the exudates of *L. hirsutum* and more sporulating bacteria observed (Mangenot and Diem, 1979). Ahmad and Baker (1988) attributed the great rhizospheric affinity in one of their strains of *Trichoderma harzianum* to its high cellulasic activity and its ability to assimilate complex hydrates of carbon: this characteristic would enable competitive utilisation of the debris of desquamous and senescent cells abundant in the rhizhosphere. It should also be mentioned that the selectivity of the Komada medium (1975), universally used for isolation of *Fusarium oxysporum* from the soil, is, among other things, due to the replacement of glucose (the common carbon nutrient in synthetic media) by another simple sugar, D-galactose.

Several observations indicate that rhizocompetence involves not only adaptation to particular environmental conditions of the rhizosphere, but also the direct relations between the microorganisms and the plants with which they come into contact. Mucilage that covers the epidermal cells contains water-soluble, high molecular weight glycoproteins endowed with agglutinating properties. When this mucilage was harvested from healthy roots of kidney-beans by simple rinsing, it provoked differential agglutination of saprophytic pseudomonads of the soil. Strains of *P. putida* generally agglutinated more often and more strongly than those of *P. fluorescens* or *P. aeruginosa* (Anderson, 1983). When the effect of the agglutinin of pea roots on one hundred bacteria of diverse origin was studied, it was found that

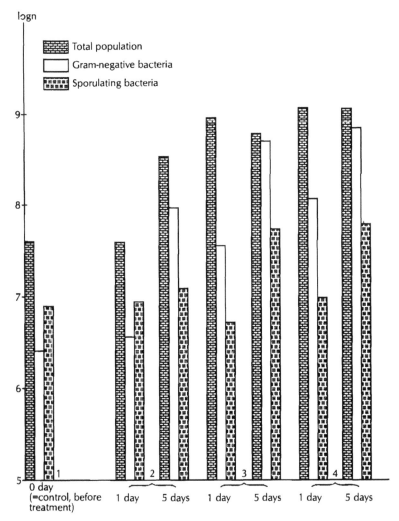

Fig. 7.9 Effect of percolation of nutritive solutions on populations of soil bacteria (after Mangenot and Diem, 1979).
1. Soil populations before treatment.
2. Populations after percolation with a mineral solution containing 1% saccharose.
3. Populations after percolation with sugar-mineral solution to which amino acid exudates of the roots of *Lycopersicon esculentum* cv. Marmande were added.
4. Populations after percolation with sugar-mineral solution to which amino acid exudates of the roots of *L. hirsutum* were added so that they contained the same concentrations of nitrogen as solution 3.
Samplings were done after 1 or 5 days of treatment. The number of bacteria was expressed in logarithms.

82% of the strains isolated from the rhizosphere agglutinated, but only 34% of the strains from non-rhizospheric soil. These isolates responded differently to agglutinins of other plants (Chao et al., 1988) and bacteria not habitual hosts in the rhizosphere, such as *Escherichia coli*, showed little or no reaction. It therefore seems that at least for some common bacteria of the rhizosphere, a recognition mechanism of the lectin-sugar type exists. Microbial sites of recognition of lectin are located in the mucous polysaccharide capsule enclosing the cells. Thus, in an experiment normal cells of *Azospirillum lipoferum* and *A. brasilense* bound to the wheat-germ agglutinin present on the roots of various Gramineae. But the same bacteria, deprived of their capsule, were incapable of agglutination (Del Gallo and Haegi, 1990).

Researchers are not in agreement about the exact role played by the phenomenon of agglutination. Considering the work already carried out with pathogenic or symbiotic microorganisms, it would appear that lectins contribute to the short-term adherence of bacteria on the roots, but do not play a decisive role in their long-term binding. In a situation wherein competition occurs, even a temporary advantage can be decisive, however.

> Purely physical phenomena, such as hydrophobic interactions among non-polar groups on the surface of roots and bacterial envelopes, also play a role in the preliminary phases of adhesion. It has further been demonstrated that membrane proteins called 'porines' are involved in adhesion of *Pseudomonas fluorescens* (De Mot and Vanderleyden, 1991) and *Rahnella aquatilis* (Achouak et al., 1995) to the roots; their function is to ensure passage of hydrophilic molecules in the bacterial cells.

Substances that have the same effect as N-acyl homoserine lactones have also been demonstrated in the root exudates of rice, tomato and several leguminous plants (Teplitski et al., 2000). These signals, which regulate expression of bacterial genes (see page 174), probably play an important role in plant-microorganism interactions and in the equilibrium among rhizospheric populations.

MICROORGANISMS THAT COLONISE THE ROOT

Parasitic and symbiotic microorganisms do not constitute a part of the rhizospheric flora as they live inside the living root tissue. But before penetration, they must pass through the same stages that authentic inhabitants of the rhizosphere do: overcome microbiostasis, grow until reaching the root while withstanding competition from the other microorganisms, fixation and recognition of compatibility.

From Overcoming Microbiostasis to Infection

Overcoming Microbiostasis

Stimulation of infective propagules is due to diffusion of exudates, either soluble in the soil solution or volatile, arising from the roots or germinating seeds. The soluble compounds are essentially made up of sugars and amino acids and help to overcome microbiostasis in a relatively barely specific manner. In the presence of this type of exudate, the parasites or symbionts therefore encounter competition with a large number of other microorganisms. This is the case, for example, with *Fusarium solani* f. sp. *phaseoli*, a fungal infector of kidney-bean, which causes elongated necrosis at the base of the hypocotyls. In the spermosphere and rhizosphere, exudates that are abundant and have a low C/N ratio, facilitate abundant germination of chlamydospores, the resting stage of this fungus. But the germinative tubes are subject to intense competition, developing with great difficulty, and most are lysed before reaching the plants. Hence very little necrosis is observed in the young parts of roots. Contrarily, in the older parts near the hypocotyl, exudates are less abundant and contain hardly any amino acids. Their high C/N ratio is barely favourable to bacterial proliferation. Germination of chlamydospores is limited but, as competition is not so stiff, the germinative tubes can reach the surface of the hypocotyls, penetrate, and cause lesions at the collar that are typical of the disease (Toussoun, 1970). Exudation of sugars by roots and seeds also results in germination of sporocysts of *Pythium ultimum* and other less specialised Pythiaceae. But the ethanol and volatile aldehydes released by these organs are capable of more extensive diffusion, resulting in a still more rapid response (Nelson, 1987).

Specificity of the stimulatory effect is more often the rule when it involves parasites or symbionts. For example, the abietic acid exuded by roots of *Pinus sylvestris* triggers germination of basidiospores of three species of *Suillus* that form ectomycorrhizae in association with this tree, whereas the acid has no effect on other mycorrhizal fungi such as *Paxillus involutus* or *Thelephora terrestris* (Fries et al., 1987). The fungus *Sclerotium cepivorum* is strictly restricted to *Allium*, in which it causes bulb white rot. The fungus can remain for many years in the soil in the form of small sclerotia that germinate only in the vicinity of the roots of *Allium* or some related species. Germination of sclerotia can be experimentally induced by introducing the crushed roots of *Allium* into the soil. The effect is felt even at a distance (a few centimetres), suggesting that it is caused by volatile compounds. Several experiments have shown, however, that neither exudates nor crushings have a direct action on *S. cepivorum*. For germination to occur, it is necessary that the non-volatile alkyl and alkenyl-cysteine sulphoxides exuded by the roots be first transformed by the rhizospheric microflora into volatile thiols and organic sulphides (Coley-Smith, 1987); these then act on the sclerotia (Fig. 7.10). Diallyl disulphide is one of the most active of these metabolites.

Hatching of eggs of most parasitic nematodes occurs by analogous processes. Thus the first stage larvae of *Globodera rostochiensis*, a parasite of potato, which encyst under the desiccated teguments of the female, are stimulated by the root exudates of their host as well as a few other Solanaceae.

In fact, in the rhizosphere of a given plant, it is generally possible to find exudates that are attractive as well as exudates inhibitory towards one given fungus or nematode. The end results observed are those of a dynamic equilibrium modulated by the environment.

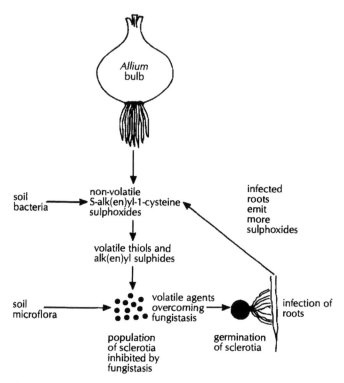

Fig. 7.10   The two steps in stimulation of *Sclerotium cepivorum* by root exudates of *Allium* under the effect of associated microflora. First step: roots excrete non-volatile sulfoxides. Second step: bacteria of the rhizosphere transform these sulphoxides into volatile thiols and organic sulphides. These vapours diffuse to the sclerotia, earlier quiescent due to fungistasis, and induce their germination, rendering it possible for them to infect the host plant. The infected roots excrete more sulphoxides, which results in amplification of the phenomenon (after Coley-Smith, 1987).

Advancement Towards the Host

Once the microbiostasis imposed on the various forms of conservation (chlamydospores, sclerotia, eggs, spores, cysts etc.) is overcome by the root exudates, the organisms resume active life. They produce germinative tubes or release motile forms, which move or grow mostly in the direction of the roots under the influence of the exudates. Amino as well as fatty acids, aldehydes and alcohols play an important role in attraction (Cameron and Carlile, 1978). Recent studies have shown the highly specific effect of flavonoids. Organisms sensitive to these compounds react even at extremely weak concentrations. Thus, to obtain a positive chemotactic response from *Sinorhizobium meliloti*, 10 nM of luteolin (secreted by seeds of alfalfa) or 7,4′-dihydroxyflavone (exuded by the roots) sufficed (Caetano-Anollés et al., 1988). The same concentration of daidzein and genistein attracted the zoospores of *Phytophthora sojae* towards the roots of soybean (Morris and

Ward, 1992). The zoospores of *Aphanomyces cochlioides* manifested a positive chemotaxis even at 1 nM of cochliophilin A, a flavonoid present in the roots of its host, spinach (Horio et al., 1992).

Zoospores also show proof of electrotaxis. Under experimental conditions they reacted to very low electric fields, comparable to those that exist in the rhizosphere (Morris and Gow, 1993).

The period between overcoming microbiostasis and penetration of the host is a critical phase during which the potential invader is especially vulnerable. Because generally poorly adapted to saprophytic life, the invader is subject to all types of antagonism and competition. Even if the organism is not destroyed before attaining its goal, the distance it has to cover or the constraints it encounters are sometimes such that its reserves are prematurely depleted. Such a situation does not necessarily result in the disappearance of the infectious organism. In fungi, autolysis of the germinative tube is often observed: the nutrients thus salvaged aid formation of a new chlamydospore or a smaller sclerotium inside the older one.

Fixation on the Host

The processes conducive to infection are still not clear and have been studied in only a small number of examples. It appears that very close contact between the host and the infectious organism, involving secretion of adhesive material, constitutes a prelude to the triggering of mechanisms of penetration. In rhizobia, finding on the host is apparently realised in two stages. According to a model still hypothetical proposed for the pea-*Rhizobium leguminosarum* pair (Fig. 7.11), initially a few bacterial cells attach to the surface of the absorbent hairs by means of an adhesine (rhicadhesine) whose secretion requires the presence of divalent cations (in fact, mainly Ca^{++} ions). In the

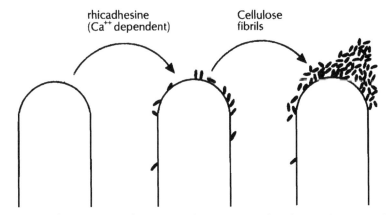

rhicadhesine
(Ca^{++} dependent)

Cellulose
fibrils

Fig. 7.11　Scheme proposed by Smit et al. (1987) to explain the attachment of *Rhizobium leguminosarum* to the absorbent hairs of pea. A small number of bacteria first affix to the hairs through an adhesine, rhicadhesine, whose secretion requires the presence of Ca^{++} ions. The bacteria then agglutinate to each other by cellulose fibrils, forming a mass at the extremity of the absorbent hair.

second stage, a mass of bacterial cells affix themselves on this first cluster of microbial cells. The long extracellular fibrils of cellulose produced by the bacteria are responsible for this agglutination. Fixation is not observed on epidermal cells that are not differentiated into absorbent hairs, which suggests that receptors, or a state of receptivity, is necessary. Fixation of the bacterial pathogen responsible for crown gall, *Agrobacterium tumefaciens*, also occurs in two stages and also involves synthesis of cellulose fibrils.

In oomycetes also, infection is preceded by an attachment phase (Deacon and Donaldson, 1993). The zoospores are attracted by chemotaxis towards the roots and accumulate around the sites where the stimulus is most intense—generally the zone of elongation and adjacent regions. Specific receptors present in the mucilage of the host trigger encystment of zoospores, a complex process occurring in two stages: secretion of adhesive glycoproteins that form the envelope of the cyst and affix it to the surface of the root and synthesis of a cell wall that transforms the protoplast, naked so far, into a cyst. The zoospores always attach by their ventral surface. This is the face that bears the flagella and there are reasons to assume that these flagella play an important role in the recognition of sites of infection as well as in orientation of the future cyst. This positioning helps the germinative tube projected from the ventral surface soon after the cyst is formed, to immediately penetrate the host tissue. These events unfold quite rapidly: it takes no more than 30 minutes between recognition of receptor sites on the root surface and formation of the germinative tube. As in the case of rhizobia, Ca^{++} ions play an important regulatory role.

At the time of encystment, the zoospores excrete Ca^{++} ions which combine with glycoproteins on the ventral side of their envelope, thereby ensuring fixation of the cyst. If Ca^{++} is chelated by a complexing agent at the time of cyst formation, the cysts lose their adhesiveness. The small adhesive cushion thus formed prevents dispersal of Ca^{++} ions into the surroundings. Because of the orientation of the cyst at the time of attachment, these ions can be rapidly resorbed by the cell and reach the regulatory centre of germination, which is now triggered off. If for some reason fixation of the cysts does not occur, Ca^{++} diffuses into the environment and cannot be resorbed in sufficient quantities. An exogenous supply of Ca^{++} is then required to effect germination. The action of calcium is reinforced in the presence of the amino acids (asparagine, aspartic acid and glutamic acid) generally present in the root exudates. These amino acids could attach to membrane receptors which control the opening of the ionic canals for Ca^{++} (Deacon and Donaldson, 1993).

In eumycetes, a mucilaginous matrix was demonstrated around the hyphae of *Verticillium albo-atrum* and two specialised forms of *Fusarium oxysporum* growing on the surface of the roots of their hosts (Bishop and Cooper, 1983). This material, polysaccharide or glycoprotein in nature, is analogous to that which covers the germinative tubes of the spores of aerial parasites. In addition to its adhesive role, this mucus seems to help in localisation of enzymes that play a role in lysis of the plant cell walls.

Infection

At the extremity of their germinative tube, aerial parasites develop a bulging appressorium, strongly attached to the surface of the epidermis. The ventral surface of this appressorium sends out a very thin penetration filament or 'infection peg' that traverses the cuticle and emerges in the underlying epidermal cell. The mycelium grows back to its normal diameter in the invaded cell, ramifies inside it, then penetrates the parenchyma. In the soil, desquamation of the absorbent hairs and cortical cells, growth of secondary roots and ruptures caused by the friction of mineral particles, constitute additional breaches in the protective envelopes of the tissues. So, fungal parasites of the root generally do not form an appressorium. They penetrate between the epidermal cells of the young roots, colonise the senescent parts of the cortex of the older roots and develop between cells in the living cortical parenchyma (*Fusarium solani*), or pass directly from cell to cell, traversing the cell wall by means of infection pegs (*Chalara elegans*). However, *Rhizoctonia solani* forms a small cushion of infection, comprising a dense mass of short and ramifying mycelial elements, on the surface of the hypocotyls and roots, which is equivalent to a compound appressorium, from which issue a multitude of filaments equivalent to infective hyphae. It has long been thought that the entry of vascular parasites into the xylem occurs passively, by aspiration, at the wound site, which would explain how saprophytic fungi or bacteria gain entry and maintain themselves for some time in the vessels (Gardner et al., 1985). Recent works have questioned this interpretation. Using a transgenic strain possessing the *Gus A* marker gene, Turlier et al. (1995) showed that *Fusarium oxysporum* f. sp. *lini* actively penetrates the immature cells encircling the subapical meristem of roots of flax. When these cells undergo division and differentiation, the *Fusarium* is eliminated from the parts that will mature into phloem and cortex but remains in the vessels and medullary parenchyma.

The site where root lesions occur is often characteristic of the host-parasite pair. On the one hand, as already seen, when a phenomenon of mutual recognition is necessary, it can only take place at specific sites. On the other hand, localisation of the 'point of impact' of a fungus depends on the time needed by the resting infective unit to start germinating, rate of elongation of the germinative tube and rate of root growth (Fig. 7.12). Thus fungi that respond extremely rapidly to the rhizosphere effect and have a very high growth rate, such as *Pythium*, generally attack the extremities of the roots. Fungi with a slow response (6 to 10 hours for fusaria) and moderate growth, mostly reach the already differentiated regions. These differences are masked in in vitro infections, in which nothing interferes with progression of the parasite, but do distinctly occur in a non-sterile soil (Fig. 7.13).

The various successive stages taken from the time that inhibition imposed by microbiostasis is overcome until the time of penetration into the living tissues of the plant are schematically depicted in Figure 7.14. It may be stated that failure of infection does not necessarily lead to disappearance of the inoculum.

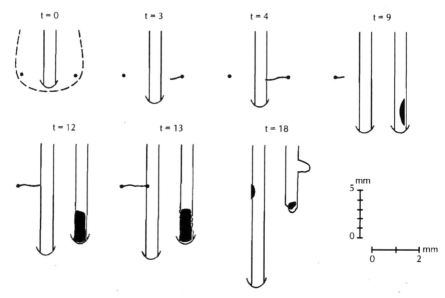

Fig. 7.12 Influence of the respective growth rates of the root and the germinative filament on the point of entry of a fungal parasite. For convenience, the unit of length is larger on the horizontal axis than on the vertical. In this example, the rate of root progression is 12 mm day^{-1}. The cfu on the left in each Figure begins germinating in 8 h and the rate of growth of its germinative tube is 6 mm day^{-1}. The cfu represented on the right side of the Figures responds to the root exudates in 2 h and the rate of elongation of its germinative tube is 12 mm day^{-1}. Time is counted in hours and the initial instant is determined as the moment when the cfu, situated 1 mm from the root, are influenced by the rhizosphere effect. A cortical lesion is observed in one case and an apical necrosis in the other.

When infection takes place, its subsequent progress depends on a very complex series of relations between the invading organism and its host. A review of the defence reactions of the plant is beyond the scope of this book. We shall give only one example (see box on page 229) which illustrates particularly well the phenomena of incompatibility between a plant and a microbial aggressor. Depending on the efficacy of the defence reactions of the plant, the infection may/may not persevere. Should the infection persist, the infectious organism develops and after a more or less lengthy time period, forms new infective units representing a higher level of inoculum compared to the initial one.

Degrees of Relation between Plant and Microorganism

This relation can be considered from the point of view of the microorganism or of the plant.

From the point of view of the microorganism, the possibility of developing inside the host tissue is always advantageous, ensuring the organism

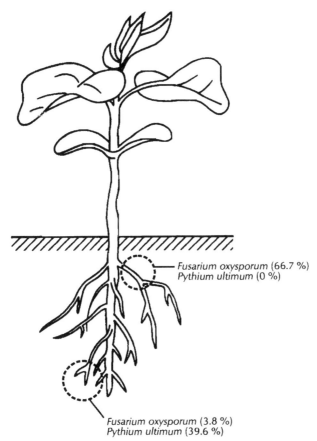

Fusarium oxysporum (66.7 %)
Pythium ultimum (0 %)

Fusarium oxysporum (3.8 %)
Pythium ultimum (39.6 %)

Fig. 7.13 Localisation of two parasitic fungi on the roots of young soybean sown in a non-sterile soil containing both *Fusarium oxysporum* (cortical strain) and *Pythium ultimum*. Fragments of equal length were taken from the extremities or the base of the roots, washed carefully and incubated on an agar medium. *F. oxysporum* was mostly present at the base of the roots and *P. ultimum* at the extremities.

abundant nutritive resources while sheltering it from competition. The relation is more or less straightforward depending on the mode of life of the microorganism. Based on the ecological type of classification applied by Garrett (1960) to soil fungi, the following categories can be distinguished:

— **Soil inhabitants:** organisms fit for competition and capable of surviving indefinitely in a saprophytic state. Some are exclusively saprophytic, others can actively colonise living tissues. These facultative parasites are generally barely specialised and present in numerous soils. Examples: *Rhizoctonia solani* and some *Pythium*.

— **Root inhabitants:** some develop perfectly on artifical substrates or in sterilised soil but, being only slightly competitive, are inhibited and incapable of colonising organic debris in an ordinary soil. Because of

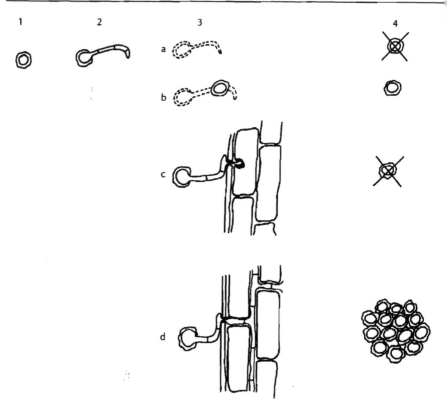

Fig. 7.14   Various stages that an infectious cfu (exemplified here by a chlamydospore) passes through from the time dormancy is overcome up to its introduction into the root.

    1. Cfu subjected to microbiostasis.

    2. Cfu, stimulated by root exudates, commences germination.

    3a. Germinative tube is lysed before it reaches the root because subjected to excessive competition;

    3b. growth of germinative tube inhibited but, after autolysis, has sufficient time to form a new chlamydospore;

    3c. germinative tube reaches the root and penetrates it, but is killed by hypersensitivity reaction;

    3d. parasite penetrates the root and grows there.

    4. Results of preceding events in terms of inoculum.

external constraints, they can only multiply in a living host. These include numerous bacterial (*Clavibacter michiganensis*) or fungal parasites (*Pyrenochaeta lycopersici, Chalara elegans,* for example) and many ectomycorrhizal fungi. Others have nutritive exigencies that, even in the absence of competition, preclude their development either in soil or in inert organic substrates. Examples: parasitic nematodes and endomycorrhizal fungi.

Example of an Incompatible Reaction: Tobacco-
Phytophthora cryptogea Interaction

In most regions where tobacco can be cultivated, the plant can be seriously affected by some strains of *Phytophthora parasitica* that cause collar rot, while strains other than the *nicotianae* variety and other species have no pathogenic effect. This difference in behaviour can be demonstrated by inoculating whole plants or by depositing a suspension of spores on a section of decapitated stems. While *P. parasitica* var. *nicotianae* invades the entire stem, a species such as *P. cryptogea*, for example, causes a hypersensitivity reaction, which just induces necrosis limited to the cells at the point of inoculation. Both in vitro and in the plant, *P. cryptogea* secretes a low molecular weight protein termed cryptogein, which is an elicitor. One nanomole of pure cryptogein suffices to induce all the symptoms of a hypersensitivity reaction in tobacco plants: these include localised necrosis, emission of ethylene, formation of phytoalexins and proteins termed PR proteins (pathogenesis-related). Work carried out at INRA in Antibes for the last fifteen years has helped in the proposal of a coherent (but still partly hypothetical) interpretation of the phenomena observed (Ricci et al., 1994). Cryptogein binds reversibly to the receptor sites of the plasmalemma (not yet identified). This bond leads to the phosphorylation of certain cellular proteins and results almost instantly in the excretion of K^+, an increase in ambient pH due to blockage of $H^+ATPase$ (which functions like a proton pump) and the appearance of superoxide radicals, which are probably responsible for necrosis. The other reactions mentioned above occur a little later. Salicylic acid apparently plays a role in transmission of the signal. This succession of events results in a long-lasting, non-specific, systemic acquired resistance.

Other species of *Phytophthora* also produce proteins which, like cryptogein, elicit hypersensitivity reactions in tobacco. These proteins are very similar and constitute a family of elicitins (Bonnet et al., 1996). Plants lacking specific receptors for elicitins show no hypersensitivity reactions and are rapidly invaded. Strains of *P. parasitica* var. *nicotianae*, which are pathogenic for tobacco, do not produce (or do not produce enough) elicitin: this is the reason they are not recognised (or not recognised fast enough) by the host plant.

It is generally considered that a close connection exists between the more or less obligate nature of the plant-microorganism relation and the specificity of this relation. But such is not always so. For example, the association between leguminous plants and rhizobia, the product of long coevolution, is extremely specific. Nevertheless, rhizobia can survive perfectly well outside their host: strains of *Bradyrhizobium* (strictly confined to soybean) were found in the soil of a field in which soybean had never been sown, 20 years after their experimental introduction. The size of the population was almost the same as at initiation of the trial (Obaton, pers. comm.). Inversely, fungi adapted to a wide range of hosts, such as endomycorrhizal fungi, are incapable of multiplying outside a living root and can only survive because they are able to infect a great variety of plants.

From the point of view of the plant, infection is an aggression against which self-defence is imperative. The study of mechanisms of resistance lies in the domain of physiopathology, which is beyond the scope of this book. Simply summarised, infection can:

- progress very rapidly and result in a generalised invasion. In the absence of effective reactions by the plant, a single point of infection theoretically suffices to obtain this result. Examples: rot caused by *Sclerotinia* and to some extent, vascular diseases.

- be slow, at least for some time, and result in local lesions of limited size. In this case, the severity of the disease depends on the number of infection points. Examples: cortical lesions of *Pyrenochaeta lycopersici* in tomato and nematodal attacks. The infection does not always result in the death of the invaded tissue. Sometimes the cells remain alive but are profoundly modified, which favours the pathogen, as in the case of galls caused by the nematodes *Meloidogyne*.

- be very rapidly arrested by the recognition of incompatibility, triggering a hypersensitivity reaction. Example: *Phytophthora cryptogea*-tobacco relation (see box immediately above).

- progress under close control by the plant. In the case of rhizobia, a nodule is the result of invasion of a single cell, the absorbent hair. In exchange for a supply of energising compounds, the host reaps the benefits of part of the activities of its invader. Symbiotic nitrogen fixation has already been mentioned. Other examples of such mutualistic associations are presented below.

Examples of Mutualism between Plants and Microorganisms

Mycorrhizae

Mycorrhizae are the result of the symbiotic association of a root and a fungus. Vestiges of endomycorrhizae have been discovered in sediments of Devonian age, over 400 million years old, which makes us think that this symbiosis is an extremely ancient phenomenon (having perhaps been previously necessary for colonisation of the Earth's surface by the first chlorophyllous plants) rather than the recent result of a long evolution in subterranean surroundings. All the terrestrial gymnosperms, nearly 85% of the angiosperms and some of the pteridophytes (ferns, horsetails etc.) representing nearly 95% of the vascular plants have mycorrhizal associations. Similar associations (mycothalli) are also known among bryophytes.

As a matter of fact, there are several types of mycorrhizae, as shall be seen later, but these associations have a few common general characteristics, clarified below.

Mycorrhizal fungi live partly inside plant tissues where their mycelia come into close contact with living cells of the cortical parenchyma, and partly outside the plant where their filaments prospect through a large volume of soil, incomparably larger than that accessible to the absorbent root hairs. While the average length of the absorbent hair is in the range of one millimetre, the mycelial network arising from the mycorrhiza extends up to at least 8 cm from the root in endomycorrhizae and even more in ectomycorrhizae. The hyphae, attached to the surface of the aggregates, are in intimate contact with the elementary soil particles and their small diameter allows them to penetrate the finest pores. Mycorrhizal fungi thus literally connect the soil to the plant.

The fungi receive simple carbon substrates from the plant (5 to 10% of photosynthesised compounds) and also growth substances, and benefit from a situation wherein they are sheltered from competitors.

The plants profit from this association, reflected in enhanced growth, all the more spectacular (compared to non-mycorrhizal plants) when the soil is poorer and less propitious for their development. The most direct and best studied benefit is improvement in phosphate nutrition: it has been demonstrated that plants possessing endomycorrhizae have a coefficient of phosphate fertiliser utilisation 3 to 5 times higher than non-mycorrhizal (control) plants (Blal et al., 1990). Phosphate, only slightly soluble, is rapidly exhausted in the soil solution in the vicinity of the roots and releasing, followed by diffusion of immobilised soluble ions, is rarely rapid enough to satisfy the total plant requirements. Mycorrhizal fungi, while exploring a large volume of soil, enable better uptake of the phosphorus available. Rousseau et al. (1992) demonstrated that the phosphorus content of young pine plants (*Pinus taeda*) was directly proportionate to the fungal biomass (*Pisolithus tinctorius*) associated with them (Fig. 7.15). By their acidifying action, mycorrhizal fungi can also contribute to solubilisation of insoluble mineral phosphates, but this effect is negligible compared to mineralisation of organic phosphorus effected by their enzymes (phytase, acid phosphatase). The phosphorus thus extracted from the soil is accumulated in the form of polyphosphates in the vacuoles of the mycelial filaments, from where it is transported towards the roots.

As for elements other than phosphorus, results vary, which probably reflects the great variability in the functioning of symbioses. Mycorrhizal plants are generally good at assimilating zinc, an element as little motile as phosphorus (Kothari et al., 1991). Several authors have observed better assimilation of nitrogen by mycorrhizal plants. In leguminous plants, this improvement is associated with an increase in number and weight of nodules and an augmented nitrogenase activity (Asimi et al., 1980). In other plants, improved assimilation can be due to a transfer brought about by the fungus. For example, it is possible to cultivate young birch with mycorrhizae in the

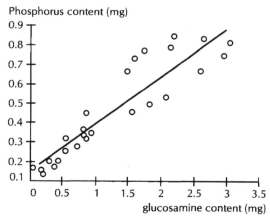

Phosphorus content (mg)

glucosamine content (mg)

Fig. 7.15 Relation between quantity of mycelium in an ectomycorrhizal fungus *Pisolithus tinctorius* (estimated by measurement of glucosamine in the cell walls) present in the fine roots of young pine (*Pinus taeda*) and the phosphorus content of the whole plant (Rousseau et al., 1992).

presence of peptides as the only source of nitrogen whereas non-mycorrhizal birch trees are incapable of utilising peptides. Symbiotic fungi hydrolyse the substrate and furnish mineral nitrogen to the plant. The benefit for the plant varies depending on the share the fungus retains in its own cytoplasm: *Amanita muscaria*, with just 1.9% mycelial nitrogen, enhances birch development better than *Paxillus involutus* with 6.7% (Abuzinadh and Read, 1989). Organic nitrogen from plant litter can thus be recycled and reassimilated very rapidly by trees.

It seems that mycorrhization also augments assimilation of water by the plant under limiting conditions, perhaps by reducing root resistance to absorption of water, but the mechanism is not yet clear (Allen and Allen, 1986). Better water absorption can, in turn, facilitate assimilation of readily mobilisable cations such as potassium or ammonium.

The rhizosphere of a mycorrhizal root is not the same as that of a non-mycorrhizal root and therefore it is more appropriate to use the word **mycorhizosphere.** The morphology of the root can be significantly altered. In addition to the products of rhizodeposition (different from those of a non-mycorrhizal root), there are also exudates of the fungus as well (organic acids, antibiotics). In the mycorhizosphere, actinomycetes are generally rarer and nitrogen-fixing bacteria more numerous. The mycorhizosphere effect often serves (but not automatically) in protection of roots against a large number of telluric parasites: fungi (*Pythium, Phytophthora, Fusarium oxysporum, Chalara elegans*), bacteria (*Ralstonia solanacearum*) and nematodes (*Meloidogyne*). The soil structure is also considerably modified in the vicinity of the mycorrhizal roots. The great abundance of mycelial filaments ensures the assembly of soil particles into stable and voluminous aggregates, which ameliorates both the circulation of gases and water retention.

Another aspect of mycorrhizal symbiosis is the possibility that a fungus, in addition to colonising the roots of its host, may colonise roots of other plants in the vicinity. This creates multiple **mycelial bridges** that bind together not only plants of the same species, but also plants of different species, genera, or even families. Through these connections passage of nutrients from one plant to another is possible (and has been demonstrated). This has two ecologically important consequences: firstly, the matter of senescent roots can be recuperated and reutilised immediately by growing roots and secondly, in an environment in which competition among species is stiff, the less competitive plants can nevertheless thrive because of the nutritional transfers assured by the mycelial bridges between the dominant and dominated species. Maintenance of a wide specific diversity thus remains possible (Grime et al., 1987).

ENDOMYCORRHIZAE WITH ARBUSCLES AND VESICLES

Mycorrhizae with arbuscles and vesicles (abridged VAM) are the most common and probably the oldest of the mycorrhizae. Present in most of the larger botanical families (except Chenopodiaceae and Cruciferae), they occur in almost all cultivated plants and in forest trees as well as herbaceous plants. They are the only mycorrhizae present on such common trees as ash, maple, wild cherry and juniper.

VAM fungi are zygomycetes (order Glomales) and efforts to culture them on artificial media have not been successful to date. Currently, nearly 120 species are known but the gamut of hosts of each species is very extended. The fungi remain in the soil in the form of thick-walled spores. These spores are fit for germination but, despite their abundant nutritive reserves, growth of the germinative tube is quickly arrested in the absence of a host root. Possibly a stimulus originating from the root, analogous to the signals which trigger the mechanisms of nodulation in rhizobia, is necessary for germination to progress normally (Bécard and Piché, 1989). Border cells could be one of the sites of emission of these signals. Anyway, a strong correlation was observed between aptitude of a plant species to form endomycorrhizae and quantity of border cells produced (Niemira et al., 1996). The germinative tube forms an appressorium on the surface of the root. An infection peg emerges and penetrates the cortical bed. Progression in the parenchyma is both intercellular and intracellular but never reaches the central cylinder (Fig. 7.16). Inside the cells, the fungi form highly ramified haustoria, the **arbuscles**, which are enveloped by a periarbuscular membrane formed by the host whose composition differs slightly from that of the plasma membrane. There is much similarity between the arbuscles and the bacteroids formed by rhizobia in the nodules. In both cases, the first stages of symbiosis are controlled by the *early nodulin* genes of the host plant, at least in leguminous plants (Gough et al., 2000). The interface formed by the periarbuscular membrane is the seat of active exchange, which is evidenced

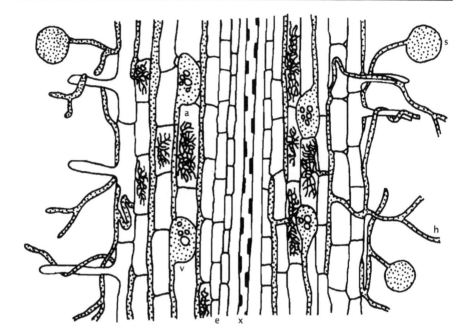

Fig. 7.16   Diagram of an endomycorrhiza with arbuscles and vesicles (Strullu, 1991).
a: arbuscle, e: endodermis, h: extramatricial hyphae, s: sporangium, v: vesicle,
x: xylem

by the existence of high ATPase activity (Gianinazzi-Pearson, 1992). The
invaded cells contain more plasts and mitochondria and their nuclei become
swollen and multilobed; their shape normalises after senescence and
disappearance of the arbuscle.

Voluminous **vesicles**, rich in lipids and calcium and similar to
chlamydospores, are differentiated along the intercellular hyphae and
sometimes within the cells (Gigasporineae, however, form vesicles only
outside the roots). Multinuclear sporangia with thick walls are also formed
outside the roots (Fig. 7.17). As sexual reproduction has not been observed
in Glomales so far, their genetic study is difficult.

Infection by a VAM fungus results in no apparent alteration of the external
aspects of the roots, which probably explains why the universality of this
association is often unrecognised. However, the presence of VAM results in
more abundant root ramification. The ability of the plant to form mycorrhizae
depends on several dominant genes presumably capable of inhibiting the
expression of normal reactions of defence against invasion or else inducing
production by the symbiont of suppressors of these defence reactions
(Gianinazzi-Pearson, 1992).

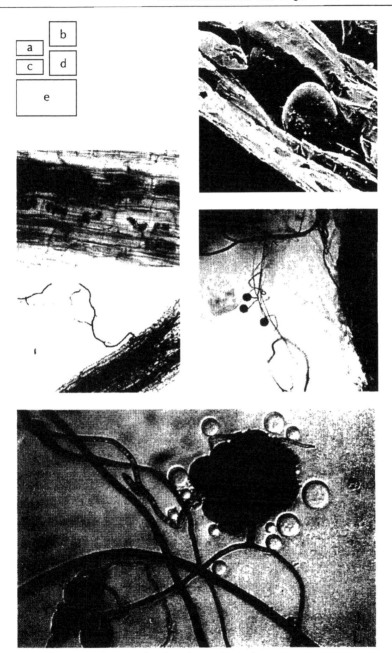

Fig. 7.17 Endomycorrhizae with arbuscles and vesicles.

    a. arbuscles inside host cells (plate, S. Gianinazzi, INRA).

    b. vesicle formed inside a cell (plate, J. Garbaye, photo library INRA).

    c. fungal mycelium extending far away from the root (Gianinazzi).

    d. spores formed outside the root (Gianinazzi).

    e. another example of external fructification (plate, L. Lopez, photo library INRA).

ECTOMYCORRHIZAE

Readily observable and universally well known, at least for the truffle-oak association, ectomycorrhizae occur only in a small percentage (3 to 4%) of phanerogams. But all the major temperate forest species are concerned and hence the paramount significance of these organisms in silviculture. However, ectomycorrhizae are not limited to trees. They are also found in herbaceous plants (Polygonaceae, Cyperaceae) in arctic regions. Ectomycorrhizae are not known to occur in monocotyledons.

Ectomycorrhizal fungal species that can be associated with roots are much more numerous than VAM fungi. A few zygomycetes (*Endogone*), ascomycetes (including truffle) and mainly basidiomycetes are found. The specificity of symbiosis may be extremely narrow or, on the contrary, very broad: Douglas fir (*Pseudotsuga menziesii*) can be mycorrhized by nearly 2000 species of fungi (Trappe, 1977). Some of these fungi can have an autonomous saprophytic existence, but can only complete their cycle if associated with a host plant. Others, such as truffle, minimally competitive, show no active development in the absence of their plant partner. Conta-mination progresses after spore germination, stimulated by root exudates, or simply from root to root. A single tree can be infected by several different ectomycorrhizal fungi. The species installed first are generally not particularly specialised; they are gradually replaced by species that are more specific. A single tree can simultaneously host ecto- and endomycorrhizae (Lopez Aguillon and Garbaye, 1989).

Achievement of mycorrhizal symbiosis can be aided by bacteria present in the hyphosphere and the mantle. These **helper bacteria** aid and abet fungal development during the presymbiotic phase when fungi search for new hosts; the mechanisms involved in these processes have yet to be elucidated and probably differ from fungus to fungus (Garbaye, 1994).

Ectomycorrhizae are constituted of short roots with characteristic bulges (Fig. 7.18) and ramifications, completely encircled by a sheath of interlaced hyphae, the fungal mantle. It is in such associations that the notion of a mycorhizosphere takes on its whole meaning. The mantle is tightly fitted to the root by hyphae that creep in-between the cells of the external layers of cortical parenchyma, without penetrating them, forming the **Hartig network** (Fig. 7.19). Exteriorly, the mantle extends into the soil by an extramatricial network analogous to that in endomycorrhizae.

Other Mycorrhizae

In **Ericaceae** (heather, for example), a special type of association is found, namely the ericoid mycorrhizae. The infected roots are filiform. The cortical tissue is limited to a bed of cells that harbour the endophyte. These are endomycorrhizae without arbuscles: the fungus (almost always *Pezizella ericae*, an ascomycete) simply forms rings or loops inside the host cells.

Orchidaceae also form unique endomycorrhizae. The particularly thick roots are infected by basidiomycetes as soon as they emerge, the curled mycelia developing inside the cortical parenchymal cells.

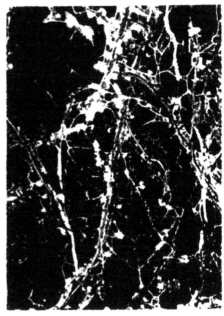

Fig. 7.18 Ectomycorrhizal associations.
Left: typical aspect of a root with short and thick ramifications (plate, Voiry, photo library INRA).
Right: dichotomous or coralloidal mycorrhizae engendered by *Rhizopogon rubescens* on *Pinus pinaster* (plate, D. Mousain, INRA).

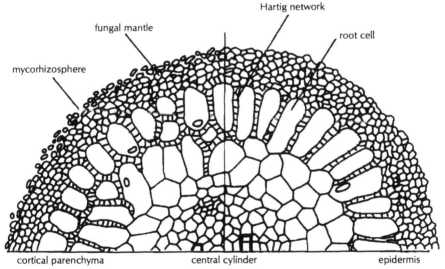

ABIETACEAE   FAGACEAE

Hartig network

fungal mantle

root cell

mycorhizosphere

cortical parenchyma   central cylinder   epidermis

Fig. 7.19 Diagram of ectomycorrhizae of gymnosperms (Abietaceae) and angiosperms (Fagaceae) (Strullu, 1991).

In yet other cases, the plants are associated with fungi which form mycelial clusters more or less developed inside the epidermal or cortical cells, but are also differentiated into an external mantle and a Hartig network. Given their mixed character, these associations are called ectendomycorrhizae. They are seen in some Ericales (arbutus or strawberry tree, Monotropaceae, Pyrolaceae) and a few gymnosperms.

Some of these associations are established between fungi and plants devoid of chlorophyll. In that case, their functioning differs markedly from that of typical mycorrhizal symbioses in which the plant supplies the fungus with carbohydrates. Here, it is the fungus that ensures the carbon nutrition of its host. Thus the non-chlorophyllous orchids, some of which can attain great dimensions in tropical zones, depend entirely on their endophyte throughout their cycle of development. The fungi involved in these associations have a high cellulolytic and ligninolytic activity; some are also active parasites of forest trees, for example *Armillaria mellea* and certain *Fomes* and *Thanatephorus*. *Monotropa*, plants without chlorophyll and close to Ericaceae, form ectendomycorrhizae with fungi, which, furthermore, form ectomycorrhizae in association with nearby trees. These fungi play an intermediary role for *Monotropa*, diverting compounds to them photosynthesised by chlorophyllous species.

Other Symbiotic Endophytes

Any organism that lives in the living tissues of a plant is an endophyte. Some endophytes, such as *Meloidogyne* or *Agrobacterium tumefaciens*, are parasites. Others seem to exercise neither a favourable nor an unfavourable effect on the plants that harbour them: this is the case with bacteria present in the vacuoles of the cortical parenchymal cells of the roots of sugar-beet (Jacobs et al., 1985). Lastly, others are symbionts that procure selective advantages for the plant. We have already covered *Rhizobium* and *Frankia*, the cyanobacteria and the mycorrhizal fungi. There are other examples of such mutualistic associations, less well known but nonetheless important.

NITROGEN-FIXING ENDOPHYTIC BACTERIA

A very unique bacterium was isolated from the roots and stems of sugar-cane in 1988 by the Döbereiner team in Brazil. This aerobic bacterium, called *Acetobacter diazotrophicus*, grows in the presence of elevated concentrations of saccharose, acidifying its environment to such an extent that the pH could go below 3, and in particular fixes atmospheric nitrogen. This fixation is not inhibited by oxygen or nitrates and is just slightly suppressed by NH_4^+. Field studies have shown that *A. diazotrophicus* can supply on average 60% of the nitrogen requirement of sugar-cane (up to 80% for certain cultivars), which represents an annual fixation of over 200 kg nitrogen per hectare. The bacterium seems to be incapable of multiplying in nature outside its host. In the soil, it is found only in the rhizosphere. It is perpetuated by stem fragments cut for cane propagation. Even though this mode of life indicates a long coevolution, *A. diazotrophicus* is not restricted to just sugar-cane. It was found in the roots and aerial parts of a wild member of Gramineae (*Pennisetum purpureum*), sweet potato and the coffee tree. By co-inoculation with a VAM fungus, it can be introduced experimentally into plants that are not natural hosts, such as sugar-beet. Such a possibility engenders considerable economic interest.

Other endophytic nitrogen-fixing bacteria were recently isolated from the roots of sugar-cane, *P. purpureum*, rice, maize, sorghum and various wild Gramineae. When reinoculated into young rice plants, the bacteria *Herbaspirillum seropedicae* and *Burkholderia brasilensis* effected a 20% increase in nitrogen assimilation and enhanced grain yield from 19 to 60% depending on the strain-cultivar combination (Baldani et al., 2000). There is no formal proof that this stimulation is due to transfer of the nitrogen fixed by these bacteria to the plant rather than their ability to produce growth factors such as indoleacetic acid, for example. However, trials with an isolate of *Azoarcus* helped to demonstrate nitrogenase activity in the elongation zone and epidermal cells of rice roots where this endophytic bacterium was present (Egener et al., 1998). However, the relations between *Azoarcus* and its host are rather coarse and could better be considered as a controlled parasitic invasion. Thus, while rhizobia and endomycorrhizal fungi form a true symbiotic entity (symbiosome) with the cells that harbour them, endophytic bacteria disintegrate the cytoplasm of the invaded cells.

ENDOPHYTES WITH STERILE MYCELIUM

A sterile binucleate basidiomycete with a red mycelium is associated with the roots of wheat and many other plants in Australia. It can invade up to 97% of the cortical parenchymal cells and part of the cells of the central cylinder with no discernible deleterious effect on the plant. On the contrary, roots of infected plants were better developed and the weights of both roots and aerial parts higher than those of controls grown under aseptic culture conditions. In wheat, the basidiomycete induced a systemic resistance to *Gaeumannomyces graminis* var. *tritici* but did not protect the plant from *Rhizoctonia solani* or *Pythium*. The fungus produced acid phosphatases and reduced manganese in the rhizosphere, thus improving assimilation of phosphorus and manganese by its hosts (Rowland et al., 1994). In vitro, the fungus excreted antibiotics and growth stimulators that could contribute to its favourable effect on its hosts (Sivasithamparam, 1998).

Other sterile fungi, with dark and partitioned mycelium, have been isolated from roots of trees and herbaceous plants, especially in regions with cold climate and barely fertile soil. The kinds of association of these endophytes with their hosts vary markedly—from a rudimentary ectomycorrhizal form with a thin mantle and a Hartig network, to relations close to parasitism. Hence much research devoted to just this aspect is needed.

Phialophora graminicola, which penetrates the cortical parenchyma of *Vulpia ciliata* (Gramineae) without forming a specialised structure, augments the phosphorus content of the roots and aerial parts (Newsham, 1999).

FUNGI ASSOCIATED WITH ORCHIDACEAE

In orchids, the seeds contain only a few undifferentiated cells and, deprived of reserves, are incapable of germinating spontaneously in nature. Thanks to research carried out by Noël Bernard at the beginning of the twentieth century, it is known that growth of the protocorm and differentiation of the embryo can only begin after invasion of the seed by a symbiotic fungus that supplies it with the glucides and growth substances it lacks. The fungus (several species of basidiomycetes have been identified) on reaching the central part of the seminal cellular mass, forms small clusters inside the cells. The degree of infection seems to be strictly controlled by the plant as the mycelial clusters are lysed after some time and new cells simultaneously invaded. Infection seems to elicit production of fungitoxic phytoalexins, such as orchinol for example.

After germination of the orchids, the same fungal species can form endomycorrhizae with their hosts. But mycorrhization does not result from passage of the mycelium from the seedling to the root cortex. The roots must be infected from the soil at the time of their formation.

Acremonium of Forage Crops Gramineae

These fungi have only been studied in the last twenty years. They have lost the ability to live outside their host plants and are transmitted from generation to generation through seeds. Hence they cannot strictly speaking be included under soil fungi. Akin to *Epichloe typhina*, a parasite responsible for a stem disease of Gramineae, they colonise the aerial parts of fescue (*Festuca*) and ray-grass (*Lolium*) as these plants grow. They ameliorate the resistance of their hosts to drought and synthesise alkaloids toxic to arthropods and vertebrates. Consequently, these fungi are often responsible for serious intoxication of livestock (Siegel et al., 1987). The toxic compounds are apparently transported down to the roots because the infected plants are better able to resist attacks by soil fungi. Another consequence is that formation of mycorrhizae is more difficult.

Associative Symbioses

Unlike the highly interdependent relations established between a plant and its endophyte, some mutualistic associations are very relaxed and referred to as associative symbioses. This badly coined term can be applied to numerous more or less free unions between plants and microorganisms of the rhizoplane or the outer cortex (PGPR bacteria for example). The term is commonly used to designate relations between Poaceae (cereals and wild Gramineae) and free-living nitrogen-fixing bacteria (*Azospirillum, Azotobacter, Bacillus, Beijerinckia, Enterobacter, Klebsiella, Rahnella* etc.). All these bacteria can live outside their host but are specifically attracted by its root exudates and show preferences that are sometimes strict: for example, *Azotobacter paspali* is linked with *Paspalum notatum*.

Azospirillum spp. have been the most widely studied. These bacteria attach to the surface of plant roots in the zone of elongation or on the absorbent hairs. This fixation implies a prior recognition of polysaccharides of the capsule by plant lectins and is accomplished in two stages, as in the case of infection by a parasite or a symbiont (see page 223). First there is reversible attachment to the host surface by the flagella, followed by firm anchorage by the exopolysaccharides. *Azospirillum* remain in the mucilaginous coat covering the surface of the roots or penetrate the cortical layers. They possess pectinolytic enzymes that enable them to penetrate the middle lamellae of the cells and descend sometimes to the endodermis. Growth substances produced directly by the bacteria or synthesised by the plant in response to colonisation, modify the aspect of the root system: the roots elongate and sprout more lateral roots and absorbent hairs, of which quite a number of the latter are curved and sometimes bifurcate. The same succession of events is seen when confrontation occurs between *Klebsiella pneumoniae* and *Poa pratensis* (Haahtela et al., 1988). *Nif* genes for nitrogen fixation and plasmidic genes analogous to certain *nod* genes of *Sinorhizobium meliloti* have been demonstrated in *Azospirillum*. However, the positive response of Gramineae inoculated with these bacteria (more fertile stems, higher biomass, higher yield) can be attributed more to modifications induced in their physiology than to nitrogen fixation. These modifications include increase in root absorption, slowdown of respiratory level expressed with reference to dry weight (Hadas and Okon, 1987) and improvement in water economy (Sarig et al., 1988). These effects are especially perceptible during the first weeks of plant development.

DISEASES OF COMPLEX ETIOLOGY

All standard research in human, animal or plant pathology implicitly assumes a distinct and bi-univocal relation between a disease and the pathogenic agent associated with it. Koch rules are the concrete and operational form of this postulate; it has long been considered impossible to circumvent them and they have markedly advanced understanding of the causes and development of maladies. However, cases occur in which it is not possible to establish a clear relation between a specific microbial parasite and a particular symptom: the same unhealthy appearance can be associated with several different parasites; or a slightly aggressive pathogen can, under certain circumstances, cause a serious disease; or the microorganism identified as a parasite is only an intermediary between the plant and the actual pathogenic agent. We shall give a few examples of all three situations.

Parasite Complexes

A plant introduced into new soil is less likely to encounter numerous and aggressive pathogenic agents. This is not so after many years of repeated cultivation, which leads to progressive accumulation of a pathogenic inoculum. In that case, very often lesions of the root system are observed which rarely are attributable to a single causative agent. Most often several fungi are associated in what is termed a parasite complex. *Pseudocercosporella herpotrichoides, Fusarium* spp. and *Rhizoctonia cerealis* on wheat (Cavelier et al., 1985); *Phoma terrestris, Pythium irregulare* and *Fusarium acuminatum* on maize (Mao et al., 1998); *Fusarium solani* f. sp. *phaseoli, Chalara elegans, Rhizoctonia solani* and *Pythium* spp. on kidney-bean (Lechappe et al., 1988); *Pyrenochaeta lycopersici, Colletotrichum coccodes, Fusarium* spp. and *Rhizoctonia solani* on tomato (Davet, 1976b); *Aphanomyces cochlioides, Pyrenochaeta fallax, Fusarium* and *Cylindrocarpon* spp. on beetroot (Didelot et al., 1994).

In some cases parasite complexes bring well-known pathogens together, in others positive identification of the individuals may be difficult. Sometimes the combination of several pathogenic agents that individually might be mild, can lead to significant weakening of the plant: this could be called 'the Lilliput effect', referring to the humiliating defeat of Gulliver by a multitude of Lilliputians. Customarily, in this case the soil is said to be 'fatigued' (see page 191).

Parasitic microorganisms sometimes merely cohabit and independently colonise different parts of the root. The symptoms observed therefore represent the sum of these individual attacks. It also happens that a primary parasite leads the way for a secondary one which, by itself, could not have overcome the defence reactions of the plant. The presence of a multitude of parasites on a single root system is particularly harmful in two particular cases:

— *When optimal conditions for growth of the parasites differ appreciably.* In that case, it may be considered that a 'buffered' system operates because, when regression of one of the parasites occurs under unfavourable conditions (natural or not), it is compensated by the development of another in such a way that the overall effect for the plant remains the same. Thus at low temperatures, *F. solani* f. sp. *phaseoli* is inhibited while *C. elegans* shows active development; but as the temperature rises *F. solani* overtakes *C. elegans* (Lechappe et al., 1988). Treatments with benzimidazoles in wheat reduced the foot rot (or eyespot) caused by *P. herpotrichoides* but *R. cerealis*, less sensitive to these fungicides, immediately occupied the place vacated by *P. herpotrichoides* (Cavelier et al., 1985). Similarly, treatment of 'soil sickness' in apple with fludioxinil or difenconazole eliminated *R. solani* and *Cylindrocarpon destructans* in the parasite complex but favoured *Phytophthora cactorum* and *Pythium* spp. (Mazzola, 1998).

— *When the parasites are not independent.* Instead they act synergistically and the effect produced is far superior to that attainable by each alone. Such synergy is evident, for example, between *F. solani* and *Pythium ultimum* on pea or kidney-bean (Fig. 7.20), or even between *F. solani* f. sp. *phaseoli* and *C. elegans* (Lechappe et al., 1988). It seems that attacks

Severity index

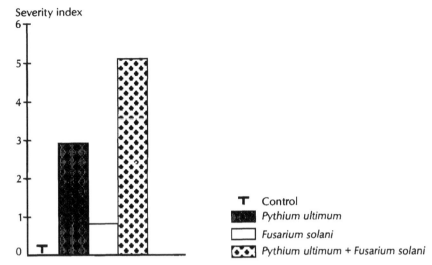

Fig. 7.20 Effect of inoculation of kidney-beans with *Pythium ultimum* and *Fusarium solani* f. sp. *phaseoli*, either singly or together, in a pasteurised and moistened soil, in a growth chamber maintained at 21°C. The severity of the symptoms was measured on a scale of 0 to 6. Perfectly healthy roots were marked 0 (after Pieczarka and Abawi, 1978).

by *P. ultimum* or *C. elegans* augment exudations and this stimulates development of *F. solani*, which is normally restricted to plant crowns. An important case is the synergy observed between the nematodes *Meloidogyne* and the vascular *F. oxysporum*. In order to obtain the symptoms of wilt in cotton, it is necessary to add an inoculum of 77,000 cfu of *F. oxysporum* f. sp. *vasinfectum* per g soil in the absence of nematodes, whereas 650 cfu associated with 50 larvae of *M. incognita* suffice to obtain the same result (Garber et al., 1979). The presence of *Meloidogyne* not only aggravates the symptoms of fusariosis in susceptible plants, but in many cases helps the fungus overcome the defences of resistant cultivars. The physiology of the plant is highly modified by gall nematodes. The rhizosphere is more attractive, moreover the gall tissues formed of undifferentiated giant cells contain large quantities of sugars, lipids and organic acids and constitute a preferential passage for the invaders. At the level of these galls, cells are incapable of differentiating tyloses that normally prevent progression of the fungus in the xylem (Mai and Abawi, 1987). Lastly, production of rishitin, a phytoalexin, is inhibited in roots carrying galls (Noguera, 1982). These interactions are not only true in the case of *Fusarium oxysporum*. *R. solani* is also attracted by the galls of *M. incognita* and causes root rot in adult plants, which it is not capable of attacking in soils devoid of nematodes (Golden and Van Gundy, 1975).

Cases are also known wherein not a nematode but a fungus facilitates entry of a secondary parasite. One then finds lesions in which a parasitic

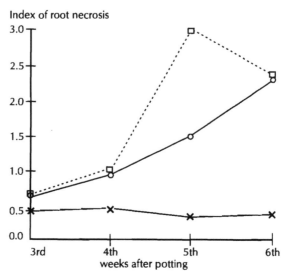

Fig. 7.21 Effect of inoculation with *Pyrenochaeta fallax* (O) and *Fusarium oxysporum* (X), either singly or in association (□), on the root system of beetroot. The fungi were incorporated in pots containing sterile soil. Necroses of the roots were estimated according to a scale of zero (no necrosis) to three for very abundant necroses (after Didelot et al., 1994).

cortege is established, often in a well-determined sequential order. Colonisation of the cortical root tissues of tomato by *Pyrenochaeta lycopersici* favours their subsequent invasion by a cortical *F. oxysporum*. *F. oxysporum* develops better in vitro on colonised roots than on healthy ones (Davet, 1976c). There is probably a similar interaction between *Pyrenochaeta fallax* and *F. oxysporum* on the roots of beetroot (Fig. 7.21). *F. oxysporum* and *F. solani* can be accompanied by *R. solani*.

Old works by Elarosi (1958) have shown that the pectinolytic enzymes of these fungi have a complementary action and caused more extensive maceration of the tissues when they were associated.

The concepts of buffered systems and complexes with synergic effects could also be applicable to colonisation of the roots by associations of benificial microorganisms. For example, specific Gram-negative bacterial communities associated with the fungus *Rhizopogon luteolus*, stimulated formation of mycorrhizae on the roots of *Pinus radiata* (Garbaye and Bowen, 1989; see *helper bacteria*, page 236). Similarly, the number of nodules formed on alder by *Frankia* was 4 to 5 times more in the presence of rhizospheric bacteria than in axenic conditions. These bacteria seem necessary for provoking curving of the absorbent hairs, the first stage before penetration of the actinomycete (Knowlton et al., 1980).

Opportunistic Parasites and Stress Diseases

Considering the infinite number of microorganisms present in the soil, only a very small percentage are capable of overcoming plant defences. Even those that reach the cortical tissues have to be content with colonising mostly dead or senescent cells. But this equilibrium is unstable and perpetually threatened. It is always likely to be toppled should lowering of root resistance occur following alteration of plant physiology. This modification can be genetically programmed and correspond to the developmental cycle of an annual plant, or can be due to gross alteration of external conditions (excessive heat or cold, drought, toxic chemicals etc.), resulting in a stress situation. In both cases, microorganisms harmless up to that point in time, can rapidly transform into aggressive invaders: they are opportunistic parasites, well known in human medicine, for example in the pathology that accompanies the acquired immunodeficiency syndrome.

The withering induced by *Macrophomina phaseolina* is a typical example of this type of disease. This fungus of tropical origin, even though quite common, never attacks plants in a state of active development in temperate region, probably because these regions constitute the limit of its thermal zone of development. But it can invade the roots and then the stems of weakened plants within a few days. In highly selected annual plants such as maize, sorghum or sunflower, whose stems bear huge inflorescences compared to the vegetal parts, the metabolic sink created at the time of seed formation constitutes a natural stress. It is at precisely this stage that *M. phaseolina* spreads in the plant (Fig. 7.22); only after depletion of reserves are the tissues colonised by the fungus (Davet and Serieys, 1987). All the factors of environmental stress accentuate these natural disequilibria and by accelerating depletion of reserves, augment the precocity of attacks. The most common stress that unirrigated cultures of sunflower are subject to is drought. Consequently, those cultivars most resistant to drought are indirectly most resistant to *M. phaseolina* (Davet et al., 1986). The effect of drought on the progress of fusariosis in winter wheat is evident as well. The causative agent, *Fusarium roseum* f. sp. *cerealis 'culmorum'*, behaves like a minor parasite when the hydric potential is higher than –3 MPa, but causes significant damage when the potential falls below this threshold (Papendick and Cook, 1974).

Vector-transmitted Diseases

Plants can be infected by viruses transmitted by telluric microorganisms. The virus multiplies only inside the cytoplasm where it can only penetrate by breaking in. It is introduced by piercing nematodes or pathogens that develop inside the cells. Only microorganisms without cell walls are vectors of viruses.

In the field, the maladies always appear in patches that are initially small. These patches grow very slowly, as the vectors are capable of only limited displacement, but persist and remain in the same place even after some years of interruption of cultivation of the susceptible crop.

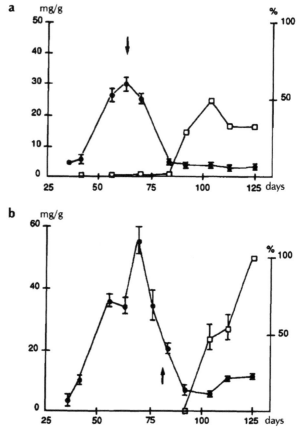

Fig. 7.22 Changes in the sugar content (●) of the roots (**a**) and bases of the stems (**b**) of sunflowers, cultivar Rodeo, and their colonisation by *Macrophomina phaseolina* (□). The time is counted as number of days after sowing. The sugar content is expressed in mg g^{-1} fresh weight and colonisation in percentage of number of fragments exhibiting colonies after culturing. The arrow in Figure **a** indicates flowering of liguled flowers. The arrow in Figure **b** corresponds to blackening of achenes and commencement of maturation of seeds. Irrespective of whether it is the roots or the stems, the tissues are invaded when their soluble sugar reserves are exhausted. Diminution of sugar reserves is accompanied by a fall in total phenols (not shown in Figure). Critical stages are attained in the stems nearly one week later than in the roots (after Davet and Serieys, 1987).

Nematodes that Act as Vectors

Only a few phytoparasitic nematodes are capable of transmitting viruses. All are free ectoparasites of the order Dorylaimida, better represented in light soils than in clayey terrains. Longidoridae, large-size nematodes (1.5 to 13 mm), have a long axial stylet, the odontostyle, ending in an odontophore (Fig. 7.23). *Longidorus* and *Xiphinema* are vectors of nepovirus (nematode-

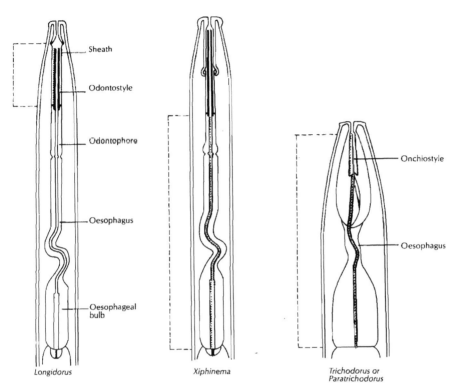

Fig. 7.23   Scheme of mouthparts of nematodes that are vectors of viruses and specific sites of adsorption of viral particles (green). The corresponding portion of the digestive tract is indicated by the hatched line (Taylor, 1980).

transmitted **polyhedral virus**). For example, the fanleaf vineyard virus found in the major wine-producing regions of the world, is transmitted by X. *index* and X. *italiae*. Trichodoridae, which are smaller (0.5 to 1.5 mm), pierce the cell walls with an onchiostyle deprived of interior aperture (Fig. 7.23). They transmit tobravirus (**tobacco rattle virus**), which are less numerous but slightly specialised. The most common virus is tobacco rattle whose vectors belong to genera *Trichodorus* and *Paratrichodorus*.

Nematodes absorb the viral particles after piercing an infected root. The viruses are retained on the surface of definite regions of the stylets, sheath or oesophagus (Fig. 7.23). This adsorption is specific, limited to a few types of viruses and apparently involves a correspondence between the structure of the viral proteinic envelope and the teguments of the animal. The nematodes can remain infectious for a long time—a few weeks to many months—in the absence of a host. Desorption occurs at the time of passage of saliva from the oesophagus into the cell. It occurs progressively, in such a way that many roots can be successively contaminated after a single infectious taking. But the viruses never penetrate the tissues of the vector. They are eliminated during moult if absorption occurred at a juvenile stage.

Fungi (sensu lato) that Act as Vectors

Only microorganisms with zoospores act as vectors of plant viruses (Fig. 7.24); they serve as the only possible mode of transmission of these viruses: roots immersed in viral suspension do not contract the infection in the absence of zoospores. These microorganisms are either chytridiomycetes or plasmodiophoromycetes.

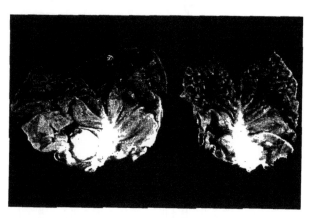

Fig. 7.24 Symptoms of Big Vein, a virus of lettuce transmitted by a chytridiomycete, *Olpidium brassicae*.
Right: note enlargement of the veins, swollen appearance of the limb and reduced size with reference to the leaf on the left taken from a healthy plant (plate, C. Martin, Agriphyto).

Some viruses remain outside the plasma membrane of the vector, as in the case of nematodes. It suffices to place the zoospores in a suspension of viral particles for them to become infectious: this happens in the case of tobacco necrosis virus, transmitted by *Olpidium brassicae*. In nature, contamination of zoospores probably occurs at the time of their liberation from sporangia in contact with infected cells. This adsorption is specific but hardly durable. But most of the viruses transmitted by these vectors are internal. It is not possible to render the zoospores infectious by immersing them in a viral suspension. Contamination of the thalli occurs during development in the infected tissues. These thalli then produce zoospores that contain the virus and can inoculate new roots. The resting form of these parasites is highly resistant and remains viable and infectious for several years: more than 10 years in the case of *Polymyxa graminis*. The viruses carried by plasmodiophoromycetes presently seem to be gaining an upper hand and are the causative organisms for economically important diseases, such as rhizomania (beet necrotic yellow vein) of beetroot (Fig. 7.25) and the mosaic diseases of cereals transmitted respectively by *Polymyxa betae* and *P. graminis*. (The distinction between these two species is questionable since tropical strains of *P. graminis* that can infect dicotyledons such as groundnut have been discovered.) The viruses transmitted by plasmodiophoromycetes belong either to the group of Furoviruses (**fungus-transmitted rod-shaped virus**), generally capable of infecting several hosts (at least experimentally) or to the Bymoviruses (**barley yellow mosaic virus**), which are limited to species of the same botanical genus (Table 7.2). Among viruses transmitted by chytridiomycetes, the specificity can be very narrow and even go beyond the level of a species (Campbell et al., 1995).

Table 7.2 Plant viruses transmitted by zoospores of soil microorganisms (after Campbell, 1996)

Vectors	Virus	Principal hosts
Plasmodiophoromycetes		
Spongospora subterranea	potato mop top (PMTV)	potato
	watercress yellow spot (WYSV)	watercress
	watercress chlorotic leafspot agent (WCLA)	watercress
Polymyxa graminis	soilborne wheat mosaic (SBWMV)	wheat, barley
	oat golden stripe (OGSV)	oat
	peanut clump (PCV)	peanut, sorghum
	Indian peanut clump (IPCV)	peanut
	rice stripe necrosis (RSNV)	rice
	barley mild mosaic (BaMMV)	barley
	barley yellow mosaic (BaYMV)	barley
	oat mosaic (OMV)	oat
	rice necrosis mosaic (RNMV)	rice
	wheat spindle streak (WSSMV)	wheat
Polymyxa betae	beet necrotic yellow vein (rhizomania) (BNYVV)	beet, spinach
	beet soilborne (BSBV)	beet
Chytridiomycetes		
Olpidium brassicae	tobacco necrosis (TNV)	tobacco
	tobacco necrosis satellite (STNV)	tobacco
	chenopodium necrosis (ChNV)	chenopodium
	lisianthus necrosis (LNV)	lisianthus
	lettuce big vein (LBVV)	lettuce
	tobacco stunt (TSV)	tobacco
	freesia leaf necrosis (FLNA)	freesia
	lettuce ring necrosis (LRNA)	lettuce
	pepper yellow vein (PYVA)	pepper
Olpidium bornovanus (= *O. radicale*)	cucumber necrosis (CNV)	cucumber
	melon necrotic spot (MNSV)	melon
	cucumber leafspot (CLSV)	cucumber
	cucumber soilborne (CSBV)	cucumber
	squash necrosis (SqNV)	squash
	red clover necrotic mosaic (RCNMV)	red clover

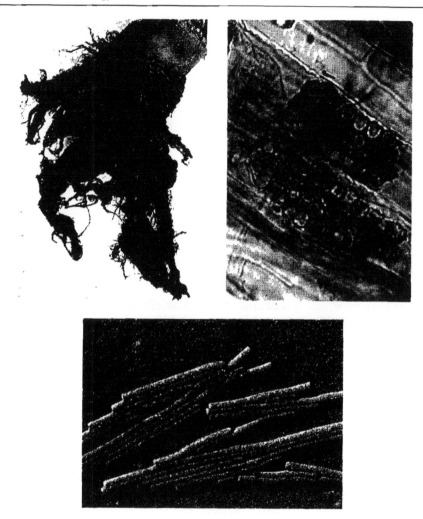

Fig. 7.25   Rhizomania of beetroot (plates by C. Putz, INRA). Above left: General aspect
of the root system of a virus-infected fodder beetroot. Among the abnormally
developed roots an entanglement of white and necrotic brown roots can be
seen.
Above right: Plasmodium of the vector agent *Polymyxa betae* which has
differentiated into cystosores in the root cells of the host.
Below: Particles of Beet Necrotic Yellow Vein virus, responsible for rhizomania
of beetroot as seen under scanning electron microscope (photo library, INRA).

Other Cases

In addition to the preceding examples, in which the vector is an obligate
intermediary between the host and its virus, there are also occasional vectors
of parasites capable of invading the roots on their own. In this case, the role
of vector is limited to accelerating dispersal of the pathogenic agent. Soil

microfauna (nematodes, collemboles and acarids) thus plays an active part in dissemination of fungi and bacteria. We have already seen an example in pod rot of peanut, which is much more severe when acarids are present (Shew and Beute, 1979). Another acarid, *Rhizoglyphus echinopus*, is responsible for the generalised infection of gladioli bulbs from a few plants contaminated by *Pseudomonas marginata*. In a soil devoid of acarids, healthy bulbs planted near diseased ones were not infected. Infection spread rapidly when acarids were introduced (Forsberg, 1959). Saprophagous nematodes ingest large quantities of bacteria and fungal spores. Jensen (1967) showed that the conidia of *Fusarium oxysporum* and *Verticillium dahliae* remain viable after passing through the digestive tube of nematodes, which suggests that these nematodes can transport infectious propagules from the rhizosphere of a contaminated plant to that of a healthy one. Soil fungi can also be spread rapidly by insects whose larvae develop in soil, sometimes at the cost of the roots, the adults of which are capable of flight. This is a particularly significant problem in sheltered cultures. *Fusarium oxysporum* f. sp. *radicis-lycopersici*, responsible for fusarial collar rot of tomato, and *Pythium aphanidermatum* can be disseminated by *Bradysia impatiens*, a small dipteran. *Scatella stagnalis* ingests and rejects in faecal pellets oospores of *Pythium* and chlamydospores of *Chalara elegans* which are perfectly viable (Stanghellini and Rasmussen, 1994). But wind and, under glasshouse conditions, air currents can act as agents of transport of contaminated soil particles, which can be just as effective as animal vectors.

THEORETICAL ASPECTS OF RELATIONS AMONG SOIL MICROORGANISMS AND PLANTS

We saw in the previous sections a few examples of the relations established when a pathogenic or symbiotic agent comes into contact with a root. Any such event is risky. For it to take place, firstly, an infectious unit must necessarily be present in the zone of influence of the root at a given moment. The probability of such an incident occurring depends on the extension of the root system and also stature of the concerned microbial population. It is therefore necessary to know what is called the **inoculum density** in order to understand and, eventually, predict the progress of the plant-microorganism relation. However, as already mentioned with reference to resistant soils, it does not suffice that the infectious agent be present in the rhizosphere for infection to take place. It is the value of **inoculum potential** (French: *potentiel infectieux*) that finally determines the intensity of the symptoms. These ideas are fundamental. Initially proposed to understand parasitic phenomena, they can also serve in the study of associative symbioses and are perfectly applicable, in the area of biological control, to the study of relations between a pathogenic agent and an antagonistic microorganism.

Inoculum

Definition

This term designates all the biological entities belonging to the same taxonomic group, present in the volume of soil under consideration and potentially capable of causing an infection in a given host. This definition, though quite long, can only be used if each of the terms is exactly and carefully defined.

The taxonomic group considered most often is the species. But it may be broader (genus *Pythium*) or on the contrary narrower (a race of a special form of *Fusarium oxysporum*, a clone of *Trichoderma harzianum*, a pathotype or a particular strain of a bacterium).

The inoculum must be quantified. It is then characterised by density, expressed in number of units per gram or per cm^3 soil. The unit is a biological entity easy to define when it refers to individualised organisms such as bacteria or nematodes. The problem is far more complex with fungi. Even when present in the soil in the form of definite countable units (spores, for example), it is not possible to exclude that mycelial hyphae present in the plant debris or occurring free in the soil (threads of *Phymatotrichum*, rhizomorphs) also play a role in infection. On the other hand, many types of spores can coexist in the same species (hyaline endospores with limited life and melanised multilocular spores with much longer life in *Chalara elegans*, for example). These considerations have led to use of an arbitrary unit of inoculum, generally referred to as a colony-forming unit (cfu) (French: 'propagule').

> A colony-forming unit represents the biological entity that gives rise to a colony on an appropriate isolation medium: a single individual cell or group of non-dissociated cells in the case of bacteria, and a chlamydospore, a microsclerotium or conidium or a mycelial fragment for fungi. It is a convenient term for designating the basal unit of enumeration or of dissemination, removing the necessity to be more precise about its nature.

But what is the relation between the cfu counted and the inoculum to be evaluated? The inoculum is defined by the target selected: the biological units that constitute it can only be distinguished from the numerous other living constituents of the soil by their ability to establish successfully on a given host. However, it is not possible to quantify the inoculum present in the soil from observations of symptoms evoked on hosts grown in this soil: this merely indicates the soil infectivity and not the inoculum density. As a result the following problem is encountered: counting the cfu without the presence of their host, knowing that the legitimacy of cfu identification is based on their confrontation with that host.

Measurement of Inoculum Density

When the size of the cfu is sufficiently large, counting can be done directly after fragmentation of the aggregates. This pertains mainly to fungi with sclerotia such as *Sclerotium rolfsii* or *Sclerotinia*. Direct counting can also be done in cases of voluminous or easily identifiable conidia (Ledingham and Chinn, 1955; Gerdemann and Nicholson, 1963). Immunofluorescence techniques and ELISA tests have revived interest in direct counting, if only an antiserum specific for the organism to be quantified can be obtained. But the threshold of detection is generally too high for the study of natural populations and the method is mainly useful in previously enriched soils.

Most often, it is necessary to 'develop' the microorganism by isolating it on a suitable culture medium before proceeding with the count. Satisfactory interpretations of results of these methods of enumeration assume that all the cfu are distributed randomly in the samples, but this is not the case if they are grouped or massed. In general, many successive dilutions are carried out and the population evaluated by counting the colonies in the dilution that is most favourable. Discs of water agar containing a known mass of sieved soil can also be incubated (Ricci, 1974; Davet, 1979) or traps that the inoculum selectively colonises can be used (Ricci, 1972; Rittenhouse and Griffin, 1985). In each case, only the presence or absence of the organism in the disc or trap is noted. These are the all-or-nothing methods wherein one is not preoccupied with ascertaining whether the positive response is due to the presence of a single or several cfu. For analysis of the results, tables are used that enable computation of the most probable number (MPN) of cfu present in the sample. Precision is comparable to that of standard methods of enumeration but counting is often much easier.

> Preparation of the soil for analysis is an operation to which generally scant attention is paid. Some microorganisms are not able to withstand desiccation. We have already seen (see page 32-33) how difficult it is to extract all the germs from soil aggregates. In the case of many fungi it is advisable to directly incorporate finely crushed soil in the isolation medium maintained in a molten state. However, crushing is not effective unless the soil is dry. Rouxel and Bouhot (1971) have shown how conditions of drying, crushing and sieving the samples can modify the results of population enumeration in *Fusarium oxysporum*.

Interest in methods that use a seeding of cfu on a culture medium is due to the fact that they help estimate, after inoculation of the host in optimal conditions, which fraction of the counted population fits to the criterion determined initially, viz, the theoretical likelihood of inducing an infection. A population may indeed appear to be very homogeneous but actually comprise a high proportion of individuals incapable of providing the expected response: in the present case, host infection. This inability can be temporary (due, for example, to dormancy of the oospores in *Pythium*) or constitutive (resulting, for example, from the absence of a gene or loss of a plasmid). Inversely, some active elements of a population can form colonies with a considerable variability in appearance, making identification difficult.

Under experimental conditions wherein one is not interested in natural populations and wherein an artificially introduced active inoculum is studied, counting can be greatly expedited by use of strains resistant to an antibiotic. Undesirable populations can thus be eliminated from the isolation medium. Reporter genes such as the *Gus A* gene of *Escherichia coli* that codes for a β-glucuronidase can also be used. The activity of this enzyme in the transformed strains results in staining the medium in the presence of an adequate substrate, which enables rapid identification of the colonies. The *gfp* gene extracted from a jellyfish, which codes for a green fluorescent protein, is also frequently used.

Distribution of the Inoculum

Determination of a single inoculum density may suffice when working on small volumes of soil under controlled conditions. But at the field scale, significance of the inoculum density becomes questionable when it is estimated from one soil sample of a few grams or a few mg. Representation of the measurement is improved if the computed value is an average of countings carried out on several samples collected from all over the field. As a result of several observations, it appears that the sampling should not be done randomly but according to a precise pattern termed 'in diamond' (Fig. 7.26).

Even under these conditions, knowing the average value is of little interest unless one is certain that the inoculum was haphazardly distributed in the field. This is applicable if the land has been divided into subunits of equal size and the quantities of cfu counted in each subunit distributed according to Poisson's law (in this case, variance is equal to average). In fact, only a small number of cases have been systematically studied because such an experiment is extremely laborious. Actually, distribution of inoculum densities very rarely obeys Poisson's law. Most often variance in sampling is higher than the average because, in addition to oscillation of enumeration around the average, there is also a fluctuation in the average level itself. This denotes an **aggregation** of the cfu which is higher when the ratio of variance to average is higher. Hence distribution either follows a negative binomial law (Punja et al., 1985; Dillard and Grogan, 1985) or obeys no statistically known law (Mihail and Alcorn, 1987), and the population is thus distributed in patches or according to a gradient. It is important to emphasise that aggregation of the cfu can go unnoticed if the size of the elementary sampling plot is too large (Mihail and Alcorn, 1987).

Inoculum Potential

Colonisation of a plant by a microorganism depends on the encounter (the probability of which augments with inoculum density) between a root and a cfu. But encounter does not suffice. It is also necessary that the cfu be capable of accessing the living tissues of the host. This invasion is opposed by the plant through resistance dependent on both its genetic patrimony

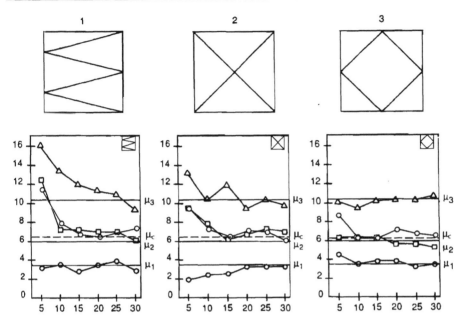

Fig. 7.26 Inoculum density of *Macrophomina phaseolina*.
Above: examples of various patterns used for soil sampling: W (1), crossed diagonals (2), diamond (3).
Below: inoculum density (number of sclerotia per g soil) obtained by computing the average of 5, 10, 15, 20, 25 and 30 samples taken along the length of each pattern. The field was divided into 3 blocks, analysed separately. O: block 1; □: block 2; △: block 3. The entire field was also sampled based on each pattern (in green). The values μ_1, μ_2 and μ_3 are the average inoculum densities calculated after dividing each block into 750 adjacent areas, each representing a sample. The average of the entire field is μ_c (after Mihail and Alcorn, 1987).

and its environment. Successful infection is thus conditioned by an interplay of forces between the plant and the microorganism. In an infested soil, the force exerted on the root and used to overcome plant resistance is the inoculum potential.

Since the time of its conception, about sixty years ago, the idea of inoculum potential has undergone considerable revision. We shall not narrate its history, despite the present interest in a study of these modifications. Due to the enterprising and reflective work of S.D. Garrett in England and R. Baker in the United States, the concept has matured into its present signification, currently accepted by most researchers. Hence we shall abide by their definition. The concept of inoculum potential has lately been enriched by works of French researchers who, as we shall see, have given the term a slightly different meaning.

Inoculum Potential According to Garrett

According to Garrett (1970), the inoculum potential represents the 'invasive force' of a parasite. It is defined as the growth energy of a parasite available for infection of the host, evaluated at the level of infection site, per unit surface of that host.

The gathering of several conidia on a given surface, or several hyphae inside a rhizomorph creates an inoculum potential that is higher than that caused by the presence of a single spore. But the hyphae that direct themselves towards the host must also have a certain vigour, determined by the nutritional status of the inoculum: the conidia possess less reserves and therefore less energy when produced on an impoverished rather than rich medium. Likewise, the infectious energy of *Gaeumannomyces graminis* var. *tritici* in soil depends on the size of the plant fragments containing the fungus (Wilkinson et al., 1985). The distance limit of an inoculum unit for initiating a lesion on a wheat root was 11 to 12 mm when the fragments were 1 to 2 mm in diameter, and less than 2 mm if they were only 0.25 to 0.50 mm in diameter (Fig. 7.27). However, the endogenous resources of the

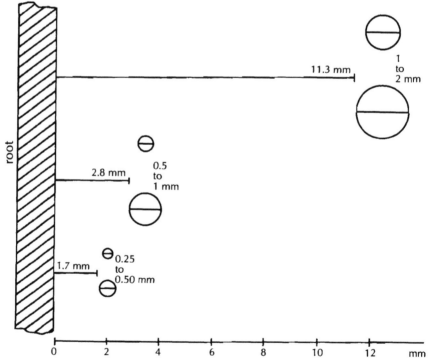

Fig. 7.27   Influence of size of infectious particles on average distance limit within which an inoculum of *Gaeumannomyces graminis* var. *tritici* has to be present in order to infect a wheat root. The fungus was grown on sterilised grains of barley which were later fragmented. The fragments were distributed into three categories depending on their size (after Wilkinson et al., 1985).

cfu can be complemented by an exogenous supply of nutrients, furnished mainly by root exudates, to the extent that they have not already been consumed by antagonistic microflora: obviously an environmental factor is also involved here. Besides biological components of the surroundings (antagonists, predators, competitors etc.), the effects of physicochemical factors (pH, temperature, water content etc.) must be added. One last condition is necessary for colonisation to be realised: the candidate for infection should have the genetic equipment that precludes its recognition by the host as an invader and that helps to surmount the host defences and successfully penetrate its tissues. The inoculum potential is thus a function of four principal variables, namely: inoculum density, resources of the parasite, its genetic patrimony and the environment.

Inoculum Potential According to French Researchers

Garrett and most Anglo-Saxon authors give emphasis mainly to the infectious agent: the inoculum potential represents the energy a microorganism should possess in order to infect a root. Without discarding the definition of Garrett, French authors have adopted a notably different view. For them the inoculum is just one factor of the surroundings and like other factors is certainly indispensable but not hierarchically paramount. What is most important is the soil: the lesions, nodules and mycorrhizae observed are all an expression of relations between a particular soil and a host plant under defined conditions.

It is therefore necessary to distinguish between the inoculum potential of an inoculum (as defined by Garrett) and the inoculum potential of the soil (better expressed in English as soil infectivity: Hornby, 1990). According to Bouhot (1980), the inoculum potential of a soil infested by a pathogen is the quantity of pathogenic energy stocked and available in the soil. The pathogenic energy is manifested by its effects on the plant. A unit of inoculum potential of the soil (UPI) can be defined as the quantity of pathogenic energy necessary and sufficient to induce an infection in a susceptible host. It is not possible, however, to count the UPI directly. But considering that the greater the inoculum potential, the greater the number of UPI of a given volume of soil, it is possible to decide that the minimum volume of soil necessary for causing death in 50% of the susceptible plants, under conditions most favourable for development of the disease, corresponds to a standard unit of inoculum potential, referred to as UPI_{50} (Bouhot, 1979b). Such volumes of soil are measurable using biological tests (see box below). However the efficacy of these tests requires the response to be rapid (a few days) and deprived of ambiguity (of the 'all-or-nothing' type). The method works very well for agents of damping-off (*Pythium, Rhizoctonia solani*) but is not applicable to parasites showing slow development such as *Pyrenochaeta lycopersici* or *Phomopsis sclerotioides*.

In order to better concretise the concept of inoculum potential of the soil (or soil infectivity), Bouhot (1979b) proposed that in the soil, a flux of pathogenic energy flowed from the inoculum to the plant. The intensity of this flux is controlled by diverse environmental components that act in a manner similar to rheostats in an electrical circuit (Fig. 7.28). This representation, albeit schematic, demonstrates well that the manifestations observed in the roots are the expression of a condition of the soil taken in its entirety: they are not the result of a simple plant-infectious agent confrontation, but rather of the functioning of the whole system.

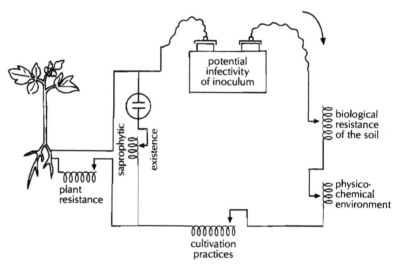

Fig. 7.28 Analogy between an electrical circuit and a 'pathogenic energy circuit' (after Bouhot, 1979b).

Method of Estimation of Inoculum Potential of
Soils Infested with *Pythium*

Soil infectivity was evaluated by observing the mortality of a trap plant very susceptible to *Pythium*, namely cucumber. Development of the *Pythium* was ameliorated by the addition of pulverised oat flakes to the infested soil, at the rate of 20 g/L. The infective mixture was sprinkled around the collars of 6-day-old cucumber seedlings. Expression of the disease was facilitated by adjusting the moisture level to 70% of the soil retention capacity and by incubating the plants at 15°C in darkness for 24 h. The seedlings were then subjected to alternation of light and temperature (day = 15 h, 20°C; night = 9 h, 18°C). Dead plants were counted on the fifth day.

To make the test quantitative, it is necessary to construct a curve expressing the relation between mortality rate and concentration of the inoculum in the soil around the collar of the plants. To accomplish this, several dilutions of the soil under study were made with sterile soil: in general, 30, 10, 1 and 0.1% in volume. If the curve obtained is not a straight line, it can be linearised by transforming the mortality level y into log y, Log (1/1–y), or into Probit y.

The concentrations are always expressed in a logarithmic scale.

In order to obtain the concentration Cx which corresponds to 1 UPI_{50}, the equation of the regression line has to be solved for y = 0.50. Cx is thereby obtained. Knowing the volume of soil V placed around the collar of the plants in each trial, it is possible to calculate how many UPI_{50} are contained in 1 g soil, using the following formula:

$$UPI_{50}/g = \frac{1}{Cx \times V \times d}$$

where d represents the apparent density of the soil (after Bouhot, 1975).

If soil infectivity varies as a function of the environmental conditions, it nonetheless remains an **intrinsic characteristic** of the infested soil, as measured by its effects on a test plant under well-defined conditions. Stating that biological manipulation of the plant, such as premunising or mycorrhization, diminishes soil inoculum potential, constitutes a misuse of language: these treatments change nothing in the response of the test plant under standard conditions.

Modelling Infection

Mathematical models give a simplified representation of complex biological phenomena. If valid, they have a predictive value: knowing the situation at a given moment t, the model should indicate how the disease will develop at an instant t + Δt, in the same or a different environment. Mathematical models can have explanatory value when based on previously verified laws. In these conditions, the parameters they contain have biological significance: such is the case, for example, for the Monod equation (see box below). But often a mathematical representation is selected preferentially simply for its good replay of data obtained in the field. It would therefore be dangerous to place great confidence in explanatory values of parameters used empirically in establishing the formula. As a matter of fact, a model is always a reducing representation and encompassing the complexities of biological phenomena is difficult. Its main interest probably lies in the fact that it necessitates a due reflection on characterisation of the variables of a system and on concatenation of causes and effects. Therefore, rather than dwelling on mathematical functions, none of which is wholly satisfactory, we shall limit our exposition below to the primary considerations that preside over the choice of these functions.

Monod Equation

The Monod equation enables calculation of the rate of growth of a microorganism in a closed medium in which a nutritive element occurs in a limited quantity:

$$\mu = \mu_{max} \cdot \frac{s}{Ks + s}$$

where μ is the rate of specific growth, expressed in grams of biomass produced per gram of substrate and per hour; μ_{max} the level of maximum specific growth, obtained in the absence of any limitation of substrate; s the concentration of the limiting factor; and Ks the saturation constant, or that concentration of the limiting factor for which the rate of specific growth of the microorganism is equal to a moiety of the maximum level attainable.

Parameter Ks is comparable to the molecular dissociation constant Km of the Michaelis-Menten equation, used for enzymatic reactions. In this equation, Km represents the substrate concentration for which the speed of reaction attains a moiety of its value limit in the presence of a given quantity of enzyme. The greater the affinity of elements of the reaction, the smaller the Km. In the Monod formula, Ks can be taken as representative of the affinity of a microorganism for the substrate: the lower the Ks value, the higher this affinity and the greater the ability of the microorganism to rapidly utilise the limited quantities of the substrate.

In a situation wherein the supply of nutrients is reduced, the organisms characterised by a low saturation constant Ks are therefore the most competitive. On the other hand, when the substrate is abundant, organisms characterised by a high maximum specific growth μ_{max} are benefited the most.

Disease Progression as a Function of Time

Infection commences from the primary sources, viz. the soil inoculum. It proceeds with colonisation of the host tissue, followed by formation of new cfu. The simplest case is that of a single cycle of infection that occurs during the plant's life: the disease is monocyclic. However, most often secondary lesions appear, either after dispersal of the cfu formed in the initial lesions, or following growth of mycelial filaments along the root, or from one root to another (Fig. 7.29). Unlike infections of the aerial parts where it is relatively

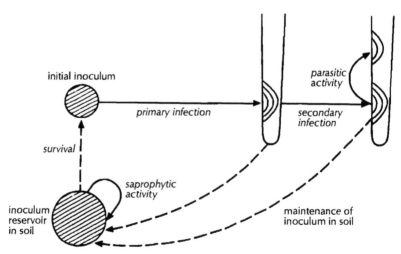

Fig. 7.29 Diagram of a polycyclic soil infection (Gilligan, 1987).

easy to distinguish primary infections from secondary ones since they occur one after the other, polycyclic infections are not easy to identify in the roots (Gilligan, 1987). Since the root system is perpetually engaged in exploration of the soil, the same plant can simultaneously exhibit both primary lesions (on new roots) and secondary lesions (on older roots). Progression of the disease therefore generally follows neither the law of simple interest (monocyclic diseases) nor that of compound interest (polycyclic diseases).

It is also important to specify how disease progression will be evaluated. Generally, the number of plants exhibiting symptoms are counted. But the same plant can be attacked by several cfu and hence the number of diseased plants counted may be less than the number of cfu that have infected them and the number of lesions resulting from the attacks. The multiple infection transformation developed by Gregory (1948) enables estimation of the average number x of infections per plant based on the proportion y of infected plants:

$$x = \text{Log } \frac{1}{1-y}$$

This transformation is currently employed without always verifying the conditions that limit its usage. It is imperative that all the underground parts of the host be equally susceptible, that all the cfu have the same infectious capability and all the infections have occurred haphazardly. This last condition is actually rarely fulfilled in natural infections because, as already seen, the inoculum is most often aggregated in more or less large masses. The most precise method of evaluating disease progression therefore would be to determine the number of points of infection per metre of root, per day, per cfu and per gram of soil, as done by Tomimatsu and Griffin (1982).

Infection-Infestation Relation

Knowing the degree of infestation in a field, anticipation of the level of infection of a crop is possible if a graphic representation of disease progression as a function of inoculum density is available. To construct such graphs, generally the rate of infection is evaluated in several fields, each of them being characterised by a particular level of inoculum. Caution must be exercised, however, to preclude hasty conclusions regarding the biology of the infectious agents based on these graphs: the fields in which these surveys are conducted differ not only in inoculum density, but also in numerous environmental conditions which are not under control.

Baker et al. (1967), considering the laws of physical chemistry, compared the cfu to points equidistant from each other, distributed uniformly in the soil around the roots at the vertices of a tridimensional tetrahedral network. According to their computations, the number of infections S is related to the inoculum density I by the formula $S = aI^b$ or even log $S = b$ log I + log a.

In logarithmic coordinates, S varies linearly as a function of I. Parameter b represents the slope of the straight line and can theoretically take only two values: if the cfu can germinate at a certain distance from the root before infecting it, b = 1; if a contact is necessary between the cfu and the roots for infection to occur, b = 2/3. Inversely, if the relation between S and I, constructed according to experimental data under controlled conditions, is represented by a straight line the slope of which is 1 in logarithmic coordinates, Baker et al. (1967) concluded that a rhizosphere effect takes place (Fig. 7.30). If the slope is equal to 2/3, a rhizoplane effect is evidenced.

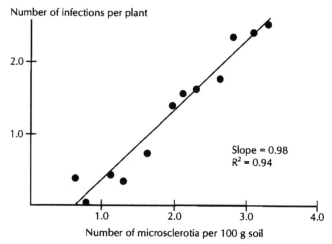

Fig. 7.30 Relation in logarithmic coordinates between inoculum density of *Cylindrocladium crotalariae* (expressed as number of microsclerotia in 100 g soil) and number of infections observed per peanut plant. The slope of the straight line is very close to 1 and according to Baker et al. (1967), this indicates that the parasite is influenced by the rhizosphere of the plant (Tomimatsu and Griffin, 1982).

This model, validated by some results but invalidated by others, has been highly criticised and given rise to a lively controversy aired in articles in *Phytopathology* for almost twenty years. Calculation of the rhizoplane effect and even the existence of this effect are particularly contested. Another school of thought, led by Gilligan (1979), has made use of a probabilistic approach to ascertain the likelihood of a cfu infecting a subterranean organ. In order for this to happen, many conditions have to be fulfilled, as can be seen from Table 7.3. The first condition is that the cfu be located in a particular portion of space that some authors call the **pathozone**. It is this probability that has been most commonly studied mathematically.

Shape and Dimensions of the Pathozone

The pathozone is the volume of soil around an underground organ in which the centre of a cfu must be situated for the infection process to be initiated. This idea enlarges the concept of the rhizosphere: it includes indeed every organ (seed, rhizome, coleoptile and of course the root) susceptible to infection and, moreover, takes into account the fact that certain large cfu,

Table 7.3 Analysis of conditions to be fulfilled for a host to be infected by a cfu. The probability of infection taking place is the product $P = P_1 \times P_2 \times P_3 \times P_4$

Elementary probabilities	Characteristic parameters
P_1: Probability of a cfu occurring in the pathozone	Survival capacity of inoculum Inoculum density Possibility of displacement of cfu Growth rate of host Dimensions of pathozone
P_2: Probability of cfu situated in the pathozone germinating	Viability of cfu Fungistasis Favourable conditions (temperature, moisture etc.)
P_3: Probability that germinative tube reaches the host	Nutritional status of cfu (resources and exogenous supply) Antagonism in a broad sense
P_4: Probability that cfu will infect host after having reached it	Nutritional status of cfu Genotype of cfu Susceptibility of host (genetic) Receptivity of host (environment) Susceptibility of site of infection

such as the sclerotia of *Sclerotium rolfsii* for example, are capable of germinating spontaneously, independent of the stimulation exercised by the host. The filaments arising from these sclerotia, however, can only infect a plant if they are close enough. For *S. rolfsii* the pathozone is broader than the rhizosphere, but remains limited in dimensions. For a fungus with rhizomorphs, on the other hand, the pathozone is much more extended.

The pathozone is generally likened to a cylinder or a sphere, more precisely to a tube or a shell if the dimensions of the root or seed inside it are taken into account. The pathozone can widen out into a truncated cone (if soil depth plays a role) or take the form of an irregular ellipsoid (if the seed has polarity).

An estimation of the probability of a cfu localising in the pathozone and infecting a host is: $\Phi = \dfrac{\text{Log}\,(N/N_s)}{P}$

where N indicates the quantity of roots present in a volume V of soil containing P cfu and N_s the number of these roots that are not infected (Gilligan, 1990). It can also be considered that $\Phi = v/V = $ with v representing the volume of the pathozone.

The equation $v/V = \dfrac{\text{Log}\,(N/N_s)}{P}$ enables calculation of the dimensions of the pathozone, especially its radius: this radius can be considered as the distance beyond which prospection by a cfu in search of a host cannot result in an infection. (All things considered, this radius is the equivalent of the critical distance from which an asteroid passing in the proximity of the

Earth can be captured and attracted by the force of gravity. This distance depends on the characteristics of the planet, but also varies depending on the mass and speed of the asteroid.) More precisely, Bailey and Gilligan (1997) defined the radius of the pathozone as the distance beyond which the probability of infection is below 5%. Several other formulations have been proposed, which either take into account or ignore root growth, compaction of the soil engendered by this growth, aggregation of the inoculum, reduction in likelihood of infection when the cfu is far removed from the host while still remaining in the pathozone and the possibility of secondary infections. Some results of the computation of the radius of this zone of influence are given in Table 7.4. As can be seen, the variability within the same series of trials is very important whenever a factor of the environment is modified.

Table 7.4 Some examples of estimation of radius r of the pathozone (the word pathozone is actually inadequate in the case of *Glomus* sp., a non-pathogenic symbiotic fungus). Computation methods vary depending on author.

Host	Infectious agent	Organs	r (mm)	Principal factor of variation	Reference
Wheat	*Gaeumannomyces graminis* var. *tritici*	roots	1.7–11.3	size of particles containing inoculum	Wilkinson et al., 1985
Wheat	*Gaeumannomyces graminis* var. *tritici*	roots	4–6.6	sowing density	Gilligan, 1990
Abies fraseri	*Phytophthora cinnamomi*	roots	0.06–5.2	inoculum density and matricial potential	Reynolds et al., 1985
Soybean	*Pythium ultimum*	seeds	0.08–0.16 0.6	live seeds dead seeds	Ferriss, 1982
Trifolium subterraneum	*Glomus* sp.	roots	2.5–13.2	inoculum density (variability depending on repetitions)	Smith et al., 1986
Egg-plant	*Verticillium dahliae*	roots	0.1–0.3		Nagtzaam et al., 1997

The limits of the pathozone also change as a function of time, which is especially true for a germinating seed, as illustrated in Figure 7.31. The difficulty in precisely characterising all the environmental parameters in a field, explains why despite the progress made, very few mathematical models are applicable as yet in the field of epidemiology of telluric microorganisms. It should also be noted that to date apparently no cross-check has been done between works on modelling of infection and reflections on inoculum potential and soil infectivity.

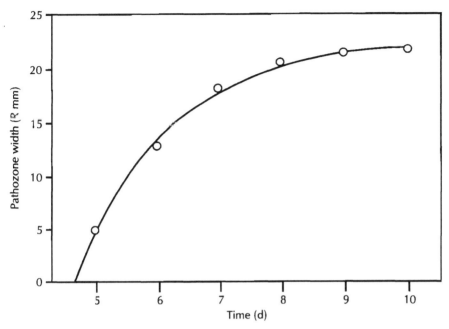

Fig. 7.31   Change in the extent (R) of pathozone influence, beyond which ≤ 5% of colony-forming units succeed in causing disease, for *Rhizoctonia solani* and germinating radish over time (adapted from Bailey and Gilligan, 1997).

For Further Information on Interactions between Microorganisms and Plants

Bacon C.W. and White J.F. (eds.). 2000. Microbial Endophytes. Marcel Dekker Inc., New York, NY.

Barker K.R., Pederson G.A. and Windham G.L. (eds.). 1998. Plant and Nematode Interactions. ASA Agronomy Monograph 36, USA.

Bolton H., Fredrickson J.K. and Elliott L.F. 1993. Microbial ecology of the rhizosphere. In F.B. Metting (ed.). Soil Microbial Ecology, pp. 27-63. Marcel Dekker Inc., New York, NY.

Dénarié J., Debellé F. and Promé J.-C. 1996. *Rhizobium* lipo-chitooligo-saccharide nodulation factors: signalling molecules mediating recognition and morphogenesis. Annu. Rev. Biochem. 65: 503-535.

Döbereiner J. and Pedrosa F.O. 1987. Nitrogen-fixing Bacteria in Nonleguminous Crop Plants. Science Technology, Madison WI.

Foster R.C. 1986. The ultrastructure of the rhizoplane and rhizosphere. Annu. Rev. Phytopathol. 24: 211-234.

Garrett S.D. 1963. Soil Fungi and Soil Fertility. Pergamon Press, Oxford, UK.

Isaac S. 1992. Fungal-Plant Interactions. Chapman & Hall, London, UK.

Kerry B.R. 2000. Rhizosphere interactions and the exploitation of microbial agents for the biological control of plant-parasitic nematodes. Annu. Rev. Phytopathol. 38: 423-441.

Lockwood J.L. 1988. Evolution of concepts associated with soilborne plant pathogens. Annu. Rev. Phytopathol. 26: 93-121.

Lynch J.M. (ed.). 1990. The Rhizosphere. John Wiley & Sons, New York, NY.

Mukerji K.G., Chamalo B.P. and Singh J. (eds.). 2000. Mycorrhizal Biology. Kluwer Acad. Publ., Dordrecht, The Netherlands/Plenum Press, New York, NY.

Peters G.A. and Meeks J.C. 1989. The *Azolla-Anabaena* symbiosis: basic biology. Annu. Rev. Plant Physiol. Plant Mol. Biol. 40: 193-210.

Pfleger F. L. and Linderman R.G. (eds.). 1994. Mycorrhizae and Plant Health. Amer. Phytopath. Soc., St. Paul, MN.

Podilla G.K. and Douds D.D. (eds.). 2000. Current Advances in Mycorrhizae Research. Amer. Phytopath. Soc., St. Paul, MN.

Schwintzer C.R. and Tjepkema J.D. (eds.). 1990. The Biology of *Frankia* and Actinorhizal Plants. Academic Press, New York, NY

Sivasithamparam K. 1998. Root cortex. The final frontier for the biocontrol of root-rot with fungal antagonists: a case study on a sterile red fungus. Annu. Rev. Phytopathol. 36: 439 – 452.

Smith K.P. and Goodman R.M. 1999. Host variation for interactions with beneficial plant-associated microbes. Annu. Rev. Phytopathol. 37: 473 – 491.

Taylor C.E. and Brown D.J.F. 1997. Nematode Vectors of Plant Viruses. CAB Intnat'l, Wallingford, UK.

Varma A. and Hock B. (eds.). 1999. Mycorrhiza: Structure, Function, Molecular Biology, and Biotechnology. Springer-Verlag, Berlin, 2nd ed.

Part III
Possibilities of Intervention

8

Why Intervene?

Cultivation of soil, even in the most primitive systems, results in the modification of natural equilibria and the imposition of new, artificially maintained ones. This disturbance is detectable at all levels, from the rural landscape right up to the microbiocenose. Cultural systems, whether traditional or modern, are all conceived with the aim of getting the best out of the space, water, effort invested as well as the fertility of the soil. This fertility is largely the result of biological processes. Management of microbial communities is thus an essential, sometimes underestimated, aspect of cultivation practices. For thousands of years the health of the soil has been maintained in an empirical manner by the addition of organic matter, essentially in the form of manure. Since the origin of agriculture, farmers have been facing losses in yield due to parasitic attacks and former agriculturists have gradually adapted their systems of cultivation to maintain risks at an acceptable overall level. Thus, in pre-Columbian Peru, the country of origin of potato (and also its pests), it was prohibited under pain of death to cultivate potatoes in the same field at less than a 7-year interval (Schumann, 1991). We now know that this precaution helped to limit the proliferation of *Globodera rostochiensis*, a nematode that produces cysts.

Diseases are only the most perceptible, sometimes even spectacular, manifestations of microbial activity, however. We have discovered that an extraordinary variety of microorganisms also have a beneficial action. The free or symbiotic nitrogen-fixing bacteria and the phosphorus-harvesting mycorrhizal fungi contribute directly to soil fertility. All the decomposers play a no less important role by ensuring plants a regular supply of mineral elements recycled from organic waste. Some microorganisms seem capable of stimulating plant growth and development by direct action on their physiology, the mechanism of which is still not clear. Many others protect them against pathogenic agents by highly varied mechanisms. Microorganisms also contribute directly to the maintenance of soil structure, necessary for good functioning of the roots, by ensuring cohesion of aggregates. Lastly, capable of metabolising most of the artificially synthesised molecules manufactured by the chemical industry, microorganisms can play a valuable role in depolluting contaminated sites.

Discovery of soil microbial life dates back to the end of the nineteenth century. It was necessary to first observe and describe the phenomena, a first step still far from over. Until recent years, intervention for modification of natural equilibria did not seem possible. At most, the methods proposed to combat parasites were so severe that they indiscriminately eliminated all the populations present, both deleterious and useful. In future, the progress made in understanding the biology of telluric microorganisms will generally help in adopting a less crude and more rational approach.

In subsequent sections we shall see how it is possible to prevent or restrict the action of unfavourable organisms and, on the other hand, how to maximise the benefits of favourable organisms and what the problems are that still limit their usage. However, before going further, it is necessary to briefly review the enemies and allies in this combat.

DELETERIOUS ORGANISMS

Parasitic Microorganisms

Parasitic soil microorganisms use the underground living parts of plants (seeds, roots or rhizomes) as substrates for their nutrition. Conquest of these special habitats allows them to escape general competition for nutritive soil resources. However, most of these parasites are also saprophytes capable of surviving and often even actively growing on organic debris. Specialised or not, these microorganisms can thrive on the plant, one after the other, from the beginning to the end of the plant cycle. The problems caused by them can be grouped into one of the following five major categories.

Damping-off

Parasites that attack seeds are generally not very specialised fungi (*Pythium, Rhizoctonia solani, Aphanomyces*) and their rapid development is stimulated by the exudations of germinating seeds. They can destroy seedlings in a few days: in 1982, damping-off resulted in the reseeding of 15,000 ha of sugar-beet in northern and eastern France (Richard-Molard, 1983). When plant tissues undergo lignification, they become resistant to the fungi causing damping-off, but the root tips made up of very young undifferentiated cells remain extremely vulnerable. The significance of such attacks on adult plants has long been largely underestimated, if not ignored. Realisation that they can seriously compromise development of the root system has only taken hold in the last few years. Thus, in the great cereal belt of the north-western

United States, 60 to 70% wheat plants may be infected by *Pythium* spp., resulting in a yield reduction of nearly 8% (Cook and Veseth, 1991). Nematodes such as *Trichodorus* can also attack root tips and arrest their growth.

Cortical Parasites

These parasites develop in the cortical parenchyma where they cause more or less extensive lesions, often colonised by secondary invaders. Necroses can remain discreet and without major incidence; numerous, they can lead to disappearance of the root; when close to the crown, plant death may ensue (see coloured Plate I, following this chapter). Many of these parasites cause serious damage to a small number of specific hosts, but can develop, without attracting attention, in a large variety of plants. Thus, *Chalara elegans* (see coloured Plate I), a major parasite of tobacco and kidney- or French bean can maintain itself on the roots of most dicotyledons. *Pyrenochaeta lycopersici*, which causes corky root disease in tomato and sometimes results in a yield loss of up to 20%, can also colonise kidney-bean, lettuce and members of Cruciferae and Cucurbitaceae without visibly disturbing their growth. But a high pathogenicity does not always induce the appearance of cortical lesions: *Pythium dissotocum*, for example, can considerably retard development of lettuce in hydroponic cultivation with no symptom visible on the infected roots.

Many of the serious parasites of cereals are fungi living in the root cortex: in France, take-all (*Gaeumannomyces graminis* var. *tritici*) causes 20 to 30% loss in yield after 2 or 3 years of repeated cultivation of wheat and foot rot (*Pseudocercosporella herpotrichoides*) reduces the harvest on average at least 5 to 10% (Fehrmann, 1988). *Cochliobolus sativus* and *Rhizoctonia cerealis* can also contribute to damage.

A large number of nematodes of the order Tylenchida can likewise be responsible for root lesions. Among the most widespread species are *Ditylenchus dipsaci* in temperate regions and *Radopholus similis* in tropical zones.

While most of the parasites mentioned above multiply rather slowly in the roots, rarely inducing death of the host, fungi with sclerotia (*Sclerotium*, *Sclerotinia*), which attack plants at the crown or above the soil (coloured Plate I), on the contrary show very rapid growth, partly aided by a strong secretion of oxalic acid that renders the cells defenceless. Damage varies depending on climatic conditions. If the autumn is moist and cool, the proportion of lettuce destroyed by *S. minor* can be over 50% in some market gardens of Roussillon. In wet soils, *Phytophthora* are often formidable invaders. A very large number of species are known, attacking annual plants as well as orchards or forest trees. In Australia, as a consequence of ecological disequilibrium of unknown origin, more than 220,000 ha of *Eucalyptus* forests in the western part of the country have been attacked and partially destroyed since 1920; the disease began spreading to the south-eastern part of the continent during the 1950s (Weste, 1984).

Bacteria such as *Erwinia carotovora* and *E. chrysanthemi* can cause rapidly progressing rots in the underground fleshy parts (roots, tubercles). They usually gain entry through a wound.

Root-and Wood-rotting Fungi (French: pourridiés)

These are lignivorous fungi capable of traversing large distances in the soil because of their rhizomorphs, which are attached to a previously colonised base. The infested zones, found in fruit orchards or in lumber forests, are thus constituted of circular patches that develop from a primary source. These fungi represent a problem with no truly satisfactory solution in orchards, in some vineyards and in silviculture. *Heterobasidion annosum* can thus cause losses of 20% in the volume of pinewood exploitable in cool temperate forests. Maritime pine is particularly susceptible to *Armillaria*: in a replantation trial in a soil in Landes naturally infested with *Armillaria obscura*, 90 and 50% of the populations of *Pinus pinaster* and *P. radiata* respectively, had been killed by the end of 6 years due to wood and root rot (Lung-Escarmant and Taris, 1988).

Vascular Diseases

Even though agents responsible for vascular diseases comprise only a very small number of species of fungi and bacteria (Table 8.1), they are widely distributed and cause great damage to a notably large number of annual and perennial plants. They develop in the xylem. Sap circulation is arrested in the infected vessels not just by the physical presence of the parasites, but also by the reaction of the companion cells of the vessels to the infection. Toxins often contribute to the spread of symptoms. In spite of the existence of resistant varieties, vascular diseases remain an important limiting factor in market gardens and in plantations. In Central America, fusariosis of banana

Table 8.1 Principal vascular parasites that can infect plants from the soil. There are other systemic parasites of the xylem (*Ceratocystis ulmi, Clavibacter michiganensis, Xanthomonas campestris* pv. *campestris*) and the phloem (phytoplasms, such as the agent causing grapevine flavescence dorée, or *Phytomonas*, a protist) but their transmission is mainly by the aerial route.

Parasites	Hosts
Fungi	
Cephalosporium gramineum	Most winter cereals and several grasses
Fusarium oxysporum	Highly numerous (high specialisation: special forms and races)
Verticillium dahliae	Highly numerous (specificity less marked)
V. albo atrum	
Phialophora gregata	Soybean
Phoma tracheiphila	*Citrus* (contamination partly by the aerial route)
Bacteria	
Ralstonia solanacearum	Highly numerous

tree, known by the name Panama disease, destroyed thousands of hectares of plantations during the first part of the twentieth century. The sensitive cultivar 'Gros Michel' was replaced by the resistant cultivar 'Cavendish' during the 1960s. But a new race of *Fusarium oxysporum* f. sp. *cubense* has surmounted this resistance and is presently spreading. Another major problem is vascular bacteriosis caused by *Ralstonia solanacearum*, which occurs in the same region. In northern Africa, bayoud or fusariosis of the date palm, progressed from Morocco (where it was known to have destroyed two-thirds of the palms by the end of the nineteenth century) towards the Algerian palm groves (reaching M'Zab in 1949) and Tunisia.

Galls and Proliferations

Unlike the attack of cortical parasites, always accompanied by a more or less extensive necrosis, some pathogenic agents do not kill the cells of the tissues they invade. But their presence leads to a hormonal imbalance that results in the formation of galls (Fig. 8.1) due to an anarchic multiplication of cells, or an abnormal proliferation of the roots. Gall-forming agents may

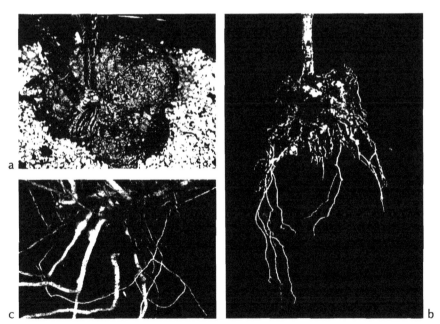

Fig. 8.1 Various types of galls: a—tumour produced at crown of a rose plant by the bacterium *Agrobacterium tumefaciens* (plate, C. Poncet, INRA); b—root system of carnation severely attacked by a gall-forming nematode (*Meloidogyne*) (plate, M. Ritter, photo library INRA); c—galls formed on tomato roots by *Spongospora subterranea*, family Plasmodiophoraceae (plate, D. Blancard, INRA).

be bacteria (*Agrobacterium tumefaciens* is the primary enemy of the rose plant in France at the present time; Aloisi et al., 1994), plasmodiophoromycetes (*Plasmodiophora brassicae*) or nematodes (*Meloidogyne*): abnormal root production is caused by bacteria (*A. rhizogenes*) or nematodes (*Heterodera*). In 1991, 100,000 ha of sugar-beet were infested in France by *H. schachtii*, with losses in yield reaching 40% in some fields (Muchembled and Richard-Molard, 1991). *H. avenae*, mainly active in southern France, can lead to a lowering of yield by more than 35% in winter wheat (Lacombe and Garcin, 1988).

The fungus *Penicillium nodositatum* is a particularly interesting case: it forms nodules on the roots of alders, which though small in size, greatly resemble nodules induced by *Frankia* with respect to formation and appearance (Sequerra et al., 1994).

Crown Gall

The bacterium *Agrobacterium tumefaciens*, closely related to *Rhizobium* and *Pseudomonas*, causes tumours at the crown and base of the superficial roots of dicotyledons. The processes of adhesion and tumour formation are initiated by low molecular weight phenolic compounds (acetosyringone, alcoholic and acidic precursors of lignin) liberated by the glucosidases of the plant cells at injury sites, even superficial ones. The genes responsible for attachment are carried by the bacterial chromosome while gall formation depends on a Ti (Tumour-inducing) plasmid. Phenolic messengers are recognised by the products of the *Vir A* gene of the plasmid, which entrains activation of the *Vir G* gene, a regulator of other *Vir* genes. In the Figure below, the principal processes in which the products of the *Vir* genes play a role are indicated by black arrows. By the action of these genes a piece of the DNA of the plasmid, the T-DNA, is mobilised, linearised and flanked by accompanying proteins (the T complex). The T-DNA then leaves the bacterium, penetrates the underlying plant cell and, guided by the Vir proteins, proceeds to enter the nucleus where it is integrated into the genome of the cell (itinerary shown by green lines in the Figure).

The plasmidic genes of T-DNA are expressed in the cell, which then starts to synthesise the plant hormones (auxin and cytokinin) that trigger the processes of tumour formation, and to produce **opines**—molecules composed of an amino acid and a sugar or a ketonic acid. The opines, of no use to the plant, serve solely as a nutritive substrate for *A. tumefaciens*.

There are several categories of opines (agropine, nopaline, octopine, succinamopine, mannopine), themselves comprising several varieties. Opines are also found in root exudates where they can be utilised by some pseudomonads and coryneform bacteria, thus modifying the composition of the rhizospheric microflora.

Crown gall is a scourge in flower cultivation and tree nurseries. It was estimated that in 1976 it caused 23 million dollars worth of damage in the United States alone.

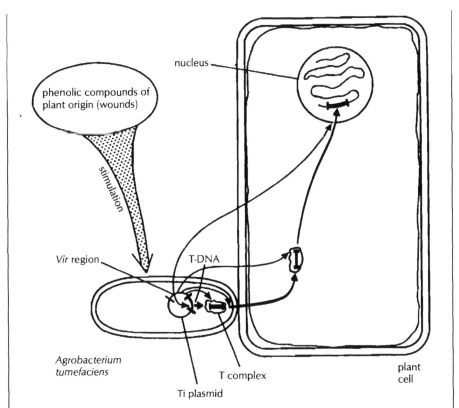

A closely related bacterium, *A. rhizogenes*, also transfers a part of its Ri plasmid to the host plant, causing abundant proliferation of the roots. These two bacteria—*A. rhizogenes* and *A. tumefaciens*—are the only known examples of exchange of DNA between organisms belonging to different systematic kingdoms.

Non-parasitic Microorganisms

Microorganisms of the Rhizosphere

It is known that many organisms living as commensals of roots in the rhizosphere may have, under certain conditions, an unfavourable effect. Nevertheless, very little work has been done to determine the economic impact of this microflora. Most of the data result from laboratory studies and are difficult to extrapolate to plants cultivated under natural conditions. For example, Alström and Burns (1989) stated that lettuce and kidney-beans cultivated for seven weeks in the presence of a strain of a cyanide-producing *Pseudomonas* reduced in dry weight by 37 and 65% respectively, with reference to control plants. Toxin-producing pseudomonads reduced the root growth of winter wheat sown in vermiculite by 70% with reference to controls, but no lesions were visible (Fredrikson and Elliott, 1985). Without undertaking a thorough demonstration, Schippers et al. (1987) attributed the lowering of yield by 10 to 15% regularly observed in potato cultivation to cyanogenic rhizospheric bacteria. Gardner et al. (1984) noted that bacteria sprayed at the foot of young *Citrus* plants could reduce their mass by more than 50% just eight months later. It should be remembered that the problems encountered during replantation of orchards are also often due to the activity of rhizospheric microorganisms.

The predatory microfauna associated with the roots can have an indirect unfavourable effect on plant growth by attacking the symbiotic flora. Several studies have shown that amoebae, nematodes and mycophagous collemboles sometimes considerably reduce the biomass of ecto- or endomycorrhizal fungi (Coûteaux and Bottner, 1994). Populations of rhizobia can also be subject to predation, such that the number of nodules formed on the roots of their hosts is significantly reduced.

Microorganisms of the Soil and Substrates

Microorganisms can enter into competition with plants when essential mineral elements are not available in sufficient quantities. On the other hand, when external conditions excessively favour their development, other microorganisms can be considered temporarily undesirable. If proteolysis releases excessive quantities of ammonia, if nitrification is too rapid, if the enhanced biodegradation of a pesticide engenders its premature disappearance, it may be necessary or at least desirable to intervene. Algae as well as cyanobacteria can pose problems in areas of soilless cultivation because of their uptake of nutritive solutions and the blockages they form in the distribution systems.

AUXILIARY MICROORGANISMS

Nitrogen-fixing Bacteria

Nitrogen is one of the principal limiting factors in plant production. Consequently, worldwide consumption of nitrogen-containing fertilisers is continuously growing. However, though nitrogen is present in the atmosphere in almost unlimited quantities, its reduction into ammonia, followed by its transport and spread, have a highly elevated energy cost. Exploitation and improvement of biological routes of nitrogen fixation therefore represent very valuable stakes.

Since the end of the nineteenth century, symbiotic bacteria of leguminous plants have been commercialised and used on a large scale for improvement of nitrogen fixation by forage crops such as alfalfa and clover. In the United States, 80-90% of alfalfa plantations are seeded each year with *Sinorhizobium meliloti*, and in France, nearly 10,000 ha, in acidic soils. But it was with the development of soybean cultivation that the market for rhizobia really expanded (Fig. 8.2). Originating in the Far East, *Bradyrhizobium japonicum* is not naturally present in soils where the plant has never been cultivated and therefore has to be introduced. Even though its reintroduction is no longer

Fig. 8.2 Effect of rhizobia on growth of leguminous plants. Above left, soybean inoculated with *Bradyrhizobium japonicum*; on the right, uninoculated control (plate, G. Sommer, photo library INRA). Below left, white lupin inoculated with rhizobia; on the right, uninoculated control with yellow leaves (plate, N. Amarger, photo library INRA).

necessary after one or two years of soybean cultivation, significant areas are still sown with *B. japonicum* each year. In the United States alone, the market for *B. japonicum* is presently estimated to be worth 10 million dollars (Glass, 1993).

Frankia spp., much lesser known, have not yet become the object of commercial production. They arouse lively interest because they are capable of forming nitrogen-fixing nodules in eight botanically different families, comprising forest species utilisable in poor or arid soils: *Casuarina*, alders, Bohemia olive trees (*Elaeagnus angustifolia*), sea buckthorn (*Hippophae*) etc. Large areas in diverse countries of the world have already been planted with young trees inoculated in nurseries, very often constituting a prelude to afforestation by other non-nitrogen-fixing species.

The high potential of endophytic diazotrophic bacteria such as *Acetobacter* and *Herbaspirillum* of sugar-cane will assuredly make them precious auxiliaries in cultivation in tropical regions. Much hope is also placed on non-symbiotic rhizospheric bacteria such as *Azospirillum lipoferum* and *A. brasilense*, *Paenibacillus polymyxa*, *Klebsiella pneumoniae* and a few others. Nevertheless, positive consequences of these associative symbioses are not due only to nitrogen fixation.

It should also be recalled that traditional agriculture in the Far East takes advantage of the associative symbiosis between a fern (*Azolla*) and a cyanobacterium (*Anabaena azollae*) in order to enrich rice fields with nitrogen. But the labour required for cultivation and harvest of ferns has led to gradual abandonment of this technique. However, there are many free nitrogen-fixing cyanobacteria and researchers are studying them in many rice-producing countries, notably India. Development of techniques favouring growth of spontaneous microflora seems more likely to produce short-term results compared to sowing rice fields with selected strains.

Mycorrhizae

Even though results vary depending on the phanerogam-fungus association under consideration, mycorrhization almost always results in an overall augmentation of plant growth. Plants with mycorrhizae more readily assimilate phosphorus and sometimes nitrogen, are more resistant to hydric stress, better withstand calcareous soil and are attacked less by root parasites.

In silviculture, plants with mycorrhizae commonly show a 20% increase in size, which translates into a 40% increment in volume of wood compared to controls without ectomycorrhizae (Garbaye, 1991). It has long been the practice in forest nurseries to inoculate plants either by empirically sowing in soil containing fragments of mycorrhizal roots or with fungi previously multiplied on artificial media. *Pisolithus tinctorius*, used on a very large scale for conifers, is produced commercially in a few countries. It has proven especially invaluable in ensuring the success of afforestation operations in poorly fertile, acidic and sandy soils, and in mine tailings and waste rock dumps rich in toxic ions. *Laccaria laccata*, *Hebeloma crustuliniforme* and *Rhizopogon vinicolor* are all largely used for mycorrhization of Douglas fir (*Pseudotsuga menziesii*). Research on controlled mycorrhization is actively carried out in Mediterranean zones where recovery of young trees after plantation is difficult due to climatic and edaphic constraints. In experimental afforestation in southern France, promising results were obtained with parasol pines mycorrhized with *Suillus collinitus* and Atlas cedars mycorrhized with *Tuber albidum*. In one of these trials the level of survival of cedars inoculated with *T. albidum* in the nursery was 79% one year after plantation versus 70% in non-mycorrhized trees (Mousain et al., 1994). Efforts are underway to select symbiotic fungi that are both efficient and harvestable for consumption. Truffle (*Tuber melanosporum*) represents an exception because here the fungus constitutes the main culture, the ligneous species being only an indispensable auxiliary symbiont. It is about 40 years since the first trail of controlled mycorrhizal association of truffles and broad-leaf trees was carried out in Italy. It is now possible to buy mycorrhized plants of oak (*Quercus pubescens, Q. ilex* etc.) and hazel (*Corylus avellana, C. colurna*) and cultivation of truffles is under expansion with more than 5000 ha of mycorrhized plants in France and average yields of 30 kg ha^{-1} on and after the 15th year following plantation. Nevertheless, much more needs to be done to understand the factors which trigger the fructification of this particular fungus and to better define the conditions for its cultivation.

It has long been thought that the addition of exogenous endomycorrhizal fungi would be useless as they are universally present in natural microflora. A study of their variability, however, showed that the efficiency of the mycorrhizal associations can vary considerably, especially in cultivated soils where the diversity of species and strains is highly reduced. Besides, some media can be completely deprived of microflora: disinfected soils in greenhouses or nurseries (Fig. 8.3), artificial substrates and culture media of vitro-plants (see box below). Production of inoculum, despite the technical problems that still persist, has started on a commercial scale. The fungus is multiplied on host plants with rapid growth (clover, ray-grass) cultivated on disinfected substrates. The mycorrhized roots or the culture substrate itself can be used as a source of inoculum. Sparingly used due to high cost, they are reserved for high-profit cultivations in which the inoculation can yield spectacular results. For example, young vines seeded with a commercial preparation at the time of planting in a vineyard in Michigan showed a growth increment of 40% over uninoculated plants during a two-year period and the yield was higher by 24% at the first harvest (Safir, 1994).

Fig. 8.3   Sterilisation of soil eliminates endomycorrhizal fungi. It is necessary to reintroduce them in order to restore biological fertility of the soil, as seen in these onion seedlings (plate, S. Gianinazzi, INRA). a—non-disinfected soil; b—disinfected soil resown with mycorrhizal fungi; c—disinfected soil not resown.

Mycorrhization of Vitro-plants

Multiplication of plants in vitro from meristem cultivation (micro-propagation) enables obtaining a large number of clones from selected plants or regenerating healthy plants from those contaminated with viruses or systemic fungi. Even though this technique is not yet applicable to all cultivated species, micropropagation is presently very commonly employed. In Europe alone, several hundred laboratories, public as well as commercial, use this method to multiply flowering or ornamental plants, widely cultivated plants (pineapple, artichoke, beetroot, wheat, strawberry, potato), fruit trees (avocado, *Citrus*, *Malus*, *Prunus*) and grapevines.

However, the moment of separation, the period when the vitro-plants are removed from the tubes and placed in a cultivation substrate, remains very dicey. It has been found that the rooting process is considerably ameliorated and plant survival excellent if endomycorrhizal fungi are added to the sterile culture substrate. If the host-fungus pair is appropriately chosen, not only is it possible to obtain more robust plants, but also to shorten by several weeks the acclimatisation period, which represents a considerable economic advantage.

In the case of vitro-plants of *Anthyllis cytisoides* and *Spartium junceum* (illustrated here), the gain in time is eight weeks. Mycorrhization enables cutting short the stay in the culture chamber and precludes maintaining the plants in saturated humidity (mist) for six weeks before transferring to the greenhouse (Salamanca et al., 1992).

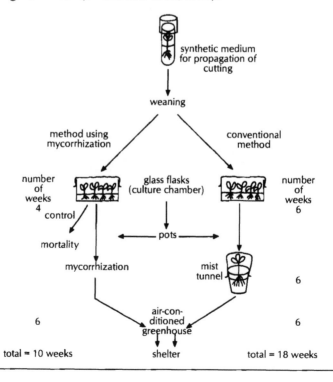

Biopesticides

By analogy with chemical pesticides, biological agents used for controlling pathogenic organisms are named biopesticides. This term, though convenient, is not readily used, probably because the word 'pesticide' is subconsciously associated with 'chemical control' and 'pollution'.

Pesticide also evokes the idea of eradication, an objective neither attained nor sought in biological control. Unlike chemical pesticides that upset natural equilibria, biopesticides on the contrary work at re-establishing the favourable equilibria of the plant disrupted by cultural practices. To better emphasise these differences, Anglo-Saxon authors tend to use the expression *plant growth-promoting microbiota* rather than biopesticides: thus the emphasis is on the positive aspect (improvement of plant growth) rather than the negative (elimination of pathogenic agents). But the qualification 'growth promoting' leads to confusion insofar as it seems to imply that the auxiliary microorganisms improve the yield because they synthesise growth factors. This is not true except in a few cases. Most often, improvement is simply due to reduction of parasite pressure under the effect of antagonists.

The list of species that can be used as agents of biological control continues to grow. It includes bacteria, fungi, protists and nematodes. Many of these will probably never progress beyond the laboratory trial stage. Others, after yielding satisfactory results on a small scale, under controlled conditions, have proven disappointing under normal conditions of application. A few species, however, have successfully passed trials of field experimentation and some of them are already commercially exploited. In England, nearly 10,000 ha of pine forests are treated each year with *Phlebia gigantea* to prevent contamination of stumps by *Heterobasidion annosum* and, in France, hypoagressive strains of *Cryphonectria parasitica* have been introduced in more than 10,000 ha of chestnut trees to combat chestnut canker. Preparations of K 84 and K 1026 strains of *Agrobacterium radiobacter* are on sale in a large number of countries to inhibit development of crown gall caused by *A. tumefaciens.*

Intensive research moreover is being conducted on *Trichoderma,* common soil fungi utilisable against a large number of fungal parasites. They assure highly effective control of damping-off and large-scale trials have also confirmed their effectiveness against various sclerotial fungi. Thus, in a carrot field infected with *Sclerotium rolfsii,* losses were reduced by 94% by a strain of *T. virens* and the commercial part of the harvest was significantly higher compared to areas treated with flutolanil, a fungicide (Papavizas, 1992). Commercialisation of biopesticides is often checked because of their specificity and therefore limited field of action. The strain T-22 of *T. harzianum,* selected by Harman (2000) after fusion of the protoplasts of two isolates with complementary abilities, has the advantage of being effective against a large number of fungal parasites: *Fusarium graminearum, F. moniliforme, F. oxysporum* f. sp. *radicis lycopersici, Pythium ultimum, Rhizoctonia solani, Sclerotium rolfsii* and others. Treatment of wheat seeds contaminated with

Pyrenophora tritici-repentis with T–22 assured a yield in experimental fields of 2166 kg ha^{-1} in lieu of 1666 kg ha^{-1} for untreated controls (da Luz et al., 1998). Other fungi are also under active research: *Coniothyrium minitans* (Fig. 8.4) and *Teratosperma sclerotivora* for combating *Sclerotinia*, *Talaromyces flavus* against *Verticillium*, *Pythium oligandrum* against various parasites. In France, theoretical and applied researches on saprophytic *Fusarium oxysporum*, antagonists of parasitic *F. oxysporum*, have been conducted in Dijon for more than 20 years. Trials carried out on several hectares of sheltered cultivation of tomatoes and melon, in soil but mostly soilless, have confirmed the efficacy of the Fo 47 strain. The protection this strain assures tomatoes against *F. oxysporum* f. sp. *radicis lycopersici* is, for example, equivalent to that provided by chemical treatment with hymexazol (Alabouvette et al., 1993).

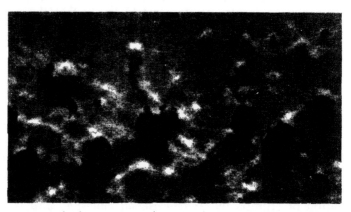

Fig. 8.4   Parasitism of *Sclerotinia minor* by *Coniothyrium minitans*. The big black masses are sclerotia of *S. minor*. The small outgrowths visible on their surface are fructifications of the mycoparasite *C. minitans* which has wholly invaded and destroyed the sclerotia (plate, P. Davet, INRA).

Research carried out since the end of the 1970s on bacteria of the rhizosphere has helped in the selection of several strains capable of ensuring efficient protection of plant roots against major parasites and/or deleterious bacteria. Most of these are fluorescent pseudomonads. Applied by coating the tubercles of potato before planting in experimental fields, they augmented yields by up to 17% (Kloepper et al., 1980). *Pseudomonas* could ensure as good an emergence in cotton sown in soil infested with *Pythium* as a treatment with metalaxyl (Loper, 1988). Other bacteria (*Enterobacter, Serratia, Streptomyces*) seem to be of equal value and a strain of *Bacillus subtilis* has even been homologised in the United States.

Rhizospheric bacteria can also exercise a general protective effect against parasitic nematodes. Research carried out thus far, however, pertains more to specific antagonists: the bacterium *Pasteuria penetrans*, for example, introduced in areas infested with *Meloidogyne incognita*, can reduce loss of yield by 24% in tobacco and between 38 and 55% in vetch (*Vicia villosa*) (Brown et al., 1985). Several fungi are under study. Experiments with predatory fungi (*Arthrobotrys robusta, A. irregularis*) have not always been conclusive and interest presently shown is centring instead on parasites of females and eggs, such as *Verticillium chlamydosporium* and *Paecilomyces lilacinus*. In a *Citrus* orchard, populations of *Tylenchulus semipenetrans* were reduced by half a few months after the introduction of *P. lilacinus* whereas, a temporary drop notwithstanding, the initial populations of *T. semipenetrans* in parts of the orchard treated with nematicides were surpassed. The total harvest and diameter of the fruits were significantly higher with biological treatment than with chemical treatments (Jatala, 1986).

Utilisation of soil fungi as **biological herbicides** is also possible. A dehydrated and crushed preparation of *Trichoderma virens* inhibited germination of amaranthus (*Amaranthus retroflexus*) without disturbing the emergence of cotton sown in the same area. The phytotoxic effect is due to a steroid, viridiol, synthesised by *T. virens* (Howell and Stipanovic, 1984). In the United States, the fungus *Phytophthora palmivora* is commercialised for eliminating the weed *Morrenia odorata* from *Citrus* orchards. But research is mainly oriented towards non-parasitic toxin-producing fungi and actinomycetes, such as *T. virens* (Jones and Hancock, 1990).

Specific deleterious bacteria could also be used as biological herbicides. It would be particularly interesting when the botanical proximity between the weed and the cultivated plant makes chemical pest control difficult, as is the case, for example, for downy brome (*Bromus tectorum*) and wheat. A treatment of seedlings of wheat with deleterious bacteria from the rhizosphere of downy brome improved the competitiveness of the cereal by reducing the emergence of brome (Mazzola et al., 1995).

Growth Stimulators

The beneficial effect of rhizospheric microorganisms, when observed in the absence of known parasites, is generally attributed to the control of minor pathogenic agents that are difficult to identify. However, trials carried out with axenic plants showed that the bacteria and the fungi can have a direct stimulatory effect, observable in the absence of any pathogenic element. This effect, already signalled in the case of *Azospirillum* but independent of their ability to fix atmospheric nitrogen, has been demonstrated in certain strains of *Pseudomonas putida*. Plants of *Brassica campestris* germinating from seeds coated with a suspension of bacteria had a more developed root system and a higher mass, and assimilated almost twice the quantity of phosphorus as control plants. The region of the bacterial DNA responsible for stimulation of root elongation could be cloned but the mechanism of action is not yet known (Lifshitz et al., 1988). The better development of the root system probably also explains, given the increased volume of soil prospected, the better resistance to drought generally observed.

Diffusible compounds of fungal origin can likewise stimulate in vitro germination and growth of various categories of plants (Gillespie-Sasse et al., 1991; Besnard and Davet, 1993) and ameliorate root assimilation of mineral phosphorus and the less soluble metallic ions such as Fe^{3+}, Mn^{4+} and Cu^{2+} (Harman, 2000). Certain strains of *Trichoderma*, introduced in trade substrates for horticulture, improve the development of numerous vegetable and floral species even in the absence of parasites. The treated plants flower earlier and more abundantly (Baker, 1988). Results of these researches are not yet highly publicised as many are patented.

Agents that Improve Soil Structure

Stabilisation of aggregates by using microbial exopolysaccharides is a solution to the problem of erosion of soils poor in organic or mineral binders. Trials on improving soil structure were carried out in Argentina by adding to the soil surface cellulose-containing waste, transformed by *Cytophaga* into a cytophagian colloidal gel. In the United States, interest is directed more towards green unicellular algae (*Chlamydomonas* and *Asterococcus*), which have quite minimal nutritive requirements as long as the water content of the soil is adequate. *Chlamydomonas* are commercialised on a small scale in the vast moist and cool plains of the Pacific north-west where irrigation systems using revolving ramps are quite common: biological treatment is less expensive than use of chemical conditioners (Zimmerman, 1993).

Biodegrading Microorganisms

Microorganisms are valuable auxiliaries in the fermentation of compost and organic waste. It should be possible to exert better control over the development of these microbial communities during the course of these processes and, for example, augment the agronomic value of the decomposed material by enriching it with nitrogen-fixing bacteria.

Usage of chemical pesticides is one of the factors in the spectacular development of modern agriculture. But were it not for the biodegrading activity of the microorganisms, their usage would have been abandoned long ago. The degree of soil sterility can well be imagined, but for the microbial decomposers that play a role in the continuous elimination of herbicides dumped on it. (In intensive agriculture, however, microbial activity has been shown to be insufficient sometimes. Thus, in Europe, groundwater layers are more and more frequently contaminated with residues of atrazine.) This ability of microorganisms to metabolise highly diverse compounds is being more and more systematically exploited for rehabilitation of soils contaminated with xenobiotic products (bioremediation) or simply for destruction of toxic compounds. Not long ago, biodegradation was currently achieved by incorporating organic materials into fields maintained at a level of mineral fertilisation and water content favourable to development of microbial life (landfarming). This procedure has now been abandoned because of the risks of pollution. Xenobiotic compounds are still treated by mixing with large volumes of soil, but only in equipment that does not allow vertical or horizontal transfer of toxic materials. It is also easier, under these conditions, to optimise development of biodegrading communities.

Workers realised that it might be useful to grow plants in soils to be depolluted because the rhizospheric microflora is more abundant and much more active than the microflora in bare soil. Mineralisation of trichlorethylene, estimated in the laboratory by measurement of $^{14}CO_2$ released from samples marked with ^{14}C, was three times more rapid in a rhizospheric soil than in the control soil (Walton and Anderson, 1990). Some plants are capable of absorbing and accumulating heavy metals or metalloids in their roots and aerial parts. It is not always necessary to uproot them to ensure decontamination of the soil: mercury and selenium, for example, get eliminated directly into the atmosphere by volatilisation from the tissues of these plants. Studies under controlled conditions, in a hydroponic culture, have shown that a selection of rhizospheric bacteria can very significantly improve absorption of Hg and Se by plants growing in wetlands, such as saltmarsh bulrush (*Scirpus robustus*) and rabbit-foot grass (*Polypogon monspeliensis*) (de Souza et al., 1999).

Other Possible Uses

Despite the progress made in the last few decades, soil microbiology is still in the stammering stage if one goes along with the concurrence of most microbiologists in estimating that nearly 90% of the soil microorganisms are still totally unknown. Among the small number already identified and partly domesticated, some are definitely of major interest for the development of agriculture: we have just seen a few examples. Others are used for the production of antibiotics or dietary proteins (for example, the cyanobacterium *Spirulina*). Still others are useful in the mining industry for extraction of metals.

The soil therefore constitutes an extraordinary reservoir of genes capable of extremely varied applications. For utilisation of these genes, the organism in which they naturally occur has been used so far. Developments in biotechnology allow envisaging their transfer from now on into other microorganisms that are easier to handle, or even directly into higher organisms. Using such technology, it was proposed to create symbiotic associations between nitrogen-fixing bacteria and plants other than legumes that are of agronomic interest, such as the cereals, or even directly introduce the *nif* gene for nitrogen fixation into the plant genomes. It is difficult to demarcate the frontier between utopia and the constraints of reality in this domain.

For Further Information on the Practical Aspect of Relations between Plants and Microorganisms

Agrios G.N. 1997. Plant Pathology. Academic Press, San Diego, CA (4th ed.).

Arshad M. and Frankenberger W.T. 1993. Microbial production of plant growth regulators. In F.B. Metting (ed.). Soil Microbial Ecology, pp. 307 – 347. Marcel Dekker Inc., New York, NY.

Balázs E., Galante E., Lynch J.M., Scheppers J.S., Toutant J.P., Werner D. and Werry P.A.Th.J. (eds.). 2000. Biological Resource Management: Connecting Science and Policy. Springer, Berlin.

Gelvin S.B. 2000. *Agrobacterium* and plant genes involved in T-DNA transfer and integration. Annu. Rev. Plant Physiol. Plant Mol. Biol. 51: 223 – 256.

Gianinazzi S. and Schuepp H. (eds.). 1994. Impact of Arbuscular Mycorrhizas on Sustainable Agriculture and Natural Ecosystems. Birkhäuser Verlag AG, Basel, CH.

Lucas J.A. 1998. Plant Pathology and Plant Pathogens. Blackwell Science Ltd., Oxford, UK.

Lynch J.M. 1983. Soil Biotechnology: Microbiological Factors in Crop Productivity. Blackwell Scientific Publ., Oxford, UK.

Ream W. and Gelvin S.B. (eds.). 1996. Crown Gall: Advances in Understanding Interkingdom Gene Transfer. American Phytopathological Society, St. Paul, MN.

Schumann G. L. 1991. Plant Diseases: Their Biology and Social Impact. American Phytopathological Society, St. Paul, MN.

Skladany G.J. and Metting F.B. 1993. Bioremediation of contaminated soil. In F.B. Metting (ed.), Soil Microbial Ecology, pp. 483 – 513. Marcel Dekker Inc., New York, NY.

White C., Sayer J.A. and Gadd G.M. 1997. Microbial solubilization and immobilization of toxic metals: key biogeochemical processes for treatment of contamination. FEMS Microbiol. Rev. 20: 503 – 516.

Zimmerman W.J. 1993. Microalgal biotechnology and applications in agriculture. In F.B. Metting (ed.), pp. 457 – 479. Marcel Dekker Inc., New York, NY.

9

Interventions Against Unfavourable Organisms

CHEMICAL METHODS

Fumigants

Fumigants are compounds with a biocidal effect that extends over all soil inhabitants. They may be either liquid or solid in form but volatilize very rapidly after application. Their volatility is often so high that it is imperative to cover the soil with plastic sheets immediately after treatment to preclude too rapid dispersion of the gases into the atmosphere. This precaution is all the more necessary as all fumigants are extremely toxic. Thus, methyl bromide and chloropicrin can be applied only by specially authorised agents. Their very large spectrum of action, which includes weeds, too often leads to consideration of fumigants as an absolute weapon whose effectiveness overshadows prior reflection on the reasons for which disinfection was deemed necessary. Fumigants often represent an easy solution and would probably be used more often if their high price did not limit usage to highly profitable cultivations.

However, it is too often forgotten that fumigants are only effective if their vapours spread throughout the volume of soil prospected by the roots. Diffusion of the gas itself depends on its density, which varies with temperature as well as preparation of the soil and its water content. The presence of recently buried and badly decomposed organic debris diminishes treatment efficacy.

Nevertheless, even carried out under good conditions, fumigation can be ineffective if basic precautions are not taken. Treatment suppresses microbiostasis, creating a **biological void** proportional to the intensity of disinfection accomplished. Under these conditions, the smallest particle of infected soil, the smallest contaminated plant debris coming into contact with the sterilised soil, constitutes a fresh inoculum of pathogenic germs whose development cannot be curbed by any amount of competition. Similarly, planting a few seedlings contaminated at the nursery stage in a disinfected soil can result in rapid spread of a disease to all the healthy plants in the vicinity. This can happen, for example, if uncertified strawberry plants harbouring *Phytophthora cactorum* in a latent state of infection are used.

Phytotoxicity of fumigants makes a delay in soil recultivation imperative. The duration of remanent toxicity can vary from a few days to a few weeks depending on the fumigant used, dosage, nature of the soil, its water content and temperature. The 'garden cress test' helps determine whether all the toxic vapours have been released.

> The test consists of taking a sample of treated soil that will one half fill a jar that can be hermetically sealed. Seeds of garden cress (*Lepidium sativum*), chosen for highly rapid germination, are deposited on the moistened soil surface. The jar is then closed. If the cress does not germinate at the end of 48 h, this means the soil still contains traces of the fumigant.

Destruction of a significant biomass by fumigation results in an accumulation of organic matter. This undergoes rapid mineralisation at the time of recolonisation of the soil by the surrounding microflora, liberating mineral elements, in particular phosphorus and nitrogen (essentially in the ammoniacal form). This must necessarily be taken into account at the time of soil fertilisation.

Fumigation indiscriminately destroys parasites and useful microorganisms. Endomycorrhizal fungi in particular are completely eliminated. This can result in reduced growth and the appearance of deficiency symptoms in crops such as *Citrus*, groundnuts and some *Allium* spp. (especially leek).

Another problem that limits, and will limit more and more the use of fumigants, is the risks they present for the environment. Among the few fumigants still permitted (Table 9.1), the most active and most commonly used is uncontestedly methyl bromide. But its usage has been strongly questioned for some years. In market gardens, copious washing of the soil after disinfection with methyl bromide is imperative to preclude accumulation of bromine in the plant tissues (especially in salad vegetables). This washing results in high pollution of groundwater in regions where the groundwater lies at a shallow depth and where fumigations are frequent, as in Belgium or the Netherlands. Methyl bromide presents yet another limitation: even when trapped under a plastic cover, part of the gas escapes into the atmosphere and can possibly lodge in the stratosphere, contributing to destruction of the ozone layer. The first directive limiting usage of methyl bromide was drafted in the early 1990s by the Montréal protocol (production was increasing then 6% each year). In 1995, a new convention decided that

Table 9.1 Broad-spectrum fumigants approved in France for soil treatment

Chemical name	Active components	Observations
Methyl bromide	Methyl bromide	Application by an authorised agent
Chloropicrin	Chloropicrin	Application by an authorised agent
Dazomet	Methylisothiocyanate	
	Methylamine	Contain
	Formaldehyde	no
	H_2S, CS_2	halogen
Metam-Sodium	Methylisothiocyanate	
Tetrathiocarbonate	CS_2	

methyl bromide should no longer be used as of 2001 in the major developed countries (Clean Air Act of the United States), phased out by 2002 in developing countries and lastly, its production halted in 2010 in industrialised countries.

The oceans are a natural source of methyl bromide. According to estimates, they produce 50,000 to 200,000 t methyl bromide per year. Industry, for its part, produces 80,000 t per year, but only part of it is released into the atmosphere. The gas has been detected in the atmosphere at an altitude of 1 km. Presumably it dissociates and the bromine oxidises:

$$CH_3Br \rightarrow Br^-, Br \text{ and } BrO$$

Study of the 'ozone hole', on the other hand, has revealed the presence of bromine in the stratosphere. Like chlorine and fluorine, bromine can destroy ozone according to the reaction:

$$Br + O_3 \rightarrow BrO + O_2$$

The present problem is to ascertain the relative proportion of industrial methyl bromide and natural methyl bromide which reaches the stratosphere. Meanwhile the producers and users of the gas propose to reduce emissions in the atmosphere by using less permeable plastic sheets. Thus, the passage across a three-layered polyamide sheet is in the range of g h^{-1} m^{-2} whereas the standard polyethylene sheets let 60 g h^{-1} m^{-2} pass through. Covers of ethylene-vinyl alcohol and plastics coated with a film of aluminium also seem to be of potential value. Furthermore, these sheets, by limiting loss of the gas, would quite appreciably reduce the doses applied.

Methyl iodide (CH_3I) could be an interesting product for substitution: it is rapidly destroyed in the atmosphere by ultraviolet rays, its decomposition products remain a short time in it and it appears to be as effective as methyl bromide. A liquid at ordinary temperatures, it is also less dangerous to handle (Ohr et al., 1996).

Selective Treatments

When the target is well defined, it is possible to make use of more specific treatments. Nematicides (Table 9.2), highly toxic, have to be used with caution; they are in fact fumigants with a limited spectrum. Fungicides are

Table 9.2 Nematicides presently approved in France. The fumigants listed in Table 9.1 also have a nematicidal action

Chemical name	Family	Observations
Aldicarb	Carbamates	Systemic
		Use under regulation
Dichloropropene	Aliphatic chlorides	Slightly fungicidal
Ethoprophos	Organo-phosphoric	Use prohibited before cultivation of carrot
		Slightly fungicidal

easier to handle and are not phytotoxic, which allows localised usage in the immediate vicinity of the plant instead of spreading over the entire surface in bare soil. Fungicides are applied at the time of sowing or at the time of planting, depending on various techniques: depositing on seedbeds, coating the seeds, incorporation into clods intended for seed plantation, soaking the plants or the cuttings. They can also be sprayed around the base of the plants during cultivation but the chances of this technique being successful are fewer.

Pesticides with a restricted spectrum have the advantage of disturbing the ecosystem far less than fumigants. As they do not create a biological void, they are safer over the short term. But over the medium term, they also reveal significant limitations due to their specificity. This is especially true for fungicides. The success of a fungicide against a major pathogenic agent allows other pathogenic agents, non-susceptible to treatment and inhibited until then by the eliminated fungus, to manifest themselves. Quintozene and benzimidazoles thus markedly favour attacks by *Pythium* and *Phytophthora*. The common use of benzimidazoles to combat fusarioses and foot rot disease (*Pseudocercosporella herpotrichoides*) of cereals seems also, for the same reasons, to be the cause of development of *Rhizoctonia cerealis* and *Typhula incarnata*, two basidiomycetes non-susceptible to these fungicides and considered until then of little importance (Cavelier et al., 1985). The imbalance is further accentuated if the natural antagonists of the new generation of pathogenic agents are inhibited by the treatment: *Trichoderma* spp., for example, are quite susceptible to benzimidazoles.

A pesticide, even when specific, can have unexpected inhibitory effects on certain microbial groups. Thus, several fungicides are toxic for rhizobia. Used for treatment of seeds of leguminous plants, they can prevent nodulation (Staphorst and Strijdom, 1976). This problem is of paramount importance in the case of soybean because the bacterium associated with it, *Bradyrhizobium japonicum*, is not naturally present in European soils and must be introduced at the time of seed sowing when this plant is cultivated for the first time. This introduction could be a momentous failure if the seeds have been treated with, for example, carboxine. Iprodione, metalaxyl and hymexazol likewise have an unfavourable effect on rhizobia (Revellin et al., 1993). Most systemic fungicides, on the other hand, are toxic for endomycorrhizal fungi. Only the anti-oomycetes are compatible; fosetyl-Al apparently has a stimulatory effect (Jabaji-Hare and Kendrick, 1987).

It should be noted that these secondary effects are not unique to fungicides. Herbicides and insecticides can also disturb microbial equilibria.

Utilisation of Non-biocidal Chemical Compounds

Chemical Baits

It is possible to obtain a reduction in the inoculum density of a parasite by inducing germination or hatching of organs of conservation when no host plant is found in the vicinity. The activated organism is then exposed to the phenomenon of competition and declines more or less rapidly. This is what happens in the case of white rot of garlic and onion caused by *Sclerotium cepivorum*. The volatile messenger that induces germination of the sclerotia is diallyl disulphide (see page 221). Injected as a solution in a contaminated soil, it reduced by 70% the quantity of viable sclerotia without disturbing the biological equilibrium (Entwistle et al., 1982; Coley-Smith, 1990). Results varied markedly depending on temperature, which should not be too high; 15°C seems to be optimal. Similarly, by watering soil containing eggs of the potato cyst nematode (*Globodera rostochiensis*) with tomato root leachates obtained from the nutritive solution of a commercial hydroponic cultivation, hatching of the cysts could be provoked, followed by death of the larvae, as they found no host on which to attach (Devine and Jones, 2000). Practical utilisation of diallyl disulphide is not yet possible because its price is too high, and trials on nematodes are still in the experimental stage. But such studies are of great interest because they show that other methods are possible besides usage of traditional pesticides.

Enzyme Inhibitors

The specific enzymes implicated in the mechanisms of parasitic invasion are still too poorly known to envisage selective inhibition of their functioning. On the other hand, it is possible in a few cases to block undesirable enzymatic reactions involving non-parasitic phenomena. This mainly concerns degradation of urea which, if too rapid, leads to accumulation of ammonia, which in turn can result in nitrogen loss by volatilisation and denitrification. There are several inhibitors of urease (Table 9.3). Applied on the soil surface, they inhibit urease activity in the superficial layers. Urea decomposes gradually in the deeper layers and ammonia can then be retained by the soil components.

To inhibit the too rapid transformation of ammoniacal nitrogen fertilisers into nitrates and thus reduce contamination of groundwater, inhibitors of nitrification are used, some of which have a broad spectrum of action, such as potassium azide or dicyandiamide, while others are more specialised. 2-chloro-6-(trichloromethyl)-pyridine, 4-amino-1,2,4-triazole, 2,4-diamino-6-trichloromethyl-s-triazine are patented.

Table 9.3 A few urease inhibitors that can be used in practice. Many other inhibitors have been experimented with but they are too costly, hardly effective in the field, or dangerous for the environment. The products listed below give bad results in rice fields, where hydromorphic conditions are particularly favourable for hydrolysis of urea. In this case, cyclohexylphosphorictriamide (CHPT) can be used. CHPT is less effective, however, in the presence of algae (Keerthisinghe and Freney, 1994).

Chemical group	Reference
Quinones	
Hydroquinone	Bremner and Mulvaney, 1978
Phosphoroamides	
Phenylphosphorodiamidate (PPD)	Martens and Bremner, 1984
N-(n-butyl)thiophosphoric triamide (NBPT)	Bremner and Chai, 1989
N-(diaminophosphinyl)benzamide (DAPB)	Martens and Bremner, 1984
N-(diaminophosphinyl)cyclohexylamine (DPCA)	Bremner and Chai, 1989
Hydroxamates	
Acetohydroxamic acid (AHA)	Pugh and Waid, 1969
Ammonium thiosulphate	Goos and Fairlie, 1988

In some cases, a too strong competition for nitrogen among soil microorganisms and plants can occur at a moment in their cycle when plants particularly need it. The first step in the transformation of ammoniacal nitrogen into organic nitrogen by the microflora is catalysed by glutamine synthetase. One can therefore imagine blocking the activity of this enzyme to prevent immobilisation of nitrogen. Trials have been carried out for this purpose using L-methionine-DL-sulphoximine (MSX), but inhibition is too short lived; the inhibitor itself is mineralised by the soil microflora (Landi et al., 1999).

Another example is the accelerated disappearance of herbicides of the thiocarbamate family observed after many successive applications, when the soil becomes excessively rich in biodegrading microorganisms. Dietholate, a compound without herbicidal action, can protect thiocarbamates from too rapid biodegradation. This probably involves an inhibition of enzymes by competition between the herbicide and the dietholate molecules (Obrigawitch et al., 1982). However, dietholate can itself be subjected to enhanced biodegradation if used several times in the same area. Another compound, S-ethyl-N,N-bis(3-chloroallyl)thiocarbamate, seems less exposed to this risk (Harvey, 1990).

Mineral Salts

We shall not discuss here the supply of metallic ions intended to compensate deficiencies of oligoelements in plants. In a few cases, mineral elements can have a beneficial effect in the absence of any deficiency. Thus, Belanger showed that the addition of silicon (in the form of potassium silicate or metasilicate) to nutritive solutions induced production of phytoalexins in cucumber in a hydroponic cultivation. These phytoalexins (flavonoids) rendered the vegetables more resistant to *Pythium ultimum* and various foliar diseases (Fawe et al., 1998).

Furthermore, addition of sodium chloride showed very good results against *Fusarium* crown and root rot of asparagus. However, very high quantities of salt were necessary (1120 kg ha^{-1}) and there was some apprehension that repeated additions could have dangerous consequences on groundwater and the soil structure (Elmer et al., 1996).

pH Modifications

By varying the pH it is possible to considerably modify the impact of a disease (Table 9.4). The effect is complex and involves the host plant, pathogenic agent and its antagonists. Elevation of pH can also be resorted to with a view to favouring useful bacteria, rhizobia for example.

Table 9.4   A few microorganisms whose expression of pathogenicity is modulated by the soil pH.

Pathogenic agents whose pathogenicity is favoured by an acidic pH	Pathogenic agents whose pathogenicity is favoured by a neutral or alkaline pH
Aphanomyces euteiches	*Chalara elegans*
Vascular *Fusarium oxysporum*	*Gaeumannomyces graminis* var. *tritici*
Plasmodiophora brassicae	*Fusarium solani* var. *coeruleum*
Pythium spp.	*Phymatotrichum omnivorum*
	Streptomyces scabies
	Verticillium albo-atrum
	Verticillium dahliae

Raising the pH is generally effected by liming, which consists of incorporating in the soil lime or material rich in calcium (accompanied sometimes by the risk of making plant assimilation of certain microelements more difficult). Lowering the pH is indirectly obtained by adding sulphur, which is oxidised into sulphate by microorganisms (Fig. 9.1). These amendments are applied to the entire field and result in relatively long-lasting modifications in soil reaction. But it is also possible to effect a change in a localised temporary manner by modifying only the pH of the rhizosphere of a particular crop. An ammoniacal nitrogenous fertiliser lowers the rhizospheric pH whereas a nitric fertiliser raises it (see page 203). The acidifying effect of the ammoniacal fertiliser can be prolonged if a nitrification inhibitor is added at the same time (Smiley and Cook, 1973).

Ozone

In the specific case of soilless cultivation, about which we shall speak a little later, the drainage fluids are very often collected and recycled. It is imperative in that case that these nutritive solutions be sterilised before re-use. Oxidation by ozone is an interesting method because it leaves behind no toxic residues. Efficiency is better when the solution is acidic, as is generally the case. The rate generally considered optimal is 10 g ozone per m^3 solution to be sterilised, with a confinement duration of 1 h.

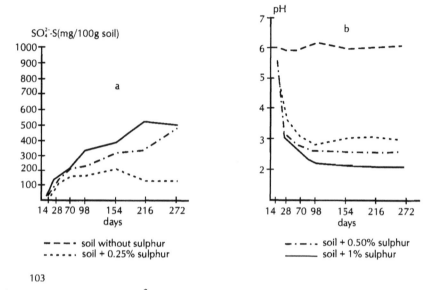

103

Fig. 9.1 Appearance of SO_4^{2-} ions (a) and lowering of pH resulting from oxidation of sulphur in the soil (b) (Adamczyk-Winiarska et al., 1975).

Chloration

The disinfectant properties of hypochlorous acid HClO are used. This acid has a strong oxidative potentiality, especially when the pH is slightly acidic. It can also combine with nitrogenous compounds to give chloramines, which likewise possess biocidal properties.

In practice, mainly the sodium salt NAOCl of hypochlorous acid is used: this is bleach, convenient to use. It can be injected into nutritive solutions using a measuring pump. In large installations, gaseous chlorine is more advantageous as it keeps better. In the presence of water, hydrochloric and hypochlorous acid accrue: $Cl_2 + H_2O \rightarrow HCl + HClO$. Chlorine dioxide, ClO_2, is very effective but dangerous to use.

In all cases, the final dose of active chlorine in the solution must be in the order of 4 ppm. This procedure is particularly useful against microorganisms with zoospores and against algae.

PHYSICAL METHODS

Heat Treatment

Heat is one of the oldest methods used for disinfection of the soil. Pressurised steam is used for this purpose and injected in the soil under metallic covers or under tarpaulins, or introduced at the base of tanks containing soil to be disinfected. The soil need not be packed and should be mostly dry. The minimum duration of treatment is about 20 minutes. Nematodes and most fungi are killed at around 60°C. Some viruses can withstand higher temperatures. The resistance of bacteria varies depending on their ability to

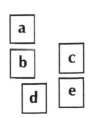

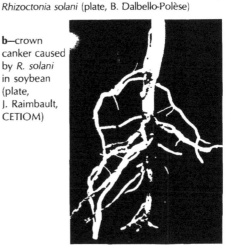

a—destruction of young soybean plants caused by *Rhizoctonia solani* (plate, B. Dalbello-Polèse)

b—crown canker caused by *R. solani* in soybean (plate, J. Raimbault, CETIOM)

c— cortical lesions (deep black) caused by *Chalara elegans* in soybean (plate, J. Raimbault)

d—basal rot in onion caused by *Sclerotium cepivorum*. Each small black dot is a sclerotium (plate, P. Bernaux, INRA)

e—attack of *Sclerotinia sclerotiorum* on soybean stem. Towards the top of the stem the white cotton-like mycelium can be seen and towards the centre of the photograph a big black sclerotium (plate, B. Dalbello-Polèse).

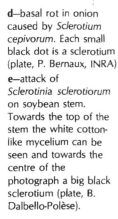

Plate I   Some examples of damage caused by soil fungi.

▲ In this greenhouse, all the plants are isolated from the soil by a plastic sheet, thus eliminating an important source of contamination (plate, M.M. Guimbard, photo library, INRA).

Plate II Importance of prophylaxis in soilless cultivation.

▼ Though cultivated in a clean horticultural mixture, the plants in the centre of the photograph show vascular fusariosis because the sacks were contaminated with soil from the greenhouse (plate, C. Martin, Agriphyto).

▼ **Plate III** Pellicled seeds of sunflower. Seeds on the left were chemically pellicled, those on the right were pellicled with the bacterium *Pseudomonas fluorescens*. The seeds retain their shape but are smoother (plate, B. Digat, INRA).

form spores: *Bacillus* spp., for example, can survive disinfection. This differential effect of steam on bacteria and fungi can be exploited by reducing the treatment temperature. By mixing air with the steam, it is possible to maintain a temperature of about 70°C in the soil, which provides a double advantage: creation of a biological void is precluded by conserving a saprophytic bacterial flora and the cost of treatment reduced. However, the cost is not that low and hence usage of steam is limited to highly profitable crops. Steam disinfection has the major advantage of not introducing toxic products into the environment. But, like chemical fumigation, it eliminates mycorrhizal fungi. If too excessive, it also creates a biological void; in moderation it could cause a temporary accumulation of ammonia because the ammonifying bacteria are better able to resist heat than nitrifying bacteria.

In horticultural exploitations, which almost always possess heating arrangements in greenhouses, the heat can also be used for disinfection of the recycled drainage liquids. An exposure of at least 10 seconds at 95°C is required to inactivate the viruses in suspension, and 30 seconds to eliminate the chlamydospores of *Fusarium oxysporum* (Runia et al., 1988). Several manufacturers have proposed boilers with submerged combustion in which the liquid to be sterilised is vaporised in the flame, then recovered by condensation at the base of the hearth (see page 299).

Trials of heating the soil by microwave emission have been attempted but as yet have not resulted in practical applications.

Solarisation

Solarisation was demonstrated by Katan in Israel in the 1970s, using the sun as a source of energy (Katan, 1981). The soil, carefully loosened and levelled, is watered in such a way that the section to be treated is copiously wetted throughout its depth (on the contrary, maintaining the deeper layers dry is preferable as loss of heat towards the subsoil through conduction is precluded). The surface is then covered with a tightly stretched transparent plastic sheet (Fig. 9.2). In Mediterranean latitudes, during periods of strong

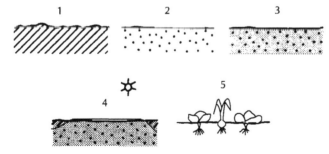

Fig. 9.2 Steps in solarisation. The soil to be disinfected (1) is carefully prepared (2). It is then copiously watered (3). A transparent plastic sheet is spread on the surface and kept well stretched (4). Then, a few weeks later, when solarisation is terminated, the plastic sheet is removed and the plants cultivated without deep tilling, to preclude raising the non-disinfected soil layers to the surface (5) (after Davet, 1991).

sunlight, temperatures of 50°C or above can be reached for a few hours each day at a depth of 10 cm. The average increase in temperature with respect to bare soil is 10 to 15°C.

Wetting the soil has three objectives: augmenting conductivity and calorific capacity, sensitising fungi and seeds and activating bacterial flora. Most of the fungi are killed when exposed to temperatures to temperatures higher than 40°C for a sufficiently long time, even if discontinuously. The critical duration of exposure reduces very rapidly when the temperature is increased: this is expressed in days at 38°C, in hours at 44°C and in minutes beyond 50°C (see Fig. 4.8, on page 103). Thus, even if the temperatures attained in the top 20 to 25 cm of soil do not seem considerable, they are generally sufficient to eliminate the pathogenic fungi if the soil is moist (the fungi resist much longer in dry soil). Even if not destroyed, the propagules that survive are weakened (enzymes and reserves denatured, physiological functions disturbed, exudations increased) and are much more susceptible to the concurrent activity of bacteria and a few saprophytic heat-tolerant fungi such as *Aspergillus terreus*, *Talaromyces flavus* and some Mucorales.

The paramount advantage of solarisation lies in this stimulation of the saprophytic microflora which remains spared by elevation of temperature. It is generally reported that the number of microorganisms having antagonistic potentialities (for example the Gram-positive antibiotics producing bacteria) increases notably. Besides the thermic, purely physical effect of solarisation, a biological effect is therefore evidenced: modification of the microfloristic equilibria confers on the soil a sort of general resistance to diseases. This phenomenon has been confirmed by numerous trials consisting of infestation of the same soil, either previously solarised or untreated, with equivalent quantities of inoculum of a pathogenic agent, then seeding or planting susceptible plants. The intensity of symptoms is always much weaker in plants cultivated in solarised reinfested soils than in control plants. This protective effect is nevertheless limited in time. It fades at the end of a few months (on average from 5 to 15) while the initial microfloristic composition gradually reestablishes from contiguous non-treated parts.

> Prolongation of the acquired resistance can be essayed by introducing auxiliary microorganisms into the soil after solarisation to reinforce its antagonisitic capability. In so doing the yield may not be conspicuously augmented but, for the same final result, duration of solarisation can be shortened appreciably. Thus, in a trial carried out in Perpignan in 1991 by C. Martin, solarisation for 7 weeks followed by the addition of *Trichoderma harzianum* had the same effectiveness against *Sclerotinia minor* as solarisation for 12 weeks. If a green manure (e.g. sorghum) is alternated with solarisation, the beneficial effect of the treatment is also extended.

The efficacy of solarisation depends linearly on the total radiation energy received on the surface of the soil (Fig. 9.3). This implies that, within certain limits, moderate radiation can be compensated by augmenting duration of treatment. The method is therefore not (as originally envisaged) limited to

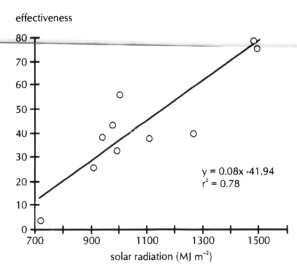

effectiveness

$y = 0.08x - 41.94$
$r^2 = 0.78$

solar radiation (MJ m^{-2})

Fig. 9.3 Relation between total quantity of energy received by a plot in the form of solar radiation and effectiveness of solarisation against crown rot of lettuce caused by *Sclerotinia minor* (Martin et al., 1995).

those regions with very strong sunlight where it was initiated—the eastern Mediterranean and California (USA). It is perfectly applicable in more temperate climates. This method was applied in Roussillon on more than 200 ha in 2000 (Fig. 9.4) and also yielded very good results in the valley of Ain and the Lyon region (Martin and Thicoïpé, 1994). Several species of nematodes as well as numerous weeds can be controlled by solarisation and the method likewise remediates the phenomenon of enhanced biodegradation, which appears in some soils treated several times successively with the same pesticide (Alfizar et al., 1992). Solarisation is less effective, however, against parasites adapted to high temperatures encountered in semi-arid regions, such as *Macrophomina phaseolina* or *Monosporascus eutypoides*. In the case of *M. eutypoides*, which has an optimal germination temperature higher than 30°C for ascospores, solarisation can even have a stimulatory effect (Reuveni et al., 1983). Similarly, nematodes that cause galls (*Meloidogyne*) and weeds such as *Portulaca oleracea*, *Cyperus* and some *Melilotus* are not well controlled.

Solarisation can be effectively combined with a chemical disinfectant of the soil, such as dazomet or metam-sodium, used in small doses. Compost or plant residues that release toxic vapours during decomposition can also be buried before covering with plastic sheets. Waste from cabbage cultivation, for example, releases volatile aldehydes and isothiocyanates with a distinct fungicidal effect. These gases do not form in a soil that has been neither wetted nor heated (Gamliel and Stapleton, 1993).

Fig. 9.4 Solarisation (plates, C. Martin, Agriphyto): a and b—application of plastic cover; c—lettuce crop grown after solarisation; d—damage caused by *Sclerotinia minor* in lettuce planted in a non-solarised part of the same plot.

Submersion

Employed on a large scale in the 1950s to combat fusariosis of banana (Stover, 1962) with varying success, submersion has fallen into desuetude nowadays. Ioannou et al. (1977) showed that it could also inhibit formation by *Verticillium dahliae* of microsclerotia in the residue of infected plants. We ourselves recorded in Roussillon that attacks of *Sclerotinia minor* on lettuce dropped to a much lower level in areas accidentally inundated in autumn following heavy rainfall. Submersion likewise strongly reduces populations of several categories of nematodal parasites, notably *Meloidogyne* (Johnson and Berger, 1972). Depending on the case, it is possible to alternate two to three rather short cycles (2 weeks) of submersion, then drying, or stagnating a sufficiently deep layer of water on the soil so as to create conditions of anaerobiosis. In countries where the soil is flat and water readily available, submersion can contribute somewhat to limiting the stock of a pathogenic inoculum to a low level in the soil. Traditional agriculture in southern China practises alternation of market gardens on non-submerged soil and submerged plants: rice, lotus, water chestnut (*Trapa natans*), etc. This assures intensive production with an acceptable level of parasitism (Williams, 1979).

Ultraviolet Radiation

Ultraviolet radiation can be utilised to sterilise recycled nutritive solutions in hydroponic cultivations. Like ozonisation, it is a 'clean' method without danger for the environment. In commercial installations, low-pressure lamps are used which emit waves around 254 nm, a value considered optimal for destroying nucleic acids. However, irradiation, lasting just a few seconds, is only effective if it reaches all the microorganisms suspended in the solution. It is therefore imperative that the liquids be filtered to clarify them before sterilisation. A sand filter can be used for this purpose, retaining already a part of the infectious agents, notably fungi.

It should be noted that in condensation boilers, sterilisation of the liquid vaporised in the hearth is assured not only by heat, but also by the ultraviolet radiation emitted by the flame of the burner (Steinberg et al., 1994).

CULTIVATION METHODS

Soil Tillage

Tillage is intended on the one hand to loosen the soil to facilitate circulation of air and water and root development, and on the other to bury harvest residues, weeds and, if required, organic and mineral fertilisers. Concomitantly, the pathogenic inoculum remaining on the soil surface and in the clods of stubble of the preceding crop is also buried. Because of this, germs which attack plants in the vicinity of the crown are distanced from their normal point of entry and the new crop can, by and large, escape infection. This is the case for many sclerotial fungi, such as *Rhizoctonia solani, Sclerotium rolfsii, S. oryzae, Sclerotinia sclerotiorum, S. minor* and *Phymatotrichum omnivorum*. But this distancing usually has little effect on the survival capacity of the sclerotia, with the result that the next tillage can bring back to the surface a highly active inoculum. However, when the fungus is susceptible to high carbon dioxide content, as is *R. solani*, the populations diminish appreciably and this reduction persists year after year (Lewis et al., 1983). When fungi capable of penetrating any point in the root system are involved, tillage contrarily tends to distribute the inoculum to a profounder soil depth, thereby augmenting the probabilities of attack.

Soil tillage also contributes, in a general manner, to horizontal spread of the inoculum with passage of the plough.

In the last few years, the cost of tillage and also concern for preserving the soil from erosion has promoted direct sowing of cereal seeds on the residues of the preceding crop still remaining on the soil surface (no-tillage, direct drilling). This method of cultivation precludes the water runoff observed on bare soil and contributes to the preservation of more important reserves in climates where rainfalls are rare and violent. The mycelial networks which ensure cohesion of the macroaggregates are not broken up by tillage and thus the soil structure is preserved.

Microbial diversity is also higher in soils that have not been disturbed by tillage. But deweeding must be done chemically, which tends to reduce plant diversity. The soil surface, protected from desiccation by the straw, also heats up less rapidly. In that case, sorghum sown after a wheat crop is much less exposed to attacks by *Fusarium moniliforme*, an opportunistic fungus governed by hydric stress and favoured by high temperatures (Doupnik and Boosalis, 1980). The same observation might surely hold true for *Macrophomina phaseolina*. On the other hand, maintaining an organic cover on the soil surface favours fungi with a high saprophytic capability, such as *Rhizoctonia* and *Pythium*, furnishing them both a substrate and moisture (Fig. 9.5).

Migratory parasitic nematodes belonging to order Dorylaimida seem to prefer tilled soils in which they move easily. *Pratylenchus*, on the other hand, being smaller, are not much affected by compaction and proliferate in no-tillage soil (Corbett and Webb, 1970).

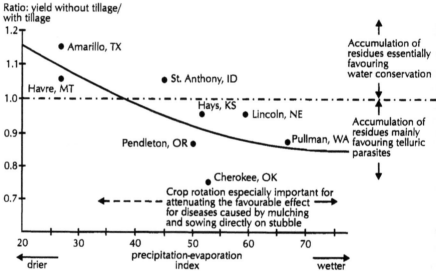

Fig. 9.5 Influence of the mode of cultivation on yield of wheat grown as a monoculture and in the absence of fungicidal treatments. Observations were carried out on 108 fields distributed over 8 cereal-growing regions of the United States. Yields were compared of crops grown with mulch tillage (seeds sown in the stubble of the preceding crop left in place) and conventional crops (clean tillage: ploughing done to bury residues of the previous crop). The average ratio between yields obtained by each cultivation method was then computed for each region. The curve represents the relation between these ratios and the climate, defined by comparing precipitation and evaporation levels. Maintaining the preceding crop in place proved advantageous in dry regions; in wet regions it led to lowering of yield, because it favoured parasitic attacks (after Zinng and Whitfield, 1957, in Cook and Veseth, 1991).

Establishment of a plough-sole at a shallow depth in conventional cultivation, or general soil compaction in no-tillage fields, contribute indirectly to development of root diseases by restricting the volume of soil explored by the plants. Loosening the soil and, if required, subsoiling, enables the roots to penetrate deeper and more rapidly and often thereby escape parasites, whether fungi (Burke et al., 1972) or nematodes (Bird et al., 1974). A plant with a well-developed root system is better able to withstand high parasitical pressure than one grown under poor conditions. The difficulties encountered by roots in properly penetrating the soil can, moreover, lead to expression of characteristic disease symptoms. Thus, in some highly clayey soils of South Africa, a particular strain of *Rhizoctonia solani* develops in the root cortex of wheat, forming spherical excrescences that cause plant stunting (Deacon and Scott, 1985). Amelioration of the soil structure attenuates the symptoms.

Water Control

The degree of soil moisture directly influences the development of microorganisms or exerts an indirect effect on them via the intermediary of their antagonists. *Chalara elegans, Gaeumannomyces graminis* var. *tritici* and *Rhizoctonia solani* are safe-havened by moist soils. Drainage of the soil induces regression. In the case of *Phytophthora, Pythium* and *Aphanomyces*, fungi with flagellated zoospores and aquatic affinity, high water content simultaneously favours their dispersion and attacks. Thus, wheat seeds sown in very wet soil (matrix potential –0.01 MPa) are infected by *Pythium* (mainly *P. ultimum* and *P. irregulare*) within two days of sowing (Fig. 9.6). Germination occurs but the plants are puny and hardly productive. Attacks are greatly reduced

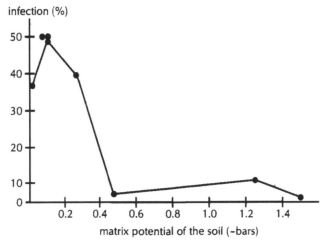

Fig. 9.6 Relation between the matrix potential of the soil, expressed here in bars (10 bars = 1 MPa) and the infection level of embryos of wheat seeds by *Pythium* spp. Readings were taken two days after sowing in pots maintained at 20°C and at fixed potentials (Hering et al., 1987).

as soon as the potential becomes lower than –0.04 to –0.05 MPa, values still quite favourable for good germination (Hering et al., 1987). In regions where the climate is sufficiently stable, damage can be minimised by sowing wheat in a soil that has just enough moisture to permit germination, before the heavy autumn rainfall or, in spring, in a soil that has drained and recovered from the period of heavy downpours (Cook and Veseth, 1991). In market gardens, where drip irrigation is common, the position of the drips with reference to the plants can be significant if the soil is infested with species of Pythiaceae. It is advantageous to keep the crowns of the plants relatively dry and hence the drips should be placed either on the surface at some distance, or deep below the plants. Nematodes also prefer moist soils and hence there is risk of their proliferation in market gardens or orchards watered by drip irrigation. If the deeper layers contain sufficient reserves from which the plant can draw water, it may be useful to let the top soil layers dry out periodically. Drying out of the arable layer can also be achieved by leaving it fallow for a few months during which several successive tillages are done.

On the contrary, in some cases it is advantageous to maintain high soil moisture, at least for a limited time, to favour the antagonistic microflora. Planning irrigation is, for example, very important in land contaminated with *Streptomyces scabies*, which causes potato scab. Risks are very high during the 5 weeks following initiation of the tubercles. If the soil is maintained moist during that period, the lenticels, entry points for the bacteria, are protected by the activity of the concurrent saprophytic bacteria and the tubers remain healthy (Lewis, 1970). It is also known that fusariosis of cereals is favoured by drought because *Fusarium roseum* f. sp. *cerealis* 'culmorum' is capable of growing at matrix potentials at which its principal antagonists are inhibited. Lowering of the hydric potential in the soil and in the plant is related to transpiration, that is to say, development of the leaves. One can thereby plan an indirect role in the demand for water by reducing the total foliar surface through reduction of seed density and nitrogenous manure (Papendick and Cook, 1974).

In other cases, watering acts both on the host and on the microorganisms. In western Africa, village rice nurseries are often established on ridges and subjected to drastic surges of irrigation, which significantly favour attacks of *Sclerotium rolfsii*. Regular irrigation, right from emergence, can itself regress the parasite and also assists in obtaining plants more resistant to it because more vigorous. After planting out, in rice fields which have no embankment and where water supply depends entirely on stream floods, withdrawal of water too soon is often accompanied by attacks of *Ophiobolus oryzinus* (related to *Gaeumannomyces graminis*) whereas inundation and maintenance of stagnant water at great depth makes the plants susceptible to *Sclerotium oryzae* (Davet and Ravisé, 1965). Soybean, sunflower, sorghum, when affected by hydric stress, are rapidly invaded by *Macrophomina phaseolina* if the temperature is likewise high enough. On the other hand, potatoes planted in a soil containing *Verticillium dahliae* need not be watered that much in the early stages of cultivation: if the aerial parts grow too abundantly, there is risk of inadequate provision of nourishment by the roots when blockage of the vascular system sets in, leading to precocious wilting. Commencement of watering after initiation of the tubers, on the contrary, leads to better survival (Cappaert et al., 1994).

It should be admitted that in practice controlling soil moisture is very difficult (it would presuppose mastery of the climate) and alone rarely suffices to ensure protection against diseases.

Sowing Conditions

The date of sowing is often of paramount importance. Thus, early sowing of winter wheat augments primary infections by *Gauemannomyces graminis* var. *tritici* and by *Pseudocercosporella herpotrichoides*: the parasites have more time to colonise the roots before the winter cold. In spring, premature sowing in land not yet heated and poorly recovered increases risk of attacks of plants susceptible to Pythiaceae through prolongation of germination. All factors that shorten this critical period diminish risk: utilisation of recent seeds with high germinative energy and deposition of fertilisers just below the seeds so that seedlings rapidly become vigorous.

Density of sowing likewise plays a role. For an inoculum distributed homogeneously, a lower seed density implies that the distance between a root and an infectious propagule is larger and therefore the probability of infection lesser. Moreover, wheat sown at low density tillers better: the tillers, formed much later than the mother-stem, are infected later and therefore attacked less by a fungus such as *P. herpotrichoides* (Colbach and Saur, 1998).

Cultivation Systems

Monoculture and its Consequences

For reasons both practical and economical, farmers commonly tend to restrict themselves to cultivation of two species, or only one, in market gardens as well as field crops. Theoretically, nothing stands in the way of the same plant species being cultivated indefinitely in the same land provided crop harvests are adequately compensated by organic and mineral fertilisers and the fields free of parasites. But a different problem is posed when a pathogenic organism is present in the soil. Three situations can then arise.

MONOCULTURE WITH NEITHER PROLIFERATION OF THE PARASITE NOR LOWERING OF YIELD

This is the situation when a resistant soil is involved. For example, the soils of Châteaurenard (see page 181) are resistant to vascular fusarioses and monoculture, or repeated cultivation of susceptible melon is possible even in the presence of *Fusarium oxysporum* f. sp. *melonis*. It is important to note, however, that a soil is generally resistant only to a given species of parasite: the soil of Châteaurenard ensures no protection against *Pyrenochaeta lycopersici*, which causes corky root disease in tomato. It is no less important to know that poorly adapted cultural practices can result in the disappearance of soil resistance: treatment of Châteaurenard soil with a fumigant intended to combat *P. lycopersici* destroyed the microflora antagonistic to vascular *F. oxysporum*; maintenance of continuous moisture in the soils of Guadeloupe resistant to *Ralstonia solanacearum* inhibited the clays from playing their protective role and suppressed the obstacles to development of the bacterium.

The rational approach of a modern farmer must therefore firstly be to verify whether the soils show resistance against one or several parasites capable of attacking the crop(s) he proposes to cultivate. If resistance is evidenced, everything possible to safeguard it must be done. If the soils are receptive to all diseases, several measures can be resorted to: use of resistant varieties, chemical or biological protection, adaptation of technical practices and, as a last resort—choice of another crop. In fact, in practice, there are still few tests that help characterise the level of resistance of soils to diseases (see box below) and when one such property is demonstrated, the mechanisms explaining it are not always clear. So ascertaining which measures should be taken up and which absolutely excluded is usually not all that easy. But, this incertitude will surely be set to rest in the near future.

INCREASE, DECLINE AND STABILISATION OF DISEASE

In the case of take-all of wheat (*Gaeumannomyces graminis* var. *tritici*), which constitutes a classic example of such a situation, regression is generally seen after a 3- to 5-year monoculture. Approximately the same length of time is needed to observe a noticeable decrease in attacks of *Streptomyces scabies* (in certain types of soils) or *Rhizoctonia solani* (in other soils) in monocultures of potato. Examples of *Rhizoctonia* decline are also known in wheat cultivation. Establishment of resistance is due to the gradual development of an antagonistic microflora. For example, after 5 to 6 years of monoculture of sunflower, reduction in attacks by *Sclerotinia sclerotiorum* was due to multiplication of the mycoparasite *Coniothyrium minitans* (Huang and Kozub, 1991). Cultivation techniques play a role in the installation of this antagonistic flora: thus, decline of take-all did not occur in no-tillage fields directly sown on stubble (Cook and Veseth, 1991). Climatic conditions must also be favourable for development of the antagonist. In the UK and northern Europe, regression of the cereal nematode *Heterodera avenae*, after several years of uninterrupted cultivation of barley or wheat was due to parasites of the females and eggs of the nematode: *Verticillium chlamydosporium* and *Nematophthora gynophila* (Kerry and Crump, 1998). *N. gynophila*, an oomycete, releases zoospores: it can only multiply readily in soils that are very wet and cannot settle in climates characterised by periods of drought.

Phytosanitary Diagnostics

The techniques used to evaluate the soil infectivity can be used to establish
sanitary diagnostics of soils. In France, the diagnostic laboratory of
pathogenic flora of the soil of Fleury les Aubrais, annexed to the Plant
Protection Service, indicates for a certain number of diseases the
importance of the damage a cultivated plant is at risk of being subjected
to in a given land. Based on test results, it is possible to estimate: no risk,
prudent to give a preventive treatment, or more profitable to envisage
cultivation of another crop. The main tests presently used are given in
the Table below.

Disease	Pathogenic agent	Test plant	Test duration
Fusariosis of asparagus	*Fusarium* spp.	asparagus	5 weeks
Vascular fusarioses (melon, tomato, flax, carnation)	*Fusarium oxysporum* (specialised forms)	melon, tomato, flax, carnation	Variable, depending on plant
Dry rot of potato tubers	*Fusarium roseum* var. *sambucinum*, *F. solani* var. *coeruleum*	half-tubers of potato	2 weeks
Gangrene of potato tubers	*Phoma exigua* var. *foveata*	half-tubers of potato	6 weeks
Damping off	*Pythium* spp.	cucumber	6 days
Cabbage clubroot	*Plasmodiophora brassicae*	Chinese cabbage (Pé Tsaï)	6 weeks
Rhizomania of beetroot	*Polymyxa betae* (vector of BNYV virus)	beetroot	3 weeks
Black foot of beetroot	*Aphanomyces cochlioides*	beetroot	8 days
Pea rots	*Aphanomyces euteiches*	pea	8 days
Root blackening of pea	*Fusarium solani* and *Phoma medicaginis* var. *pinodella*	pea	5 weeks

Sanitary diagnostics requires careful sampling: the land must be divided
into homogeneous subplots (at least 10 per ha) and for each subplot, 8 to
12 elementary samples should be grouped together to constitute a
representative selection for laboratory analyses. Carrying out these tests
therefore constitutes considerable work. That is why they are presently
limited to technical institutes or to groups of producers.

But if the antagonists do not have a high saprophytic capability, their
populations risk rapid diminishment when the susceptible plants (if the
antagonists are associated with their rhizosphere) or the pathogenic agents
(if mycoparasites or nematophages) are no longer present in the environment.
These equilibria are therefore unstable. For example, a one-year interruption
of a monoculture of wheat suffices to observe again virulent attacks of take-
all during the next crop. Three successive cultivations of oats, alfalfa, potato
or soybean after 7 years of wheat monoculture destroyed the antagonistic
potentiality acquired by the soil, to almost the same extent as a disinfection

with methyl bromide (Cook, 1981). It is therefore not possible to change from one system of cultivation to another without due precautions.

PROLIFERATION OF ROOT PARASITES

Proliferation of root parasites is unfortunately the most frequent occurrence. The parasite multiplies abundantly in its host and, after its death, remains protected from competition in the organs it has colonised, in such a way that the quantity of inoculum increases year by year. The rate of infestation of the roots of tomato by such nematodes as *Meloidogyne* or such fungi as *Pyrenochaeta lycopersici*, thus rises rapidly and attains a level such that unless the soil is disinfected, cultivation has to be abandoned. It is possible, however, that continuation of cultivation of the susceptible plant could lead to a state of equilibrium. This is suggested by a trial carried out by Ebben and Last (1975), who found that the percentage of roots infected by *P. lycopersici* stabilised after seven years of tomato monoculture. It remained at a value slightly higher than 60% until the ninth and last year of the trial. But such a rate of attack is not economically supportable.

In this situation, **rotation** becomes imperative, as it precludes cultivation of susceptible plants continuously in the same land for several years.

A monoculture sometimes also has undesirable effects even in the absence of pathogenic agents. It leads to accumulation in the soil of compounds of plant origin which, beyond a certain threshold, can become phytotoxic: this seems to be the case in old asparagus plantations (Young and Chou, 1985). Monoculturing, generally speaking, depresses diversity of microbial populations. One notices in particular a drift of endomycorrhizal fungal populations, characterised by a selection of strains with rapid growth and reproduction, with parasitic tendencies, and elimination of less vigorous but more significant strains from the agronomical point of view (Johnson et al., 1992).

Crop Rotation

Practised for ages, rotation was initially performed empirically as a response to problems of fertility and parasitism. In the classic legume-cereal succession, the legume supplies part of the nitrogen required by the cereal that succeeds it. In the Egyptian berseem clover (*Trifolium alexandrinum*)-rice rotation practised for several centuries in Egypt, the savings realised on nitrogenous manure are due not only to mineralisation of the nitrogen-rich residues contributed by the berseem clover. Yanni et al. (1998) have shown that the symbiotic bacterium of berseem clover, *Rhizobium leguminosarum* bv. *trifolii*, behaves like an endophyte in the roots of rice and ameliorates growth of this plant and efficiency of nitrogen assimilated. This is probably the result of a long coevolution. A properly studied rotation, on the other hand, generally ensures a gap, which may be considered passive, in the development cycle of the pathogenic agents. But in certain special cases, the alternately cultivated plants contribute actively to regression of the parasitic populations.

DIMINUTION OF PARASITIC POPULATIONS DUE TO INTERRUPTION IN DEVELOPMENT CYCLE

The effectiveness of an interruption introduced by cultivation of non-host plants depends on the degree of specialisation of the targeted pathogenic agent and its capability for self-maintenance in the soil. Multiplication of an unspecialised parasite such as *Verticillium dahliae*, capable of invading most herbaceous or ligneous dicotyledons, is only interrupted if the rotation includes cereals. But the microsclerotia of the fungus ensure long persistence, which makes eradication by cultivation techniques problematical. *Macrophomina phaseolina*, most *Pythium* species and gall nematodes (*Meloidogyne*) also have a very wide host range. The situation is more favourable when specialised parasites such a *Gaeumannomyces graminis* var. *tritici* are involved. Rotation in which wheat or barley is grown once in 3 or 4 years suffices to hold the disease below the harmful threshold. Nevertheless, in a situation of this kind, the inoculum does not always diminish to the extent expected. Many plants can, in fact, harbour parasitic microorganisms without manifesting symptoms of the disease. These 'healthy carriers' do not augment the populations, but at the least contribute to their maintenance: thus the vascular forms of *Fusarium oxysporum* and some varieties of *F. solani* can persist, despite their high degree of specialisation, in the absence of a susceptible crop (Armstrong and Armstrong, 1948; Schroth and Hendrix, 1962).

Rotations therefore have a preventive effect. They constitute a means of precluding a too high accumulation of the pathogenic inoculum; only rarely can they be considered a method of curative control. Interruption of the development cycle of the parasites can be assured in a more efficient manner by the practice of bare fallow, in which the soil is hoed or chemically weeded. But maintenance of barren soil for a long period of time considerably augments the risk of erosion and leaching of nitrates; hence this practice is not recommended and sometimes even prohibited.

ACTIVE INTERRUPTION OF PARASITE DEVELOPMENT CYCLE

A certain number of plants act as natural baits against parasitic organisms. They provoke germination of the parasite's organs of conservation or hatching of its eggs and then, depending on the species, penetration of the larvae, progress in the life cycle or reproduction of the adults is inhibited, or the larvae are poisoned after ingesting the contents of a few cells. As a result, in all cases a significant decrease in number of infectious propagules takes place. Several Compositae (Indian carnation or *Tagetes, Chrysanthemum morifolium, Gaillardia*) and legumes (*Concanavalia ensiformis, Crotalaria spectabilis, Styzolobium deeringianum*), graminaceous (*Avena sativa, Brachiaria decumbens, Eragrostis curvula, Panicum maximum*) as well as a few oleaginous plants of economic significance such as castor oil plant, sesame and groundnut instigate reduction in inoculum density of numerous species of phytophagous nematodes of order Tylenchyda (Panchaud-Mattei, 1990).

Varieties of white mustard and forage radish stimulate eclosion of cysts of *Heterodera schachtii* without permitting development of the larvae. Grown alternately with soybean, bahia grass (*Paspalum notatum*) and velvet bean (*Mucuna pruiens*) induced an increase in yield by 85% and 98% respectively, with respect to continuous cultivation of soybean in fields infested with *Heterodera glycines* and *Meloidogyne* spp. (Weaver et al., 1998). The plant-nematode pair must be appropriately chosen because the effect can be very specific. *Tagetes minuta*, for example, is effective against *Meloidogyne incognita* but not against *M. arenaria* (Belcher and Hussey, 1977). It is also possible to employ a host plant that is destroyed by an herbicide after the juvenile parasites have invaded its roots, but before the larvae have had time to complete their cycle. Employing this method, carrot seedlings can break the dormancy of cysts of *Heterodera*, or potato can reduce infestation of the soil by *Globodera pallida*. Wild Solanaceae, such as *Solanum sisymbriifolium* and some cultivars of *S. nigrum* are even more effective (Scholte, 2000).

Some plants can also play the role of a bait against soil fungi. Thus gladioli, even though not members of genus *Allium*, provoke germination of the sclerotia of *Sclerotium cepivorum* (see page 221) but are not colonised. Cultivated before a susceptible plant such as garlic, gladioli could effect an increase in yield of 67% compared to a preceding crop lacking a biological effect, such as potato (Table 9.5). Similarly, oats can constitute a significant crop in soils infested with oomycetes (*Aphanomyces*, *Phytophthora* and *Pythium*). The roots of oats (and *Arrhenatherum elatius*) attract zoospores that accumulate around the zone of elongation. But instead of encysting, the zoospores are immobilised some tens of μm away, dilate, then disintegrate under the effect of a diffusible toxic saponin, avenacine (Deacon and Mitchell, 1985).

Table 9.5 Effect of the preceding crop on the health of garlic in a soil infested with *Sclerotium cepivorum*. Gladioli cause a reduction in inoculum density by stimulating germination of sclerotia without being colonised. Potato has no stimulatory effect and does not modify the inoculum density (Montegano and Vergniaud, 1987).

Preceding crop	Percentage of diseased bulbs	Average weight of a bulb (in g)	Yield (in t ha^{-1})
Gladioli	6.4	114	9.5
Potato	30.6	91.8	5.7

INDIRECT EFFECT OF CROP ROTATION

Small-scale trials have shown that crop rotation can also be a remedy in cases of 'apple replant disease'. Recovery of young apple trees planted in 'diseased' apple orchard soils was considerably improved by prior cultivation of wheat (Mazzola and Gu, 2000). It seems that wheat leads to a qualitative modification of the soil microflora.

Even though a succession is not involved, only an association, cover plants can be included here. Leguminous plants are used as soil cover in tree cultivation in tropical and equatorial regions to preclude soil erosion as well as to augment nitrogen and organic matter content of the soil. These plants also induce modifications in the soil microflora. Thus, cultivation of *Pueraria javanica* and *Calopogonium coeruleum* in oil palm plantations (*Elaeis guineensis*) raised the degree of resistance of palms to vascular fusariosis. This effect was due to a notable increase in populations of the *Fusarium* species, most particularly *F. oxysporum*, which led to an increase in competition for the pathogenic agent (Abadie et al., 1998).

Mineral and Organic Soil Amendments

Mineral Elements

Manipulation of **nitrogenous fertilisers** makes rather often a modification of microbial equilibria possible. This occurs in multiple ways. Nitric or ammoniacal nitrogen can be used to increase or decrease (by almost one unit) the pH of the rhizosphere as already seen (see page 293). In case of accidental anaerobiosis, the nitrates can also serve to raise the redox potential and preclude reduction of sulphates into sulphides that are toxic for the roots. Nitrogen by itself sometimes also has a direct depressive effect on soil fungi. Thus, a nitrogenous manure, especially if supplied in the ammoniacal or ureic form (which liberates ammonia), inhibits germination and growth of *Sclerotium rolfsii*. Fractionated supply ensures better protection (Messiaen et al., 1976). *Verticillium dahliae* is highly inhibited by ammonium sulphate (Dutta and Isaac, 1979). However, numerous studies have shown that ammoniacal nitrogen also augments the severity of vascular fusarioses, not only because it lowers rhizospheric pH, but also because it increases parasitic aggressiveness: an isolate of *Fusarium oxysporum* f. sp. *lycopersici* was much more aggressive against tomato when cultivated in the presence of an ammonium salt than in the presence of a nitrate (Jones et al., 1989). We have also seen, on the other hand, that *Gaeumannomyces graminis* thrives nicely in residues of cereal crops (highly cellulosic) when the soil is nitrogen-rich. Manipulation of nitrogenous fertilisers must therefore take into account the dominant parasitic problems. Nitrogenous fertilisation can be complemented, if necessary, by inhibitors of urease or inhibitors of nitrification.

Potassium- or **phosphate**-containing fertilisers do not seem to influence the development of diseases much. Nevertheless, it seems confirmed that a high phosphate content aggravates vascular fusarioses (Jones et al., 1989).

Independent of its effect in raising the pH when added in the form of lime, **calcium** can also have a direct effect on microbial equilibria: certain soils in Hawaii became resistant to *Pythium splendens* when calcium was added, both in the form of $CaCO_3$ and $CaSO_4$, which does not modify the pH (Kao and Ko, 1986).

Microelements sometimes play a role, either because they are immobilised and made unassimilable for the plant by the action of a pathogenic agent or because they constitute a limiting factor for development of the parasite. The first situation is illustrated by oxidation of Mn^{++} (soluble) into Mn^{4+} (insoluble) under the action of *Gaeumannomyces graminis* var. *tritici* in the rhizosphere of wheat, resulting in a localised deficit even in soils where the manganese content is satisfactory. This phenomenon can be aggravated by rhizospheric bacteria that oxidise manganese. Deficient plants become more susceptible to the disease. Other rhizospheric bacteria that reduce manganese have an opposite effect. Treatment of seeds at the time of sowing with such bacteria together with a small dose of $MnSO_4$, could help to remedy this problem (Huber and McCay-Buis, 1993). Vascular fusarioses are examples of the second situation. *Fusarium oxysporum* has high requirements for iron, zinc and manganese. The low solubility of these elements when the pH exceeds 6.5 could explain the lessened incidence of fusariosis in limed or naturally alkaline soils. Furthermore, it is also known that competition for iron is one of the mechanisms of biological resistance of soils.

Organic Amendments

Organic matter has multiple beneficial effects on soil quality: amelioration of soil structure and, by stabilising the aggregates, protection of the land from erosion; increasing water capacity and exchange capacity, thus reducing leaching and pollution of the subterranean layers; enriching the soil with nutritive elements gradually liberated and, lastly, as a support for intense microbial life. The principal sources of organic matter are plant residues or green manure, dung (less and less used because of separation between agriculture and animal husbandry) and compost (in the process of development; see box below). It is important to know the composition of the buried organic matter because its influence on microbial equilibria depends on its nature.

The supply of organic matter to a soil very often has beneficial effects on the health of the roots. Thus, dung and compost decrease attacks of *Rhizoctonia solani* on radish and kidney-beans (Voland and Epstein, 1994) and attacks of *Pyrenochaeta lycopersici* and *Phytophthora parasitica* on tomato (Workneh et al., 1993). But the reasons for this amelioration are still not clearly understood. In the small number of examples studies so far, three principal mechanisms have been identified: quantitative and qualitative regulation of available nitrogen, overall or selective stimulation of antagonistic microflora and liberation of inhibitory compounds.

AVAILABILITY AND FORM OF NITROGEN
When a fungus requires an external supply of nitrogen to germinate, as is the case with *Fusarium solani* f. sp. *phaseoli*, burying a substrate with a high C/N ratio (straw, green manure mown at maturity) plays an inhibitory role in immobilising mineral nitrogen, whereas unmatured green manure (with a low C/N ratio) has no positive effect (Huber and Watson, 1970). Jouan and Lemaire (1974) also demonstrated the inhibitory effect of amendments rich in cellulose on *Rhizoctonia solani*. Fresh dung or dried blood which, on

the contrary, rapidly liberate ammoniacal nitrogen, are unfavourable to *Pythium*, *Chalara elegans* and *Sclerotium rolfsii*, as well as nematodes.

Composting

Composting consists of controlled biological decomposition of organic waste of agricultural or urban origin (household refuse: unfortunately, the frequent presence of heavy metals often hinders recuperation of urban waste). This transformation is mostly realised through aerobic microflora, which distinguishes composting from rotting or putrefaction. Accumulation of an abundant substrate, moist and aerated, leads to very rapid microbial proliferation and intense metabolic activity, accompanied by a rise in temperature. By the end of a few days, a temperature of 50 to 60°C is generally attained but to let the temperature rise further serves little purpose since the metabolic activity then decreases considerably. Cooling the fermenting mass, generally by overturning in the presence of an air current, is therefore necessary.

During composting the readily decomposable matter is consumed and transformed into a microbial biomass. The final product is rich in colloids and contains mineral elements stabilised within slowly biodegradable compounds. The pH generally lies between 7 and 8 due to the process of ammonification. Bad aeration of the fermenting mass lets anaerobic fermentation take place, leading to organic acids that are often phytotoxic and lower the pH.

Bacteria (*Bacillus* and actinomycetes) are preponderant during the first phase of composting whereas fungi dominate during the cooling phase. Most pathogenic germs are eliminated after rise in temperature.

The general antagonistic activity of the soil is reinforced by a supply of compost.

STIMULATION OF ANTAGONISTS

Along with the compost and the farmyard manure, considerable quantities of exogenous microorganisms enter the soil, while addition of fresh organic matter stimulates instead the development of indigenous microbes. In all cases a strong increase in active biomass is evidenced, which rapidly leads to accrued microbiostasis and the phenomena of competition. This is probably the main reason for the positive effect of organic manures, observed with respect to *Pythium* (Bouhot, 1981), *Sclerotinia minor* (Lumsden et al., 1986), *Pyrenochaeta lycopersici* (Workneh and Van Bruggen, 1994) and some gall nematodes (Johnson et al., 1967). Sometimes it is possible to assign a preponderant role to a particular group of the total microflora: Mucorales in the compost used by Bouhot (1981), *Trichoderma* and bacteria (*Flavobacterium balustinum*, *Pseudomonas putida*, *Xanthomonas maltophilia*) in composts made up mainly from the bark of broad-leaf trees (Hoitink and Fahy, 1986). But sometimes the activity of the total microbial biomass, along with the resultant nutritional competition, is wholly responsible for lessening severity of the disease (Chen et al., 1988; Serra-Wittling et al., 1996).

TOXIC EFFECTS

Fresh compost of the bark of broad-leaf trees contains specific soluble fungitoxic and nematotoxic compounds. Esters of hydroxyoleic acids, for example, prevent *Phytophthora* from forming sporocysts and zoospores but do not inhibit *Rhizoctonia solani* (Hoitink and Fahy, 1986). Some oilseed cakes (cotton, castor, sesame) useable as manure after extraction of their oil are toxic to several species of nematodes. A green manure constituted of Sudan grass (*Sorghum sudanense*) significantly reduced populations of *Meloidogyne hapla*, provided it was buried early enough (8 weeks at the most after sowing): the fresh leaves of *S. sudanense* contain a cyanoglucoside, dhurrin, which after hydrolysis, ultimately releases p-hydroxybenzaldehyde and toxic HCN into the soil (Widmer and Abawi, 2000).

It was seen earlier (see page 97) that residues of cruciferous crops are significant because they release various toxic sulphur compounds when undergoing decomposition. The protective effect of wastes from lettuce cultivation against fusariosis of tomato crown seems to be due to a unique mechanism. Lettuce tissues contain orthodiphenols and in particular esters of caffeic acid, capable of chelating iron. They would inhibit assimilation of iron by *Fusarium oxysporum* f. sp. *radicis lycopersici* in the same manner as the siderophores of antagonistic pseudomonads (Kasenberg and Traquair, 1989).

It is also possible to obtain a toxic effect by converting the soil to a generalised anaerobic state. To do so a green manure is buried and, after irrigation, the soil covered with a plastic sheet, barely permeable to oxygen, such as that used for ensilage. The redox potential drops rapidly (down to –200 mV). Under these conditions populations of pathogenic soil fungi are drastically reduced due to the combined action of anoxia and the products of anaerobic decomposition of the organic matter (Blok et al., 2000). This procedure is equally effective against nematodes of genera *Meloidogyne* and *Pratylenchus* provided that confinement is maintained for about 8 weeks.

Soilless Cultivation

In vegetable and flower gardens, intensive cultivation rapidly leads to accumulation of root parasites, inevitably necessitating sterilisation with steam or regular use of fumigants. To get rid of the soil and find a solution to the phytosanitary problems, artificial substrates were conceptualised, which would additionally enable better control of plant nutrition. These substrates may be constituted of organic matter (peat, compost, ligneous fibre), alone or mixed with sand or perlite, or of an inert material (pozzolan, rock wool, polyurethane). In many exploitations, there is even no substrate whatsoever. The roots develop in a flattened channel in which a nutritive solution is running: this is the hydroponic system or NFT (Nutrient Film Technique).

These media are characterised by a markedly poor buffering capacity, both biologically and chemically. Initially, they are deprived of parasitic germs but yet capable of supporting microorganisms: humid, aerated, rich n mineral elements and root exudates, they constitute contrarily an ideal substrate wherein no competition is exerted (except in the case of compost). They are therefore at the mercy of accidental contamination and necessarily require surveillance and very strict rules of hygiene. The risk is particularly high in hydroponic cultivations and installations where the drainage liquids are recycled since a single point of infection suffices to contaminate the entire nutritive solution, which circulates in a closed circuit. Pathogenic agents can be introduced right at the outset of cultivation, from plants grown in poorly maintained nurseries. Contamination can also occur during the course of cultivation through accidental pollution. In soilless systems, irrigation is localised and the ground in the greenhouses and shelters therefore remains dry. Movements in paths between beds raise dust particles which, deposited on the plant substrates, may contaminate them with pathogenic germs (see coloured Plate II). Cortical and vascular *Fusarium* spp., *Verticillium dahliae, Pyrenochaeta lycopersici, Didymella lycopersici, Phomopsis sclerotioides* are rather frequent on solid substrates. They are also seen in NFT but this mode of culture mainly favours development of a particular pathogenic flora constituted of parasites well adapted to liquid media: bacteria (*Agrobacterium tumefaciens, Erwinia* spp., *Ralstonia solanacearum*), plasmodiophoromycetes (*Spongospora subterranea* on tomato), chytridiomycetes (*Olpidium* on lamb's lettuce, lettuce and melon), Pythiaceae (various *Pythium* and *Phytophthora, Plasmopara lactucae-radicis* on lettuce). Installations for disinfecting the nutritive solutions (by heat, UV, ozone or chloration) tend to become common due to extension of parasitic problems in soilless cultivations. Slow filtration on pozzolan or rock wool allows rather good decontamination of solutions. A few weeks after implementation, the filtration system is colonised by a saprophytic microflora: its antagonistic action completes the retention effect exerted by the filter and by adsorption of the microorganisms on the colloids secreted by the microflora. Direct supply of fungicides (or in some cases, Cu^{++}) in the solutions is likewise possible. In this case, the concentrations must be adjusted in accordance with the nature of the culture substrates to preclude phytotoxic effects and the accumulation of residues in the plants. Surfactants may also be used against zoosporic parasites: they destroy the permeability of the membrane, which provokes lysis of the zoospores. Many surfactants have been synthesised by the chemical industry but a large number of bacterial species also produce them naturally and trials are under way to introduce such bacteria into the rhizosphere of plants via the intermediary of nutritive solution (Stanghellini and Miller, 1997).

BIOLOGICAL METHODS

Any action that makes use of living organisms to shift microbial equilibria in a manner favourable to the plant can be included in the panoply of methods of biological control. As per this definition, most of the cultivation techniques can be considered biological methods because they modify the biocoenoses by acting on environmental conditions. In a stricter sense, biological methods can usually be qualified as those utilising particular well-characterised living organisms. Some use auxiliary microorganisms to oppose the action of pathogenic agents: this is **biological control** in the true sense. Others try to inhibit development of diseases by action on the host plant itself: sometimes the result can be obtained thanks to auxiliary microorganisms; this type of intervention therefore belongs under biological control in the strict sense. More often, agronomists attempt to make the plant more resistant by acting on its genome: this sometimes qualifies under **genetic control**.

Biological Control

Plant pathogens can be neutralised by highly varied mechanisms that result in their destruction, their exclusion or their modification. We shall merely mention a few of the better known models.

Destruction

This can occur either near the host plant or far from it. Often, it can even happen in its absence: in this case, the auxiliary microorganisms are introduced before planting the susceptible plant, to reduce the inoculum density of the pathogen. *Trichoderma harzianum, T. hamatum, Teratosperma sclerotivora, Coniothyrium minitans*, for example, attack the sclerotia which constitute the forms of conservation of *Sclerotium cepivorum* or *Sclerotinia* spp. Their hyphae insinuate in the cracks of the pigmented envelopes of the sclerotia and then penetrate the non-melanised reserve tissues and destroy them by enzymatic lysis. The sclerotia are reduced to hollow shells and the mycoparasites sporulate abundantly inside and on their surface.

Nematodes contract infection while moving in the soil. The adhesive propagules of *Pasteuria penetrans* (an actinomycete) or of *Paecilomyces lilacinus* (a fungus) attach to the cuticle and release a germinative tube that penetrates the animal's body and forms a thallus while gradually digesting it. The nematode, after death, liberates a large quantity of new infectious propagules. Fungi with adhesive loops such as *Arthrobotrys*, and those which parasitise the eggs, such as *Verticillium chlamydosporium*, arouse less interest at the moment. All these parasites generally thrive well in the soil but show a rather narrow specificity which manifests from the first steps of fixation on the host.

Exclusion

Pathogenic microorganisms can be inhibited from penetration of the host plant without necessarily being destroyed. This involves a proximity protection that is exerted in the rhizosphere. The antagonists must be present in the immediate vicinity of the sites of penetration. They have to be rhizocompetent in order to do this, i.e., capable of rapidly colonising the roots as and when they grow, or their distribution in the substrate must be homogeneous and sufficiently high for all the susceptible zones to have a chance for protection. Exclusion can be due to highly different mechanisms which, often, are complementary.

COMPETITION

We have already seen that competition for carbon sources is the basis for the protective effect of some saprophytic *Fusarium oxysporum* against vascular and cortical *F. oxysporum*, and that competition for iron exercised by their siderophores explains in part the action of the fluorescent pseudomonads. It is generally considered that protection against *Heterobasidion annosum* ensured in pine forests by *Phlebia gigantea* is an example of site competition. *H. annosum*, which is in fact only a parasite of weakened trees, initially settles on the stumps of freshly cut pines, then descends towards the senescent roots and, profiting from the energy furnished by this mass of moribund plant tissue, then attacks living trees in the vicinity. Control against this wood-rotting fungus consists of preventing stumps from being colonised and was demonstrated by Rishbeth in the UK more than 30 years ago. For this, after felling the trees, the stumps are daubed with a suspension of spores of a saprophytic fungus, *P. gigantea*, capable of rapidly occupying the substrate and defending it against intruders (we have already discussed hyphal interference in this connection; see page 162).

At the rhizosphere level, competition for infection sites is often evoked without being properly demonstrated: only after having eliminated all other possible explanations did Sneh et al. (1989) arrive at the conclusion that only competition for sites of infection could be responsible for the protection exercised by non-pathogenic isolates of *Rhizoctonia solani* against pathogenic strains in young radish and cotton plants. More recently, Olivain and Alabouvette (1997) supported arguments in favour of this type of competition between strains of *Fusarium oxysporum* on tomato roots. Research on recognition mechanisms between plant and symbiotic or parasitic invader has shown that designated points of fixation do exist: young absorbent root hairs in the case of rhizobia, for example. Occupancy of these sites by an auxiliary microorganism should contribute therefore to protection of the roots against invasion by pathogenic microorganisms.

ANTIBIOSIS

The implication of antibiotics in the mechanisms of biological control, long controversial, was confirmed by studies on directed mutagenesis carried out on fluorescent rhizospheric pseudomonads. Strains of *P. fluorescens* that produce phenazine, phloroglucinol, pyoluteorine and pyrolnitrine were selected and are capable of protecting wheat against *Gaeumannomyces graminis* var. *tritici*, tobacco against *Chalara elegans* and cotton against *Pythium ultimum* and *Rhizoctonia solani* respectively.

Control of crown gall caused by *Agrobacterium tumefaciens* called for a special category of antibiotics, the **bacteriocins**. These bacteriocins are low molecular weight antibiotics released by certain strains of bacteria, and specifically toxic for strains of the same species that do not synthesise them. A strain of *A. tumefaciens* (K 84) that produces a bacteriocin, agrocine 84, has thus been found among the non-pathogenic *A. tumefaciens* (presently grouped under the species *A. radiobacter*). This agrocine is toxic for the more common forms of *A. tumefaciens* that induce production of nopaline and agropine in the host plant (Fig. 9.7 and box 'Crown Gall', see page 274). To protect the rooted plants and cuttings, soaking in a suspension of K 84 suffices. The strain K 84 is itself non-susceptible to the bacteriocin it secretes. However, the plasmid responsible for both the production of agrocine 84 and immunity of the K 84 strain against its own bacteriocin is easily transferable in vitro to pathogenic strains. It is likely that this transfer can also occur under natural conditions. To preclude this risk, a new plasmid was constructed, carrying a deletion in the portion of the DNA which controls the transfer, but is identical to the original plasmid for all the other genes (Jones et al., 1988). The transformed strain, denoted by the name K 1026, is now commercialised in several countries.

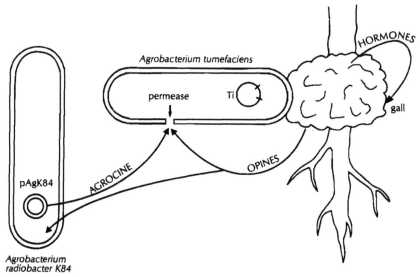

Fig. 9.7 Interactions between the K 84 strain of *Agrobacterium radiobacter* and *A. tumefaciens*. The K84 strain has a plasmid (p Ag K 84) which codes for the synthesis of agrocine, an antibiotic acting specifically on strains of *A. tumefaciens* which induce production of two opines, nopaline and agropine, in the transformed cells of the host plant. Agrocine 84 penetrates the target cells via courtesy of the same permease which allows the opines to cross the membrane and enter into the bacterial parasite. The K 84 strain can itself utilise the opines synthesised by the transformed cells.

Species of *Trichoderma* produce a very large variety of antibiotics, which are either a complementary element or, in the case of strains antagonistic to *Pythium*, an essential element of their properties as biological auxiliaries.

Many ectomycorrhizal fungi likewise secrete inhibitory compounds. *Leucopaxillus cerealis* var. *piceina*, for example, synthesises diatretyne nitrile, capable of completely inhibiting germination of zoospores of *Phytophthora cinnamomi* at concentrations of 2 µg ml^{-1}, thereby protecting the roots of various species of pines from infection (Marx and Davey, 1969). The inhibitory effect can generally be considered as a consequence of the symbiotic association as itself: thus, the synthesis of oxalic acid, toxic for *Fusarium oxysporum* f. sp. *pini*, by *Paxillus involutus*, is 5 times greater when the fungus is associated with the roots of pine rather than in pure culture (Duchesne et al., 1989). In most of the models studied, the inhibitory effect is selective and characteristic of a pathogenic fungus-ectomycorrhizal fungus pair, but often extends to the ordinary roots near the mycorrhized roots.

PHYSICAL BARRIERS OR CAMOUFLAGE

The protection assured by ectomycorrhizal fungi has sometimes been attributed to a solely mechanical effect: the mantle with which the fungi encircle the roots would physically inhibit accessions of microorganisms. Though likely, this effect (limited to the mycorrhizal region, if it exists) has never been properly demonstrated.

Growth of hyphae of *Pythium ultimum* is stimulated and oriented by the volatile exudates released by the seeds when they germinate. Coating the seeds with a strain of *Pseudomonas putida* capable of metabolising these exudates (especially ethanol) strongly diminished their diffusion and thus improved emergence in contaminated soils (Paulitz, 1991). This involves some kind of masking of the spermosphere effect. Gilbert et al. (1994) analysed studies in which the rhizospheric flora of plants subjected to parasitic attacks was compared to that of plants protected by a biological agent, or genetically resistant, or cultivated in a resistant soil. They found that in every case except one, the greater the rhizosphere effect (which results in the presence of abundant microflora differing from the soil flora), the greater the severity of the symptoms. The authors concluded that the better protected plants are those in which the rhizosphere is the most difficult to distinguish from the rest of the soil: in other words, roots thus camouflaged would no longer be detectable by the parasites. Going against the general view, Gilbert and colleagues consequently suggested that to realise this camouflage, it would be preferable to research the biological auxiliaries outside the zone of root influence rather than select rhizocompetent microorganisms.

Modification

In a few special cases, aggressive strains of the parasite can be subjected to a modification which transforms them into hypoaggressive, non-pathogenic strains. The best studied example is that of chestnut canker caused by *Cryphonectria* (=*Endothia*) *parasitica*. The fungus insinuates itself in trees through wounds and develops beneath the bark, causing the formation of cankers which eventually encircle the trunk, leading to death of the distal parts. However, some trees recover spontaneously. Strains of *C. parasitica* isolated from the lumpy scars formed around the cankers during recovery differ from other strains. They are less pigmented, sporulate poorly and produce less oxalic acid, laccase and various other toxic proteins. They are capable of penetrating the bark but subero-phellodermic barriers have enough time to come into play and fungal growth is rapidly arrested (Grente, 1965). The cytoplasm of these hypoaggressive strains contains fragments of double-stranded RNA (dsRNA) that resemble viral particles. However, no one has ever observed proteinic capsids around these nucleic acids. The hypo-aggressive strains are contagious and anastomose with the normal strains and transmit their dsRNA to them. The strains thus contaminated in turn become hypoaggressive. In practice, this result is obtained by introducing the hypoaggressive strain at the periphery of the cankers. The contaminated strains are then disseminated naturally from the cured trees by rain and insects. It suffices to treat 10% of the cankers for 4 consecutive years to obtain a good diffusion of the hypoaggressive character in the chestnut plantation. This method presents a major constraint however: the dsRNA can be transmitted from a contaminated strain to a healthy one only after anastomosis and the latter is only possible between compatible strains. Vegetative compatibility is regulated by at least 5 genes in *C. parasitica*. Prior study of the compatible groups is therefore imperative before embarking on treatment.

Japanese research has further shown that strains of *Rhizoctonia solani* can lose their pathogenicity after anastamosis with compatible hypoaggressive strains (Hashiba, 1987). The infectious agent responsible for the transformation in this case is a linear DNA plasmid.

Premunition

Premunition is a method of biological control wherein the auxiliary microorganism is utilised not to directly oppose the action of pathogenic microorganisms, but to stimulate preventive plant defence reactions. It is now more commonly termed induced systemic resistance. The presence and active participation of the host plant are therefore indispensable: it follows that in the same species, certain genotypes are more apt to be immunised than others. The protection induced by the auxiliary microorganism extends over the entire plant. It results from the activation of several different defence mechanisms: for that reason, the phenomenon ensures a wide range of resistance (the same treatment can immunise against bacteria, fungi and viruses) and great stability (a mutant bypassing the resistance induced is less likely). But duration is variable. In some plants premunition can persist for the entire development cycle and sometimes even be transmitted by grafting. Salicylic acid could constitute one of the signals which transmit the stimulus. But according to other hypotheses, the presence of salicylic acid in plants can simply result from activation by an unknown factor of pheny-lalanine-ammonia lyase, an enzyme which, moreover, intervenes in tissue lignification and synthesis of phytoalexins.

Several fluorescent pseudomonads that do not produce antibiotics, classified under the PGPR group, are capable of inducing a systemic resistance to pathogenic agents in their hosts. It is possible to protect carnations, for example, against tracheomycosis caused by *Fusarium oxysporum* f. sp. *dianthi*. Van Peer et al. (1991) showed that this involves neither competition nor antagonism, by precluding all contact between the pseudomonads and the *Fusarium*: the bacteria, in suspension, were added to the culture substrate which contained the rooted plants and the fungus was introduced directly into the xylem of the stems one week later. The plants were protected although no bacteria were detected in the stems. Phytoalexins accumulated in the stems of the bacterised carnations in response to infection by *Fusarium*. However, no phytoalexins were present in the bacterised but uninfected carnations. The role of rhizospheric *Pseudomonas* is therefore to place the plant 'on the alert'. The lipopolysaccharides of the bacterial cell walls seem to be mediators in this activation. PGPRs other than pseudomonads can also induce systemic resistance to diseases: *Serratia marcescens, Flavomonas oryzihabitans, Bacillus pumilus,* applied as seed treatments, protected cucumber against foliar parasites in an open field trial (Wei et al., 1996).

In France, studies have been undertaken on *Ralstonia solanacearum,* a vascular parasite that causes severe wilt in Solanaceae in tropical regions. *Hrp⁻* mutants were obtained by genetic manipulation. These mutants had lost their pathogenicity against susceptible cultivars and no longer caused a hypersensitive response in resistant cultivars. They did, however, remain capable of invading the roots of tomato and sustaining themselves in the base of the stems. Totally inoffensive, they induced a defensive reaction in the plants, which protected the latter from eventual invasion by pathogenic *R. solanacearum* (Trigalet, 1994). Trials are underway in several tropical countries to ascertain their conditions of use and to select mutants best adapted to the natural environment.

Plants can also be cross-protected by fungi. Thus, cotton grown from seeds treated with *Trichoderma virens* was protected against *Rhizoctonia solani*: the plants contained more phytoalexins (in this case, gossypol and related products) and peroxidase activity in their roots was higher than in controls (Howell et al., 2000). Induced systemic resistance has also been recently considered one of the mechanisms that could be responsible for wilt control induced by some non-pathogenic strains of *Fusarium oxysporum* (Alabouvette et al., 1998).

Endomycorrhizal fungi induce attenuated defence reactions in their hosts, which could contribute to improving their resistance to parasites. Thus, in roots of soybean infected with *Glomus mosseae* or *G. fasciculatum*, small but significant quantities of phytoalexins, potentially toxic for fungi (glyceolline I), bacteria and nematodes (coumestrol) were detected (Morandi et al., 1984). Mycorrhization causes other changes in plant physiology. For example, the composition of the root exudates is quantitatively and qualitatively modified. Production of chlamydospores by *Chalara elegans*, possible in ordinary roots of tobacco, was inhibited in mycorrhized roots due to the high concentration of arginine present in them (Baltruschat and Schönbeck, 1975)

Genetic Control

Induced systemic resistance activates the defence mechanisms of the plant. These mechanisms occur spontaneously in plants provided with suitable genes. Obtaining varieties resistant to diseases constitutes what is sometimes called genetic control.

A single dominant gene sometimes suffices to confer resistance. For example, the *Fom 1* gene ensures a high level of resistance against *Fusarium oxysporum* f. sp. *melonis* on the melon cultivar 'Doublon'. But **monogenic resistances** are often bypassed by new races of the parasite. This is what happened in the case of melon. The generalised cultivation of cultivar 'Doublon' and its derivatives resulted in the appearance of a race of *F. oxysporum* f. sp. *melonis* capable of overcoming the resistance procured by *Fom 1*. Another gene, *Fom 2*, was later found in oriental varieties and introduced into commercial cultivars. However, races 1-2 which can overcome resistance arising from gene *Fom 1* and gene *Fom 2* are presently known.

Polygenic resistances, which are due to the additive effects of several genes, dominant or recessive, are less strong than monogenic resistances but are not threatened by the appearance of new races of the parasite. Resistances of this type to vascular fusariosis were also shown in melons of the Far East. They are longer and more difficult to introduce in commercial cultivars.

Resistance genes can sometimes be identified in individuals belonging to cultivated populations. More often, they are present in wild species and it is necessary to cross these species with the one to be improved. Several retrocrossings are then necessary to produce a cultivar with an agronomically satisfactory quality. It may happen that acquisition of resistance is linked with shortcomings that retrocrossing cannot wholly eliminate or that hybridisation is still not possible. Thus the *Ve* gene of resistance to verticilliosis is easily transferable to tomato from *Lycopersicon pimpinellifolium* but has not yet been successfully introduced into egg-plant. This obstacle can be overcome, however, by grafting susceptible plants onto resistant stock; egg-plants grafted on tomatoes possessing the *Ve* gene escape verticilliosis and concomitantly are not attacked by *Chalara elegans* as it does not infect tomato. Thanks to grafting, it is therefore possible to manage 'as though' resistant cultivars were available. The procedure, though easy, is nonetheless costly in terms of manpower and hence reserved for highly profitable crops (Table 9.6). **Grafting** must be considered a temporary solution while awaiting resistant varieties.

Genome mapping, with determination of molecular markers attached to the principal genes enables more rapid identification of significant progeny and reduction in number of retrocrossings necessary for obtaining an agronomically valuable plant. But the constraints imposed by incompatibility between species considerably limit the possibilities of improvement by traditional genetic methods. With the development of molecular biology, it is now theoretically possible to overcome these compatibility problems and the need for retrocrossings. Several techniques enable introduction of a precise gene, irrespective of its origin, into the genome of plant cells. However, for certain species under experiment, it is not yet known how to regenerate an entire plant from transformed cells.

Table 9.6 A few examples of stocks sometimes used to protect susceptible market produce against telluric parasites

Grafted cultivated species	Stock	Parasites combated
Tomato or egg-plant	Hybrid *Lycopersicon esculentum* x *L. hirsutum*	*Pyrenochaeta lycopersici* *Fusarium oxysporum* f. sp. *lycopersici* *Fusarium oxysporum* f. sp. *radicis lycopersici* *Verticillium dahliae* *Meloidogyne incognita*
Egg-plant	Tomato (*Ve*)	*Verticillium dahliae* *Chalara elegans*
Melon	*Benincasa cerifera*	*F. oxysporum* f. sp. *melonis*
Cucumber	*Cucurbita ficifolia*	*F. oxysporum* f. sp. *cucumerinum*
Watermelon	*Lagenaria siceraria*	*F. oxysporum* f. sp. *niveum*

It is well to remember, however, that a resistance based on the functioning of a single gene, whether introduced by transgenesis or by traditional pathways of Mendelian genetics, is always susceptible to being circumvented by a simple mutation of the pathogenic agent.

Transgenetic techniques enable introduction of foreign genes provided by a microorganism, plant or animal, into the genome of a plant (or any other organism). The most commonly used method since the early 1980s utilises the Ti plasmid of *Agrobacterium tumefaciens*. The gene to be transferred is included in that part of the plasmid which becomes integrated into the genome of the host plant, the T-DNA, previously cleared of the genes coding for synthesis of the phytohormones, to preclude anarchic development of the transformed cells (Caplan et al., 1983). A marker gene that helps in recognising the transformed cells is also added. It is likewise possible to use the Ri plasmid of *A. rhizogenes*. Unfortunately, this method is not applicable to cereals because they are not infected by agrobacteria. In this case, another vector can be employed, namely a fungus, *Olpidium brassicae*, a vector for several plant viruses. A plasmid containing the gene to be transferred is constructed and encapsulated in the proteinic envelope of the Tobacco Necrosis Virus, a virus normally transmitted by *O. brassicae*. Roots of wheat have thus been transformed by simply soaking the plants in a suspension containing these pseudovirions and the zoospores of *O. brassicae* (Zhang et al., 1994). For cereals, it is also possible to directly transfer DNA fragments in protoplasts ('naked' cells) by electroporation: by using electric shocks, pores are created through which the transgene is introduced. This technique has also been applied to soybean and cotton. It is equally possible to bombard the cells with microprojectiles constituted of gold or tungsten balls covered with the DNA that needs to be incorporated. In all these cases, it is necessary to add a marker gene to the gene required to be expressed in the plant, as it allows rapid sorting of the transformed cells in the laboratory. Bacterial genes with resistance to antibiotics have long been used as markers.

There are currently a few transgenic plants resistant to insects or viruses. But only a very small number of genes of resistance to bacteria or fungi have been cloned or even characterised in plants.

Identification in certain pathogenic bacteria of avirulence genes and genes controlling hypersensitivity reactions could help in determining the molecules whose synthesis they govern and thereby in finding their receptors in plants, and thereafter the genes that code for these receptors. Likewise for fungi, attempts are underway to utilise the elicitors known, to identify the receptors and corresponding genes.

However, the functioning of a dominant gene of resistance is triggered by a corresponding specific avirulence gene of the pathogenic agent. It is therefore questionable as to what practical interest it would be to introduce one such gene in plants genetically widely separated from the species of origin and therefore probably parasitised by races of the pathogenic agent possessing different genes of avirulence.

Evaluation of the risks and advantages of transgenic plants is largely beyond the scope of this book. Limiting ourselves to the domain of soil biology, we shall merely remember that a slight risk does exist, which should not be ignored, of transformation of microorganisms by plant DNA in the soil, especially if this DNA itself contains sequences homologous to the bacterial genes (see page 329).

For Further Information on Biological Control of Harmful Microorganisms

Allmaras R.R., Kraft J.M. and Miller D.E. 1988. Effects of soil compaction and incorporated crop residue on root health. Annu. Rev. Phytopathol. 26: 219-243.

Anonymous. 2001. Induced Resistance to Plant Diseases. Papers presented at an international symposium held in Corfu 22-27 May, 2000. Eur. J. Plant Pathol. 107: 1-146.

Azcón-Aguilar C. and Barea J.M. 1997. Applying mycorrhiza biotechnology to horticulture: significance and potentials. Scientia Hortic. 68: 1-24.

Barker K.R. and Koenning S.R. 1998. Developing sustainable systems for nematode management. Annu. Rev. Phytopathol. 36: 165-205.

Boland G.J. and Kuykendall L.D. (eds.). 1998. Plant-Microbe Interactions and Biological Control. Marcel Dekker Inc., New York, NY.

Campbell C.L. and Benson D.M. (eds.). 1994. Epidemiology and Management of Root Diseases. Springer-Verlag, Heidelberg, Germany.

Cook R.J. 2000. Advances in plant health management in the twentieth century. Annu. Rev. Phytopathol. 38: 95-116.

Cook R.J. and Baker K.F. 1983. The Nature and Practice of Biological Control of Plant Pathogens. Amer. Phytopath. Soc., St. Paul, MN (2nd ed.).

Engelhard A.W. (ed.) 1989. Soilborne Plant Pathogens: Management of Diseases with Macro- and Microelements. Amer. Phytopath. Soc., St. Paul, MN.

Hall R. (ed.). 1996. Principles and Practice of Managing Soilborne Plant Pathogens. Amer. Phytopath. Soc., St. Paul, MN.

Hoitink H.A.J. and Boehm M.J. 1999. Biocontrol within the context of soil microbial communities: a substrate-dependent phenomenon. Annu. Rev. Phytopathol. 37: 427-446.

Hoitink H.A.J. and Keener H.M. (eds.). 1993. Science and Engineering of Composting: Design, Environmental, Microbiological, and Utilization Aspects. Renaissance Publ., Worthington, UK.

Jeger M.J. and Spence N.J. (eds.). 2001. Biotic Interactions in Plant-Pathogen Associations. CAB International, Wallingford, UK.

Klee H., Horsch R. and Rogers S. 1987. *Agrobacterium*-mediated plant transformation and its further applications to plant biology. Annu. Rev. Plant Physiol. 38: 467-486.

Ogoshi A., Kobayashi K., Homma Y., Kodama F., Kondo N. and Akino S. (eds.). 1997. Plant Growth-promoting Rhizobacteria. Proc. 4th Internat'l., Workshop on PGPR, Japan-OECD Workshop, Sapporo, Japan, 5-10 Oct. 1997.

Palti J. 1981. Cultural Practices and Infectious Crop Diseases. Springer-Verlag, Berlin.

Rees R.M., Ball B., Watson C. and Campbell C. (eds.) 2000. Sustainable Management of Soil Organic Matter. CAB International, Wallingford, U.K.

Whipps J.M. 1997. Developments in biological control of soil-borne plant pathogens. Adv. Bot. Res. 26: 1-134.

10

Utilisation of Auxiliary Microorganisms

A large number of cultural practices favour directly or indirectly the development of microbial populations naturally present in the environment, on which agriculture has been dependent since its origin: nitrogen-fixing bacteria, communities antagonistic to parasites and predators, mycorrhizal fungi etc. We have seen a few of these examples in the previous chapter. Having long been used empirically, collaboration of these microorganisms has been systematically researched ever since advances in microbiology clearly demonstrated their role. A novel step was achieved with conceptualisation of possible reinforcement of natural communities or remediation of their insufficiency by introducing into the soil microbial populations selected and multiplied on an industrial scale. Several problems must nevertheless be resolved before satisfactory results can be achieved. One is the choice of criteria which would allow selection of efficacious individuals posing no danger for the environment. Other problems concern production, formulation and conditions of application of these biological products.

CHOICE OF MICROORGANISMS

Microorganisms are first selected for their capability of fulfilling the function assigned to them. But for a microorganism whose behaviour is satisfactory under laboratory conditions to be retained, it must necessarily remain effective in the complex environs where it will be used. Subjected to intense competition, it has to survive. Obviously, its success under this competition should not be due to toxins dangerous to public health and excessive competitiveness not result in an upset of the natural equilibria. Lastly, use of a biological preparation must find a place in the technical route without jeopardising the ensemble of other practices.

Efficiency Criteria

The selection process involves exploitation of the variability of species and choosing the most promising strain from a collection as large as possible. Hence it is necessary to first prospect intensively for microorganisms preferably in places where chances of finding them are maximal: biodegraders in polluted soil, antagonists in the rhizosphere of healthy plants, nitrogen-fixing symbionts in nodules etc. The trickiest problem is defining the criteria according to which a given strain is declared superior to others. The ideal practice would be to compare all the isolates in their actual conditions of use but this is generally impossible for various reasons of place, time and labour. It is therefore necessary to resort to miniaturised tests (inoculation of small vitro-plants, for example) or tests representative of an activity recognised as primordial (synthesis of enzymes, antibiotics, siderophores, degradation of a particular substrate). This is possible only where the mechanisms of action are already well understood. Tests based on chemical measurement are valuable because they enable precise classification of a large number of strains in a relatively short time. But recourse to one unique criterion can lead to loss of important information and result in elimination of candidates of great value. As a matter of fact, especially in matters of biological control, the effectiveness of an antagonistic microorganism lies in the simultaneous work of several different mechanisms. To compensate this inconvenience, tests at several levels can be designed: one batch of strains retained after selection based on the first criterion, is screened by a second criterion and so forth. The difficulty of sorting operations increases rapidly with the number of levels involved.

Resorting to mutagenesis permits creating artificial variability, thus increasing the possibilities of choice. However, mutagenic agents often cause undesirable mutations which can reduce the competitiveness of the strains in a hostile environment. Techniques of genetic engineering henceforth allow for envisioning less traumatic modifications of the genome. The field of variability can be considerably enlarged by resorting to genes identified in other groups of microorganisms.

Irrespective of the methods used to obtain them, strains which showed efficiency in the laboratory must again be tested under conditions which, at one and the same time, simulate reality but still give good reproducibility of results. This new series of tests, conducted in a greenhouse or in air-conditioned enclosures, seeks primarily to study adaptability to the environment, in order to make an ultimate choice among the material selected and to specify its conditions of use. If a genetically modified microorganism is involved, it is further necessary to show that its use presents no ecological risks.

As a final step, multiple trials in an open field must be conducted to verify that the biological agent selected responds well to what is expected of it and the added value it helps to achieve is not lower than its cost.

It is important to remember that in most cases auxiliary microorganisms are associated more or less intimately with the plant. Irrespective of whether this involves nitrogen-fixing symbiotic or associated bacteria, mycorrhizal fungi, growth-promoting or resistance-inducing bacteria, microorganisms will completely realise their potentialities only if the plants correctly play their role as partners. Definitive interactions between bacterial strains and plant cultivars have been demonstrated in most of these associations. Yet for decades selection of high-yielding plants used by modern agriculture has been based on their ability to respond to high doses of fertiliser and not on capability for cooperation with soil microorganisms. This has sometimes resulted in a sort of counter-selection: the most productive cultivars generally respond very badly to an inoculation with endomycorrhizal phosphorus-assimilating fungi or to nitrogen-fixing bacteria. To return to an agriculture less demanding in inputs, it is hence necessary not only to make available good strains of auxiliary microorganisms, but also to completely review the conditions of selection of cultivated plant varieties.

Competitiveness

A microorganism that proves highly antagonistic in a petri dish or in sterilised soil can be completely inhibited by microbiostasis in a natural environment, or even be rapidly destroyed by parasites or predators. Selected strains must therefore have a high capability for saprophytic life. Such a capability depends on a large number of factors and therefore cannot be determined by simple criteria. In some cases, however, it is possible to conceive tests than can be rather rapidly executed in a laboratory. Thus clones of *Trichoderma* can be classified by evaluating their growth on a nutritive medium from discs constituted of a mix of a non-sterilised reference soil and an inoculum added in stepped concentrations. Trichodermas are readily distinguishable from other fungi growing on the medium because of their characteristic green sporulation and their level of occupancy on the substrate can be rapidly evaluated according to an arbitrary scale of notation (Davet and Camporota, 1986).

When the organism introduced is intended to destroy pathogenic agents present in the soil, the presence of other strains of the same species is generally not a hindrance: their effects are added to those of the selected auxiliary organism. But when that organism must be present at the surface or inside the roots of a host plant to act, it is extremely important that it be able to occupy the maximum possible sites. Competition with indigenous strains of the same species can then become a limiting factor. Trials carried out in pots using different *Glomus* species (endomycorrhizal fungi) showed, for example, that an introduced species can be completely excluded from the root system by indigenous *Glomus* if less competitive. On the other hand, if the foreign species is highly competitive, it suffices to place the inoculum under the seedbed to obtain good colonisation of the roots after germination (Hepper et al., 1988). The presence of competitive indigenous strains of rhizobia in the soil could compromise the inoculation of a leguminous plant by a high-performing but less competitive strain. Field trials have helped establish that to obtain a positive response to an inoculum supply, it is necessary that at least two-thirds of the total number of nodules be colonised by the strain introduced; for a foreign strain of *Rhizobium* to colonise more than half the nodules, application of a quantity 1000 times higher than the natural inoculum density is required (Thies et al., 1991). For the antagonistic strain of *Fusarium* Fo 47, the concentration has to be 100 times higher than that of pathogenic *F. oxysporum* (Alabouvette et al., 1993). In such situations, it is therefore necessary to use a highly concentrated inoculum, placing it in such a way that it comes into contact with the roots before competitors do. This condition can be accomplished either by coating the seeds, thanks to which the infection occurs right at the time of germination, or by inoculation in the nursery if plants requiring transplantation are involved. Since estimation of rhizospheric competence of microorganisms is difficult, as a clear understanding of what it is attributable to is lacking, recourse to the host plant remains imperative for judging the capability of the strains.

Innocuousness

Except for biological herbicides, auxiliary microorganisms should not cause any damage to plants. Utilisation of certain strains of *Trichoderma virens*, very efficient against *Rhizoctonia solani* and *Pythium ultimum* because of their production of gliotoxin and gliovirin, is thus limited because they also synthesise a phytotoxic steroid, viridiol. Such problems must be taken into account right from the first application studies.

Furthermore, before marketing a biological agent its non-toxicity to man as well as animals must be demonstrated, together with its incapability of growing at the human body temperature and, in general, that no risk to the environment is posed. When this involves **microorganisms selected** from natural populations using classic methods, legislation wavers between two attitudes: whether the agent should be considered simply an improved variety just like a new wheat cultivar or a new race of rabbit and homologation done essentially on the basis of criteria of effectiveness (the attitude generally adopted for microorganisms considered growth improvers), or whether a complete dossier of homologation needs to be constituted, comprising the same toxicological studies as though the agent were a new chemical compound (the fate of biological control agents, probably because of their qualification as 'biopesticides'). The latter suspicion, though legitimate, can have adverse effects: the market for biopesticides being extremely small, the effort and cost of toxicological studies constitute a considerable obstacle to the development of these methods of biological control (thus indirectly favouring continuing use of chemical pesticides); on the other hand, it prevails on firms that nevertheless desire to commercialise auxiliary microorganisms to present them systematically as growth stimulators even though their action is essentially one of protection against pathogenic agents (toxicological studies are thus skipped). An annexe to the European regulations on pesticides was drafted in 2001 to particularly take into account the case of biopesticides.

Genetically modified organisms (GMO), defined as 'organisms whose genetic material has been modified by ways other than through natural multiplication or recombination', are subject to special rules in most countries. Legislations of the countries of the European Union were harmonised in 1990. All studies carried out on GMO must receive prior permission from the Commission of Genetic Engineering. Experiments can only be carried out under confined conditions in specially equipped laboratories and greenhouses, subjected to very precise norms. Experimentation outside a confined environment and marketing of GMO are subject to authorisation by the Study Commission for Dissemination of Products Arising from Biomolecular Engineering, after examination of risks to the general public's health and to the environment.

Given the present stage of our knowledge, knowing what risks dissemination of GMO pose for the environment is very difficult. While it is possible to study and to foresee to a certain extent the behaviour of a microbial population introduced into the soil, to appreciate the risks of horizontal transfer of new genetic information it possesses is most challenging. That genes can be effectively exchanged among bacteria of taxonomically different groups is known and transfers are probably possible between prokaryotes and eukaryotes (see box). However, for these transfers to occur it is necessary that sufficiently abundant populations of the two partners be present at the same moment in the same ecological niche. The exchanged DNA must then be replicated and transmitted to subsequent generations. Even though quite slender, the probabilility of diffusion of genes outside the transgenic populations, especially in the rhizosphere where microbial activity is high, cannot be rated zero. Let us note, however, that this possibility of dissemination exists even for 'natural' genes. The risks that need to be taken into account must therefore mainly concern the 'exotic' genes which originate in surroundings that differ completely from those into which they are introduced.

It goes without saying that the commercialised preparations should not be contaminated by other microorganisms. This appears difficult, however, judging from recently published results of an enquiry into the quality of inoculums of rhizobia formulated on peat (Olsen et al., 1995). Of a total of 40 samples from non-expired lots sold in Canada by North American companies, 39 contained more (sometimes up to 1000 times more!) contaminants than cells of rhizobia. Several of these contaminants were strong inhibitors of rhizobia in vitro and one among them, *Pseudomonas aeruginosa*, a bacterium potentially pathogenic for man. French and Canadian regulations are stricter and impose quality control.

Other problems, not really taken into account by existent regulations specifically concern biopesticides. Some of these microorganisms are, in fact, closely related to the parasites destined to be overcome: the Fo 47 strain, antagonistic to *Fusarium oxysporum* f. sp. *radicis lycopersici*, is also an *F. oxysporum*; the strains of *Ralstonia solanacearum* which protect tomatoes from bacterial wilt are directly derived from the pathogenic bacterium and *Agrobacterium radiobacter* is extremely close to pathogenic *A. tumefaciens*. One can legitimately question the eventuality of these auxiliaries one day transforming into parasites. These risks are taken into account and evaluated during predevelopmental studies. For Fo 47, for example, the chances of this saprophytic strain becoming a parasitic special form are not greater than that of billions of other saprophytic *F. oxysporum* present in all soils. GMO modified by deletion of genes, such as *A. radiobacter* K 1023, are a priori less likely to become dangerous.

Do Transfers of Genes Occur in Nature among Organisms Genetically Distanced?

Numerous analogies exist not only in the succession of reactions that lead to the production of antibiotics of the β-lactames group in streptomycetes (cephamycins) and in fungi (penicillins and cephalosporins), but also in the genes that play a role in these reactions. For example, a 61% homology is observed between genes coding for isopenicillin-N-synthetase (catalysing the synthesis of an intermediary metabolite) in *Streptomyces clavuligerus*, a bacterium, and *Aspergillus nidulans*, a fungus. This seems to constitute a solid argument in favour of the hypothesis of a horizontal transfer of this gene (as well as other genes involved in the same metabolic chain) from a prokaryotic microorganism to eukaryotic organisms (Peñalva et al., 1990). Besides, it is known that during the course of the infectious process, genes of *Agrobacterium tumefaciens* are integrated into the genome of the parasitised plant.

Inversely, the transfer of DNA fragments of a eukaryote to a prokaryote is possible by transformation of competent bacteria (see page 26). One such passage was demonstrated in vitro in an *Acinetobacter* from pulverised transgenic beetroot leaves: in this case, the transfer was facilitated because the transgenic beetroots had received a DNA of bacterial origin, possessing sequences homologous to those of the competent bacteria (Gebhard and Smalla, 1998). Right now it is impossible to predict the frequency with which such exchanges are likely to occur in the soil nor what their consequences would be.

Hence it is always better to exercise prudence when introducing foreign genes into a genome.

Compatibility

To be accepted by the users, new biological products should not result in upsetting the methods and schedule of work. These products have to be formulated and conditioned in such a way that they can be applied using ordinary agricultural equipment. Lastly, insofar as possible, they have to be compatible with other treatment products.

The last point sometimes gives rise to a few problems. Thus, pesticides used to coat seeds for improving germination and emergence are not always compatible with rhizobia. Certain treatments with fungicides are toxic for endomycorrhizal fungi or for biopesticides: trichodermas, for example, are strongly inhibited by benzimidazoles. It is sometimes possible to rectify these constraints. If the seeds of legumes are impregnated with a toxic product, inoculation of rhizobia by the coating process is not carried out; instead microgranules are placed below the seeds at the time of sowing. Besides, it is rather easy to select strains of fungi resistant to a family of fungicides: for example, trichodermas resistant to benzimidazoles. In such a situation, supplying an antagonist in combination with a fungicide can be envisaged: aside from its direct action on the target parasite, the chemical treatment confers a selective advantage on the auxiliary organism by eliminating part of the competing microflora.

Association of Several Auxiliary Microorganisms

It is sometimes profitable to utilise a mixture of several microorganisms. If they are compatible, their effects are often additive and even synergetic sometimes. Thus, some strains of fluorescent pseudomonads and certain isolates of saprophytic *Fusarium oxysporum*, each in its own way, assure good protection against vascular fusarioses; but numerous trials have shown that the protection may be even more effective when there is an association of a pseudomonad with an antagonistic fusarium. The siderophores of the bacterium sequester the iron, resulting in penury that exacerbates the sensitivity of the pathogenic fusarium to competition for carbon due to the presence of the saprophytic *Fusarium oxysporum* (Lemanceau and Alabouvette, 1993). The synergetic effect can be observed even between strains belonging to the same microbial species. In trials carried out at Pullman on take-all, wheat protected with a mixture of 4 strains of *Pseudomonas fluorescens* produced yields augmented by 20.4%, while none of these strains used individually improved the harvest appreciably (some even reduced it). Unfortunately, the same mixture did not suit all the experimental sites and so far the criteria that would assist in choice of associations most effective as a function of the characteristic parameters of the environment have yet to be determined (Pierson and Weller, 1994).

In soils poor in assimilable phophorus, infection of the roots of soybean by the endomycorrhizal fungus *Glomus mosseae* resulted not only in augmented absorption of phosphorus, but also of the number of nodules formed by *Bradyrhizobium japonicum* and the total nitrogenasic activity, which eventually led to a marked increase in yield. However, a strong phosphate fertiliser eliminated the effect of *G. mosseae* on nodulation and nitrogenasic activity (Asimi et al., 1980). *Glomus* spp. can likewise behave as vectors of diazotrophic endophytic bacteria: for example, they seem to favour penetration of *Acetobacter diazotrophicus* into the roots of *Sorghum bicolor* (Isopi et al., 1995).

Root infection by a symbiotic microorganism, and thereafter the functioning of symbiosis, is very often ameliorated by the presence of rhizospheric bacteria. For example, some strains of *Pseudomonas putida* stimulated nodule formation in kidney-beans either in the presence of indigenous populations of rhizobacteria or after the addition of an extraneous inoculum of *Rhizobium etli* bv. *phaseoli* (Grimes and Mount, 1984). *P. fluorescens* contributed to augmentation of the number of nodules formed by *B. japonicum* on the roots of soybean (Nishijima et al., 1988). A phosphate solubilising fungus such as *Penicillium bilaii* also seemed to have a positive effect (Rice et al., 1995). Suspensions of *Frankia* applied under axenic conditions to roots of alders (*Alnus rubra*) did not provoke deformation of the absorbent hairs, which constitute the first step in penetration, and hardly any or no nodules formed. On the other hand, a mixture of bacteria of the rhizosphere or specific strains of *Burkholderia cepacia* added to the suspension, led to curvature of the root hairs in all the inoculated plants and a large number of nodules formed (Knowlton et al., 1980). Similarly, formation of endo-or ectomycorrhizae could be considerably improved by the participation of helper bacteria. Thus, in trials carried out with Douglas fir (*Pseudotsuga menziesii*) in production nurseries, the rate of mycorrhized roots was about 60% when the fungus *Laccaria laccata* alone was added and 83 to 88% when associated with bacilli or pseudomonads. Evaluated five months later, the overall dry mass of the roots and the aerial parts of the fir saplings were much higher in three (Fig. 10.1) of the five experimental bacterial and fungal combinations. It thus seems possible to envisage production of a mixed inoculum at the commercial level (Duponnois and Garbaye, 1991).

A similar study was carried out on the relations that come into play between a plant (*Trifolium subterraneum*) and three categories of microorganisms: rhizobia, indigenous endomycorrhizal fungi and a strain of *Pseudomonas putida*. One-and-a-half times more nodules were counted on the roots of clover when the endomycorrhizal fungi or a strain of *P. putida* were added separately, but twice the number of nodules when the two were introduced together. These increments accompanied similar variations in the dry mass of the plants. Colonisation of the roots by the endomycorrhizal fungi increased from 7 to 23% in the presence of *P. putida* (Meyer and Linderman, 1986).

Bacteria belonging to several species common in the rhizosphere (pseudomonads, bacilli...) can thus have an overall ameliorative (helper) effect on the establishment of symbiotic relations between plants and other bacteria or fungi. This is apparently a rather general phenomenon, even though this property is limited to some specific strains among the whole

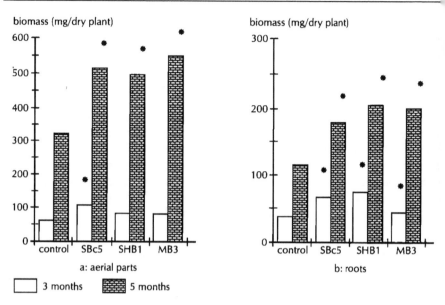

biomass (mg/dry plant)

biomass (mg/dry plant)

a: aerial parts

b: roots

☐ 3 months   ▦ 5 months

Fig. 10.1 Positive effect of a mixed inoculation (ectomycorrhizal fungus *Laccaria laccata* and selected bacteria) on the increase in weight of the aerial parts (a) and roots (b) of young Douglas fir plants (*Pseudotsuga menziesii*) measured 3 months and 5 months after cultivation in a nursery. Control: *L. laccata* only; SBc 5: *L. laccata* + *Pseudomonas* sp.; SHB 1: *L. laccata* + *Bacillus* sp.; MB 3: *L. laccata* + *B. amyloliquefaciens*. Asterisks indicate that the values differ significantly from controls at the 5% threshold (Duponnois and Garbaye, 1991).

rhizospheric populations. The mechanisms are not yet clearly understood but are probably highly varied. It can be presumed that the stimulatory effects are exercised before the establishment of symbiosis (modification of the exudations and the microflora associated with the roots, production of useful metabolites or detoxification of inhibitors), at the moment of infection (deformation of absorbent hairs, enzymatic action on the cell walls) and perpetuated later (synthesis of chelating agents and growth regulators).

PRODUCTION AND USE OF AUXILIARY MICROORGANISMS

Production of Inoculum

The technology of fermentation in a **liquid medium** was mastered long ago by the pharmaceutical industry which uses it for the production of antibiotics. Using fermenters that can easily contain volumes of 100 to 200 m^3, it is possible to precisely control the primary parameters which influence the development of cultures: pH and composition of nutritive fluids, temperature, oxygen content, speed of brewing of the medium etc. So, it is by fermentation in a liquid medium that mass production of most of the biological auxiliaries of agricultural usage is achieved: *Rhizobium, Pseudomonas, Trichoderma, Gliocladium, Fusarium, Hebeloma, Pisolithus* and others. At the end of the production cycle, the biomass must be separated from the culture medium, which is achieved by filtration or by centrifugation. But, whereas in the pharmaceutical industry it is the culture medium containing the metabolites that is of interest, in our case residual liquids represent only waste that has to be retreated before dispatching into the external milieu. Drying the biomass is a tricky operation: done too drastically, it can diminish the viability of the inoculum; carried out too long, it augments the risks of contamination. In all cases energy consumption is high. On the other hand, while fermentation in a liquid medium is perfectly suitable for bacteria, it is less so for the production of fungi: some species sporulate with difficulty under these conditions.

In such a case it is possible to resort to fermentation in a **solid state**. The substrate can be constituted of residues of low market value (bagasse of sugar-cane, bran, finely chopped straw, beetroot pulp etc.) watered with a nutritive solution containing mainly nitrogen. The risks of bacterial contamination are considerably reduced by the absence of free water and the constraints of asepsis are less rigorous than in liquid fermentation. It is not necessary to separate the biomass from its substrate: the product obtained can be used directly after limited drying, which is much less expensive in terms of energy than biomass harvested after fermentation in a liquid medium. Despite these advantages, fermentation on a solid substrate is not widely used in industry. *Talaromyces flavus* and *Teratosperma sclerotivora* are produced on a small scale using this type of procedure. An inoculum of *Trichoderma harzianum* titrating 10^9 to 10^{10} conidia per gram of dry product was produced repeatedly in a prototype of 1.6 m^3 constructed on the platform of predevelopment in biotechnology at INRA in Dijon (Durand, 1983). The strain of *Fusarium* Fo 47 readily reached concentrations higher than 10^7 cfu g^{-1} on enriched peat (Alabouvette et al., 2001).

Fermentation on solid subtrates can be utilised for artisanal production of inoculum by, for example, a group of producers or within the cadre of an agricultural economy with a low level of inputs. Arboriculturists in Roussillon have used this method for over twenty years to produce an inoculum of *Trichoderma harzianum* intended for biological control of *Armillaria mellea* in peach, apricot and kiwi plantations. Boiled grains of barley are drained, then sown with a suspension of conidia and distributed as a 10 cm layer over the soil of a clean, closed room. After initiating the culture, the grains are stirred once or twice to ensure good colonisation. The annual production of Catalan arboriculturists varies between 12 and 25 tons.

Organisms that do not develop on artificial media, such as endomycorrhizal fungi, pose a special problem: a living host is necessary to multiply them. Once colonised, the roots are harvested, chopped and eventually conditioned. The host plants, planted priorly in pots of sterile soil, are now cultivated in a soilless medium, which enables much better control of production conditions. Techniques of hydroponic (NFT) or aeroponic cultivation (the plants are suspended and a mist of nutritive solution vaporised on the roots) supply an inoculum of excellent quality and high concentration: at least 10^5 cfu per g dry roots (Sylvia, 1999).

Conditioning

Once produced, the biomass must be conditioned in a form that maximally favours survival while concomitantly allowing for easy usage. The presentation is not the same depending on whether the preparation is destined to coat seeds or to be introduced into the soil.

For treating seeds, the inoculum is generally in the form of a wettable powder constituted of dried and crushed or lyophilised biomass. The fresh biomass is often first mixed with an inert support (peat or lignite) which facilitates drying and handling. Lyophilisation is costly and not always a guarantee of good conservation: many fungi do not withstand this treatment very well. Adherence of the powder to the seed teguments can be ensured by an adhesive such as gum arabic or methyl cellulose.

For incorporation in the soil or in horticultural substrates, the inoculum may be in the form of a wettable powder or granules, or encapsulated in polymers. *Fusarium oxysporum* Fo 47, for example, is harvested by filtration after fermentation in a liquid medium, mixed with talc, then dried at 20°C under a current of filtered air (Alabouvette et al., 1993). Microorganisms produced on solid substrates of plant origin or on vermiculite imbibed with nutritive solution can be utilised directly or after rough crushing. Bacterial suspensions can likewise be used to impregnate granules made up of peat or clay. The supply of nutritive elements is not imperative for good conservation of the biological agents but helps in their becoming active much more rapidly at the time of their incorporation in the soil. Their effectiveness was found to be notably improved (Fig. 10.2).

Encapsulation of biopreparations in an organic polymer is a solution that simultaneously permits good conservation and ease of handling. The organic matrix most commonly used is alginate gel. The bacterial or fungal biomass is put in suspension in a solution of sodium alginate. This suspension, eventually added to inert or nutritive adjuvants, is introduced drop by drop into a solution of calcium chloride. The drops are rapidly coagulated by the calcium and if the depth and concentration of the solution are suitably adjusted, they deposit in the form of solid pellets at the bottom of the container. The alginate pellets, highly porous, can eventually be covered with a pellicle that ameliorates their mechanical resistance and aids in better conservation of the organisms they contain.

These preparations can remain active for many months if maintained in a cold chamber (around 5°C). But the number of viable cfus generally decreases rapidly when conserved at ambient temperature. This problem of conservation of viability and characteristics of the inoculum constitutes one of the drawbacks in employment of biological auxiliaries. *Bacillus* spp. are particularly interesting because they form spores which assure better conservation than most other bacterial species.

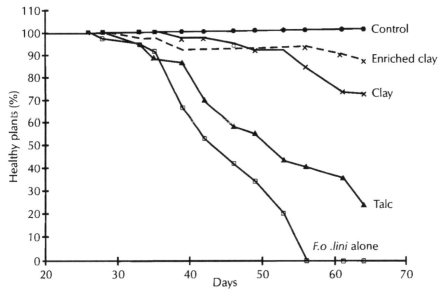

Fig. 10.2 Antagonistic activity of the non-pathogenic Fo 47 strain of *Fusarium oxysporum* against a strain of *Fusarium oxysporum* f. sp. *lini*, causing flax wilt. Fo 47 was introduced in an infested natural soil using different modes of formulation: on talc, in microgranules (clay) or in microgranules enriched with nutritive elements (enriched clay). Each treatment brought in 10^5 cfu g^{-1} soil. Antagonistic activity is expressed as the percentage of healthy plants (Alabouvette et al., 2001).

Application

Inundative additions in which the biological auxiliary is distributed or buried in the entire plot to be treated, help achieve a homogeneous infestation of the soil. This is possible when the volumes to be treated are low, as would be the case in a nursery, market gardens or substrates for soilless cultivation. They are also often used during experiments carried out in small plots. But when applied to hectares of land, the quantities of the inoculum required would be so high that implementation of this procedure on a commercial scale would be excluded in almost all cases. It is therefore necessary to think of other application techniques for large-scale field cultivations.

The quantities of inoculum required are considerably reduced if it suffices to place the auxiliary (generally in the form of granules) in furrows at the time of sowing. The operation can be automated and the sowing-machine adjusted in such a way that the inoculum is placed just beneath the seed. The young seedling is thus in contact with a very abundant population of auxiliaries right from the moment of germination. This technique is often employed for biopesticides. It is sometimes also utilised for rhizobia and endomycorrhizal fungi. The technique of sowing on gel (more commonly known under the name of fluid drilling) is extremely suitable for utilisation of biological auxiliaries. Pregerminated seeds are placed in suspension in an aqueous gel before sowing; emergence is thus very rapid and the plants obtained more vigorous; since they can be sown one by one they need not be thinned out. Chemical fungicides as well as antagonistic fungi such as *Laetisaria arvalis* (Conway, 1986) or *Trichoderma* can be added to the gels. The water contained in the gel allows for rapid germination of conidia and chlamydospores. The gel-forming agent itself intervenes directly by its pH (magnesium silicates, too alkaline, are not suitable for *Trichoderma*) and by its composition (organic polymers can serve as a carbon source for the fungi). Sowing radish seed on a hydroxyethyl cellulose gel containing a strain of *T. hamatum* regularly yielded better results than chemical treatment with quintozene in a soil infested with *Rhizoctonia solani* during trials carried out in several different sites (Mihuta-Grimm and Rowe, 1986).

It is possible to further reduce the quantities of inoculum required by directly applying the auxiliaries to the surface of the organs to be treated. These treatments most often involve the seeds and a **coating** or **pellicle** is applied in such cases (see coloured plate III).

Coating consists of depositing relatively thick layers of inert substances (peat, clay, cellulose matrix) around the seeds, alternating with biological auxiliaries or chemical agents. The coated seed resembles a sugar-coated almond. This procedure, rather expensive, is mainly used to augment the volume of very small seeds (tobacco, lettuce) or for regularising the shape of irregular grains (beetroot). Sowing is easier and more precise.

When a pellicle is employed, one or several thin layers of a porous and water-soluble polymer are applied. The microorganisms are protected by the pellicle whose wettable properties ensure a highly homogeneous distribution around the seed. The grain retains its shape but is smoother and circulates more easily in the sower.

It is certainly possible to treat seeds directly at the farm by soaking them in a suspension of auxiliary microorganisms with or without adjuvants. This is currently done for inoculation of leguminous seeds with rhizobia or for small-scale experiments with biopesticides. The most commonly used fixators are gum arabic and methyl cellulose but sometimes polyvinylic alcohol or a vegetable oil is employed.

Whatever the procedure used, it is necessary to initially wet the seeds, then dry them. Drying is a tricky operation to carry out on an industrial scale. It must be done very gradually to preclude alterations in the germinative faculties of the biological material, both vegetal and microbial. Conservation of seeds coated or pellicled with microorganisms also poses problems which have yet to be properly resolved.

A preconditioning technique for germination was used in the UK in the 1960s to facilitate emergence. Initially, the technique consisted of placing the seeds before sowing in an osmotic solution, the hydric potential of which had been calculated such that the tissues would imbibe it but not so much that the radicle emerged (*seed priming*). However, osmotic solutions are costly and this type of conditioning tricky to implement. More simply, it is possible to moisten the grains and mix them with a quantity of lignite or finely pulverised coal, calculated to absorb part of the moisture and yet to maintain the hydric potential just below the level required for them to germinate: this is *solid matrix priming*. As in fluid drilling, this preconditioning can be combined with a biological treatment. Thus, primed seeds of tomato and cucumber, coated with a suspension of spores of *Trichoderma harzianum* germinated and emerged better in a soil infested by *Pythium ultimum* than primed or non-primed grains, protected or not protected by a treatment with thiram (Harman and Taylor, 1988).

When cuttings or rooted plants are involved, it is also possible to supply the inoculum locally by submerging the cuttings or roots in suspensions obtained from microbial cultures or wettable powders. The adherence of bacterial or fungal propagules can be considerably improved by adding 0.5 or 1% sodium alginate to the suspension that can subsequently be hardened by dipping the roots in a solution of calcium chloride of the same concentration. The gel keeps the inoculum in contact with the roots and, further, ensures protection against microorganisms present in the external environment (Deacon and Fox, 1988). The supply of auxiliaries can also be accomplished in vitro for plants multiplied by micropropagation. This technique is utilisable for some mycorrhizal associations or for inoculation of nitrogen-fixing endophytes into sugar-cane.

Irrespective of whether the additive is generalised or localised, to be effective the biological auxiliaries must escape the antagonistic effects of other microbial populations in the soil. Competition is low in composted soil and almost nil in horticultural substrates such as rock wool, vermiculite or polyurethane but very intensive in ordinary soil. So it is preferable that auxiliaries be installed after prior disinfection with steam or a fumigant. If complete soil disinfection is not possible, incorporation of a fungicide to which the biological auxiliary is only slightly susceptible can be done concomitant with introduction of the inoculum. Properly chosen, the fungicide can also reinforce the action of the auxiliary if it is a biopesticide (Abd el Moity et al., 1982; Cole and Zvenyika, 1988). Solarisation also facilitates development of populations if they are introduced soon after removal of the plastic covers: for the same quantity of inoculum added, populations of *Trichoderma* (counted three months after their introduction) were 2 to 7 times higher in solarised field plots than in non-solarised plots (Martin et al., 1995).

Several other solutions can confer a selective advantage to the microorganisms whose introduction into the soil is desired. For example, the pH of the preparation can be altered: acidic pH values favour *Trichoderma* and hence preferably formulations with peat or lignite base should be used. For bacteria, to the contrary, an environment close to neutral is preferable. Conditioning in the form of alginate pellets and techniques of seed coating and fluid drilling allow a large variety of adjuvants. Some can be simply sources of carbon less accessible to other organisms, others may have more complex effects. Thus, polysaccharides (starch, inuline, laminarine, xylane) and mannitol induce the production of lytic enzymes by *Trichoderma harzianum*; germination of conidia of *T. koningii* is stimulated by certain fatty acids and its antagonistic activity is strongly increased in the presence of p-hydroxybenzoic acid, albeit the reasons for this effect are not understood (Nelson et al., 1988). It is also possible to imagine some chemical mediators intervening in the recognition phenomena between plants and symbiotic microorganisms. For example, it could be possible to procure a selective advantage for rhizobia introduced on seeds of legumes by adding infinitesimal quantities of Nod factors to the coating: it is known that they stimulate the infection and accelerate formation of nodules (Denarié and Joly, 1994). Even if it has no direct advantage for the bacterium, the adjuvant can render its action more effective: this happens when manganese salts are added to a treatment of wheat seeds with manganese-reducing bacteria.

Is Creation of Artificial Associative Symbioses Possible?

We saw in Chapter 8 that cells of galls and root proliferations formed on plants parasitised by agrobacteria produce opines, molecules which constitute a nutritive substrate specifically used by the parasitic bacterium. Similarly, rhizopines, suspected of playing a comparable role with reference to rhizobia, have been detected in nodules of leguminous plants. Production of nutrients for the exclusive use of a particular bacterial strain most probably confers a considerable advantage to that strain in the competition it encounters from other soil bacteria. Trials carried out with transformed *Lotus corniculatus* that became opine producers after incorporation of plasmidic genes of *A. rhizogenes*, have confirmed that strains of *A. tumefaciens* capable of catabolising these opines were effectively advantaged with respect to other populations in the rhizosphere of these plants (Guyon et al., 1993).

Trials of the same type were carried out, this time with transformed tobacco and two almost isogenic strains of *Pseudomonas fluorescens* differing solely in their ability to catabolise or not the agropine present in the root exudates of tobacco. Results confirmed that when the two strains were present together in the rhizosphere, the modified strain capable of utilising agropine colonised a larger portion of the roots than the ordinary strain (Savka and Farrand, 1997).

It therefore appears possible to establish privileged relations between cultivated plants and auxiliary microorganisms. Creation of a special trophic relation should favour the strain wanted in the rhizosphere even though that strain was not initially rhizocompetent.

Utilisation of transgenic endophytic bacteria and the creation of artificial symbioses could thus enable placing groups of useful genes inside a plant without having to modify it profoundly. Risks for the environment would apparently be limited since, in the absence of an adequate host, the manipulated microorganisms lose their selective advantage.

The rather recent discovery of the role of N-acyl-homoserine lactones (AHL) as regulators of gene transcription intervening in the functioning of bacterial communities of the rhizosphere has opened a new vista. Preliminary experiments have shown that it is possible to make the genes of AHL synthase express themselves in a plant (via the intermediary of its chloroplasts). This could lead to the production of transgenic plants capable of synthesising and excreting AHL, thus modifying the structure and behaviour of the microbial communities of their rhizosphere (Pierson, 2000).

It is nonetheless important to remember the rule of the limiting factor: even if the potentialities of the plant are excellent and the microorganisms associated with it are perfectly adapted, they will only fulfil their role if the environmental conditions themselves are favourable.

For Further Information on the Production and Utilisation of Auxiliary Microorganisms

Anonymous. 1995. Safety Considerations for Biotechnology: Scale-up of Micro-organisms as Biofertilizers. OECD, Paris, France.

Fisher T.W. and Bellows T.S. (eds.). 1999. Handbook of Biological Control: Principles and Applications of Biological Control. Academic Press, New York, NY.

Hall F.R. and Menn J.J. (eds.).1999. Methods in Biotechnology, vol. 5: Biopesticides: Use and Delivery. Humana Press, Totowa, NJ, USA.

Jarstfer A.G. and Sylvia D.M. 1993. Inoculum production and inoculation strategies for vesicular-arbuscular mycorrhizal fungi. In F.B. Metting (ed.). Soil Microbial Ecology, pp. 349–377. Marcel Dekker Inc., New York, NY.

Kennedy G.G. and Sutton T.B. (eds.). 2000. Emerging Technologies for Integrated Pest Management: Concepts, Research, and Implementation. American Phytopath. Soc., St. Paul, MN.

Lucas P. and Sarniguet A. 1998. Biological control of soil-borne pathogens with resident versus introduced antagonists: should diverging approaches become strategic convergence? In P. Barbosa (ed.). Conservation Biological Control. Academic Press, San Diego, CA.

Mathre D.E., Cook R.J. and Callan N.W. 1999. From discovery to use. Traversing the world of commercializing biocontrol agents for plant disease control. Plant Dis. 83: 972-983.

Mulongoy K., Gianinazzi S., Roger P.A. and Dommergues Y. 1992. Biofertilizers: agronomic and environmental impacts and economics. In E.J. Da Silva, C. Ratledge and A. Sasson (eds.). Biotechnology: Economic and Social Aspects, pp. 55–69. Cambridge Univ. Press, Cambridge, UK.

Taylor A.G. and Harman G.E. 1990. Concepts and technologies of selected seed treatments. Annu. Rev. Phytopathol. 28: 321–339.

Wajnberg E., Scott J.K., Quimby P.C. and Scott J. (eds.). 2000. Evaluating Indirect Ecological Effects of Biological Control. CAB International, Wallingford, UK.

Walter J.P. and Paau A.S. 1993. Microbial inoculant production and formulation. In F.B. Metting (ed.). Soil Microbial Ecology, pp. 579–594. Marcel Dekker Inc., New York, NY.

Whipps J.M. and Lumsden R.D. (eds.). 1989. Biotechnology of Fungi for Improving Plant Growth. Cambridge Univ. Press, Cambridge, UK.

Conclusion

We saw at the beginning of this book how even the existence of telluric microorganisms is dependent on organic matter synthesised by plants. We also saw subsequently that the microorganisms in turn exercise considerable influence on plant development. First, they exert a paramount, although indirect, influence via the medium in which the plants anchor their roots for fixation and nourishment: the pH, redox potential, chemical composition and soil structure depend largely on microbial activity. There is also a direct influence since the roots associate with or even shelter a multitude of microorganisms whose role is often significant in the development of plants, sometimes even from the time of their germination, as in the case of orchids. In this domain, the best mixes with the worst: some microorganisms furnish the plants with essential mineral elements, growth substances, protection against aggression that may be of biotic or abiotic origin; others attack them and limit their development. Their pathogenic effect is sometimes manifested by characteristic symptoms; most often their action is insidious. After comparing (to equivalent amounts of fertiliser) the yields obtained on land either disinfected or not with methyl bromide and protected from contamination of external origin, it was only recently realised that concealed parasitic attacks could result in up to 25% loss in production.

Despite this, the underground biological environment of plants is still too often neglected. How many phytopathologists take into account, in their trials, the interactions between the parasites they study and the omnipresent symbiotic microorganisms they ignore? How many plant-breeders care about the modifications of the microflora (sometimes considerable, as we have seen) in the rhizosphere caused by the new genotypes that result from crossings carried out by them? And how many consider the role of the mycorrhizae while creating plants more and more dependent on phosphate-containing manures? Nature is a whole. The plant world and the microbial world are only elements of the same entity in which we also participate. We earlier believed this fact could be ignored and, for many years, treated agriculture as an industrial activity like any other and considered the soil to be a chemical reactor in which it sufficed to introduce nitrogen, phosphorus and potassium in order to obtain continually increasing yields. In such a system only the capability of plants for biosynthesis was taken into account; the activity of microorganisms was mostly ignored and, as a matter of fact, completely masked by the addition of fertilisers and chemical treatments.

This concept eventually showed its limits: fertilisers are expensive, the soil is carried away by erosion, parasitic attacks have obligated more numerous treatments with more and more pesticides, groundwater layers have become polluted and the diversity of microflora and microfauna steadily suppressed in areas of large-scale farming.

A new concept, introduced by the Anglo-Saxons under the term **sustainable agriculture**, has thus become popular. This sustainable agriculture proposes to produce without wasting energy, exploit the soil without depleting or exhausting it, and intervene without causing pollution. It is aware that agricultural production depends on complex biological equilibria and that their disruption can have disastrous consequences. These equilibria cannot be considered naturally perfect, however. We have seen several instances wherein it is possible to modify them to obtain more favourable plant growth. But such modifications often take time. Whether the evolution of the level of organic matter is involved, or nitrogen fixation, mycorrhizal symbiosis or biological control, the effects of interventions are rarely manifested in a short time period, which makes their recognition difficult in the commonly used experimental procedures. Supply of fertiliser or treatment with fungicide to the contrary, produces an immediate result. However, modifications in microbial equilibria, albeit slow, are sustainable as well: when a population of rhizobia is well installed in a soil, it is redundant to regularly renew the inoculum supply. It is foreseeable that better understanding of the **ecology** of auxiliary microorganisms would permit selection from each of them of strains best adapted to the medium for which they are destined, or adjustment of environmental conditions which will propitiously orient the equilibria in the microbial communities, or else development of plant varieties capable of maintaining in their rhizosphere microbial populations most favourable for them. Thus it should be possible to emplace systems which are self-sustaining and require minimal renewable inputs.

Understandably, one can hardly expect industrial establishments which have to date responded to an annually increasing demand for fertilisers and pesticides, to show much interest in the research and production of microbial inoculums that do not require constant renewal. Such a market is none too attractive from a commercial point of view. Also, given the present state of affairs, research in the area of soil biology is mainly financed by public subsidies. Progress, stagnation or abandonment depends to some extent on society's choice and more particularly on political choice.

All things considered, the relatively recent notion of sustainable agriculture is only an extension to all the agricultural practices of what specialists involved in plant protection have long known under the name of integrated control. This involves creation of an environment that allows for installation of a new biological equilibrium favourable to crops as well as durable, using all the methods available (including chemicals so long as they have no harmful effects). This is a difficult approach because it entails a large number of techniques to manipulate the primary parameters that determine the environment and because it necessitates evaluation of the consequences of these interventions, at least in the medium term. The farmer thus becomes a manager. Is this attitude really anything new? **'Agriculture is management of the land'** people said during the times of Olivier de Serres.

Olivier de Serres (1539–1619) may be considered the first agronomist in the modern sense of the word. He rationalised usage of green manure and farmyard dung, started a system of crop rotation that allowed for more intensive agriculture without recourse to land fallow, and was the first to propose the idea of artificial meadows. He introduced the idea that land behaves like a living organism. He carried out research in his model farm at Pradel in Vivarais during the course of a long life beclouded with religious conflicts. He published the results of his thoughts and observations in a book continually republished until the nineteenth century: 'Le Théâtre d'Agriculture et Mesnage des Champs' [The Agricultural Scene and Land Management]. This book could be considered the first treatise furnishing the bases for a system of sustainable agriculture.

References

Abadie C., Edel V. and Alabouvette C. 1998. Soil suppressiveness to *Fusarium* wilt: influence of a cover-plant on density and diversity of *Fusarium* populations. Soil Biol. Biochem. 30: 643–649.

Abbass Z. and Okon Y. 1993. Plant growth promotion by *Azotobacter paspali* in the rhizosphere. Soil Biol. Biochem. 25: 1075–1083.

Abd El Moity T.H., Papavizas G.C. and Shatla M.N. 1982. Induction of new isolates of *Trichoderma harzianum* tolerant to fungicides and their experimental use for control of white rot of onion. Phytopathology 72: 396–400.

Abuzinadh R.A. and Read D.J. 1989. The role of proteins in the nitrogen nutrition of ectomycorrhizal plants. IV. The utilization of peptides by birch (*Betula pendula* L.) infected with different mycorrhizal fungi. New Phytol. 112: 55–60.

Achouak W., De Mot R. and Heulin T.1995. Purification and partial characterization of an outer membrane protein involved in the adhesion of *Rahnella aquatilis* to wheat roots. FEMS Microbiol. Ecol. 16: 19–24.

Adamczyk-Winiarska Z., Krol M. and Kobus J. 1975. Microbial oxidation of elemental sulphur in brown soil. Plant Soil 43: 95–100.

Adams P.B. 1971. *Pythium aphanidermatum* oospore germination as affected by time, temperature, and pH. Phytopathology 61: 1149–1150.

Ahmad J.S. and Baker R.1987. Rhizosphere competence of *Trichoderma harzianum*. Phytopathology 77: 182–189.

Ahmad J.S. and Baker R. 1988. Growth of rhizosphere-competent mutants of *Trichoderma harzianum* on carbon substrates. Can. J. Microbiol. 34: 807–814.

Akkermans A.D.L., Abdulkadir S. and Trinick M.J. 1978. N_2-fixing root nodules in Ulmaceae: from *Parasponia* or (and) *Trema* spp.? Plant Soil 49: 711–715.

Alabouvette C. 1986. Fusarium-wilt suppressive soils from the Châteaurenard region: review of a 10-year study. Agronomie 6: 273–284.

Alabouvette C., Lemanceau P. and Steinberg C. 1993. Recent advances in the biological control of *Fusarium* wilts. Pestic. Sci. 37: 365–373.

Alabouvette C., Schippers B., Lemanceau P. and Bakker P.A.H.M. 1998. Biological control of *Fusarium* wilts. Toward development of commercial products. In G.J. Boland and L.D. Kuykendall (eds.). Plant-Microbe Interactions and Biological Control, pp. 15–36. Marcel Dekker, Inc., New York, NY.

Alabouvette C., Edel V., Lemanceau P., Olivain C., Recorbet G. and Steinberg C. 2001. Diversity and interactions among strains of *Fusarium oxysporum*: application to biological control. In M.J. Jeger and N.J. Spence (eds.). Biotic Interactions in Plant-Pathogen Associations, pp. 131–158. CAB Internat., Wallingford, UK.

Alef K. and Kleiner D. 1986. Arginine ammonification, a simple method to estimate microbial activity potentials in soils. Soil Biol. Biochem. 18: 233–235.

Alfizar A., Martin C. and Davet P. 1992. Appearance, persistence and potential control of enhanced biodegration of iprodione and vinclozolin in the field. Agronomie 12: 733–738.

Allen E.B. and Allen M.F. 1986. Water relations of xeric grasses in the field: interactions of mycorrhizas and competition. New Phytol. 104: 559–571.

Allen R. N. and Newhook F.J. 1973. Chemotaxis of zoospores of *Phytophthora cinnamomi* to ethanol in capillaries of soil pore dimensions. Trans. Br. mycol. Soc. 61: 287–302.

Aloisi S., Pionnat S., Jacob Y., Pelloli G., Botton E., Bettachini A., Hericher D., Antonini C., Simonini L. and Poncet C. 1994. Crown gall du rosier: stratégies de lutte. Quatrième Conf. Internat. Maladies des Plantes, Bordeaux, 6–8 décembre 1994. Annales, A.N.P.P., Paris, pp. 791–797.

Alström S. and Burns R.G. 1989. Cyanide production by rhizobacteria as a possible mechanism of plant growth inhibition. Biol. Fertil. Soils 7: 232–238.

Amin K.S. and Sequeira L. 1966. Phytotoxic substances from decomposing lettuce residues in relation to the etiology of corky root of lettuce. Phytopathology 56: 1054–1061.

Amir H. and Alabouvette C. 1993. Involvement of soil abiotic factors in the mechanisms of soil suppressiveness to *Fusarium* wilts. Soil Biol. Biochem. 25: 157–164.

Anderson A.J. 1983. Isolation from root and shoot surfaces of agglutinins that show specificity for saprophytic Pseudomonads. Can. J. Bot. 61: 3438–3443.

Anderson J.P.E. and Domsch K.H.1978. Mineralization of bacteria and fungi in chloroform-fumigated soils. Soil Biol. Biochem. 10: 207–213.

Anderson J.W. 1978. Sulphur in Biology. Edward Arnold, London.

Anderson R.V., Gould W.D., Woods L.E., Cambardella C., Ingham R.E. and Coleman D.C.1983.Organic and inorganic nitrogenous losses by microbivorous nematodes in soil. Oikos 40: 75–80.

Archer S.A. 1976. Ethylene and fungal growth. Trans. Br. mycol. Soc. 67: 325–326.

Armstrong G.M. and Armstrong J.K. 1948. Non susceptible hosts as carriers of wilt Fusaria. Phytopathology 38: 808–826.

Arshad M. and Frankenberger W.T. 1990. Ethylene accumulation in soil in response to organic amendments. Soil Sci. Soc. Am. J. 54: 1026–1031.

Asimi S., Gianinazzi-Pearson V. and Gianinazzi S. 1980. Influence of increasing soil phosphorus levels on interactions between vesicular-arbuscular mycorrhizae and *Rhizobium* in soybeans. Can. J. Bot. 58: 2200–2205.

Bailey D.J. and Gilligan C.A. 1997. Biological control of pathozone behaviour and disease dynamics of *Rhizoctonia solani* by *Trichoderma viride*. New Phytol. 136: 359–367.

Baker R. 1988. *Trichoderma* spp. as plant-growth stimulants. CRC Crit. Rev. Biotechnol. 7: 97–106.

Baker R., Maurer C.L. and Maurer R.A. 1967. Ecology of plant pathogens in soil. VII. Mathematical models and inoculum density. Phytopathology 57: 662–666.

Bakken L.R. and Olsen R.A. 1989. DNA-content of soil bacteria of different cell size. Soil Biol. Biochem. 21: 789–793.

Baldani J.I., Oliveira A.L.M., Guimarães S.L., Baldani V.L.D., Reis Jr. F.B., Silva L.G., Reis V.M., Teixeira K.R.S. and Döbereiner J. 2000. Biological nitrogen fixation (BNF) in non-leguminous plants: the role of endophytic diazotrophs. In F.O. Pedrosa, M. Hungria, M.G. Yates and W.E. Newton (eds.). Nitrogen Fixation: from Molecules to Crop Productivity, pp. 653–658. Kluwer Acad. Publ., Dordrecht, The Netherlands.

Baltruschat H. and Schonbeck F. 1975. Untersuchungen über den Einfluss der endotrophen Mycorrhiza auf den Befall von Tabak mit *Thielaviopsis basicola*. Phytopathol. Z. 84: 172–188.

Barber D.A. and Lee R.B. 1974. The effect of micro-organisms on the absorption of manganese by plants. New Phytol. 73: 47–106.

Barclay W.R. and Lewin R.A. 1985. Microalgal polysaccharide production for the conditioning of agricultural soils. Plant Soil 88: 159–169.

Barron G.L. 1992. Lignolytic and cellulolytic fungi as predators and parasites. In G.C. Carroll and D.T. Wicklow (eds.). The Fungal Community, pp. 311–326. Marcel Dekker, New York, NY.

Barron G.L. and Dierkes Y. 1977. Nematophagous fungi: *Hohenbuehelia*, the perfect state of *Nematoctonus*. Can. J. Bot. 55: 3054–3062.

Bateman D.F. and Beer S.V. 1965. Simultaneous production and synergistic action of oxalic acid and polygalacturonase during pathogenesis by *Sclerotium rolfsii*. Phytopathology 55: 204–211.

Bécard G. and Piché Y. 1989. New aspects on the acquisition of biotrophic status by a vesicular-arbuscular mycorrhizal fungus *Gigaspora margarita*. New Phytol. 112: 77–83.

Becker J.O., Hedges R.W. and Messens E. 1985. Inhibitory effect of pseudobactin on the uptake of iron by higher plants. Appl. Environ. Microbiol. 49: 1090–1093.

Beckman C.H., Halmos S. and Mace M.E. 1962. The interaction of host, pathogen and soil temperature in relation to susceptibility to Fusarium wilt of bananas. Phytopathology 52: 134–140.

Beemster A.B.R., Bollen G.J., Gerlagh M., Ruissen M.A., Schippers B. and Tempel A. (eds.). 1991. Biotic Interactions and Soil-borne Diseases. Elsevier, Amsterdam.

Belcher J.V. and Hussey R.S. 1977. Influence of *Tagetes patula* and *Arachis hypogaea* on *Meloidogyne incognita*. Plant Dis. Rep. 61: 525–528.

Bell A.A. and Presley J.T. 1969. Temperature effects upon resistance and phytoalexin synthesis in cotton inoculated with *Verticillium albo-atrum*. Phytopathology 59: 1141–1146.

Besnard O. and Davet P. 1993. Mise en évidence de souches de*Trichoderma* spp. à la fois antagonistes de *Pythium ultimum* et stimulatrices de la croissance des plantes. Agronomie 13: 413– 421.

Bird G.W., Brooks D.L., Perry C.E., Futral J.G., Canerday T.D. and Boswell F.C. 1974. Influence of subsoiling and soil fumigation on the cotton stunt disease complex, *Hoplolaimus columbus* and *Meloidogyne incognita*. Plant Dis. Rep. 58: 541–544.

Bishop C.D. and Cooper R.M. 1983. An ultrastructural study of root invasion in three vascular wilt diseases. Physiol. Plant Pathol. 22: 15–27.

Bishop P.E. and Joerger R.D. 1990. Genetics and molecular biology of alternative nitrogen fixation systems. Annu. Rev. Plant Physiol. Plant Mol. Biol. 41: 109–125.

Bishop P.E., Jarlenski D.M.L. and Hetherington D.R. 1982. Expression of an alternative nitrogen fixation system in *Azotobacter vinelandii*. J. Bacteriol. 150: 1244–1251.

Blake D.R. and Rowland F.S. 1988. Continuing world-wide increase in the tropospheric methane, 1978 to 1987. Science 239: 1129–1131.

Blal B., Morel C., Gianinazzi-Pearson V., Fardeau J.C. and Gianinazzi S. 1990. Influence of vesicular-arbuscular mycorrhizae on phosphate fertilizer efficiency in two tropical acid soils planted with micropropagated oil palm (*Elaeis guineensis* Jacq.). Biol. Fertil. Soils 9: 43–48.

Blok W.J., Lamers J.G., Termorshuizen A.J. and Bollen G.J. 2000. Control of soil-borne plant pathogens by incorporating fresh organic amendments followed by tarping. Phytopathology 90: 253–259.

Bolton H. and Elliott L.F. 1989. Toxin production by a rhizobacterial *Pseudomonas* sp. that inhibits wheat root growth. Plant Soil 114: 269–278.

Bonnet P., Bourdon E., Ponchet M., Blein J.-P. and Ricci P. 1996. Acquired resistance triggered by elicitins in tobacco and other plants. Eur. J. Plant Pathol. 102: 181–192.

Bottner P. and Billès G. 1987. La rhizosphère: site d'interactions biologiques. Rev. Ecol. Biol. Sol 24: 369–388.

Bouhot D. 1975. Technique sélective et quantitative d'estimation du potentiel infectieux des sols, terreaux et substrats infectés par *Pythium* sp. Ann. Phytopathol. 7: 155–158.

Bouhot D. 1979a. Un test biologique à deux niveaux pour l'étude des fatigues de sol. Application à l'étude des nécroses des racines de céleri-rave. Ann. Phytopathol. 11: 95–109.

Bouhot D. 1979b. Estimation of inoculum density and inoculum potential: techniques and their value for disease prediction. In B. Schippers and W. Gams (eds.). Soil-borne Plant Pathogens, pp. 21–34. Academic Press, London.

Bouhot D. 1980. Le potentiel infectieux des sols (soil infectivity). Thèse de Doctorat ès Sciences, Université de Nancy.

Bouhot D. 1981. Induction d'une résistance biologique aux *Pythium* dans les sols par l'apport d'une matière organique. Soil Biol. Biochem. 13: 269–274.

Bourquin A.W. 1990. Bioremediation of hazardous waste. Biofutur 93: 24–35.

Bowen G.D. and Rovira A.D. 1976. Microbial colonization of plant roots. Annu. Rev. Phytopathol. 14:121–144.

Bowers J.H. and Parke J.L. 1993. Colonization of pea (*Pisum sativum* L.) taproots by *Pseudomonas fluorescens*: effect of soil temperature and bacterial motility. Soil Biol. Biochem. 25: 1693–1701.

Braun P.G. 1991. The combination of *Cylindrocarpon lucidum* and *Pythium irregulare* as a possible cause of apple replant disease in Nova Scotia. Can. J. Plant Pathol. 13: 291–297.

Bremner J.M. and Mulvaney R.L. 1978. Urease activity in soils. In R.G. Burns (ed.). Soil Enzymes, pp. 149–196. Academic Press, London.

Bremner J.M. and Chai H.S. 1989. Effects of phosphoroamides on ammonia volatilization and nitrite accumulation in soils treated with urea. Biol. Fertil. Soils 8: 227–230.

Brisbane P.G. and Rovira A.D. 1988. Mechanisms of inhibition of *Gaeumannomyces graminis* var. *tritici* by fluorescent pseudomonads. Plant Pathol. 37: 104–111.

Brown M.E. 1974. Seed and root bacterization. Annu. Rev. Phytopathol. 12: 181–197.

Brown S.M., Kepner J. and Smart G.C. 1985. Increased crop yields following application of *Bacillus penetrans* to field plots infested with *Meloidogyne incognita*. Soil Biol. Biochem. 17: 483–486.

Buresh R.J., Samson M.I. and De Datta S.K. 1993. Quantification of denitrification in flooded soils as affected by rice establishment method. Soil Biol. Biochem. 25: 843–848.

Burke D.W., Miller D.E., Holmes L.D. and Barker A.W. 1972. Counteracting bean root rot by loosening the soil. Phytopathology 62: 306–309.

Burns R.G. 1978. Enzyme activity in soil: some theoretical and practical considerations. In R.G Burns (ed.). Soil Enzymes, pp. 295–340. Academic Press, London.

Burns R.G. 1982. Enzyme activity in soil: location and a possible role in microbial ecology. Soil Biol. Biochem. 14: 423–427.

Bushby H.V.A. and Marshall K.C. 1977. Water status of *Rhizobia* in relation to their susceptibility to desiccation and to their protection by montmorillonite. J. Gen. Microbiol. 99: 19–27.

Caetano-Anolles G., Crist-Estes D.K. and Bauer W.D. 1988. Chemotaxis of *Rhizobium meliloti* to the plant flavone luteolin requires functional nodulation genes. J. Bacteriol. 170: 3164–3169.

Callot G., Chamayou H., Maertens C. and Salsac L. 1982. Les interactions sol-racine. Incidence sur la nutrition minérale. 'Mieux comprendre'. INRA, Paris.

Cameron J.N. and Carlile M.J. 1978. Fatty acids, aldehydes and alcohols as attractants for zoospores of *Phytophthora palmivora*. Nature 271: 448–449.

Campbell R.N. 1996. Fungal transmission of plant viruses. Annu. Rev. Phytopathol. 34: 87–108.

Campbell R.N. and Ephgrave J.M. 1983. Effect of bentonite clay on the growth of *Gauemannomyces graminis* var. *tritici* and on its interaction with antagonistic bacteria. J. Gen. Microbiol. 129: 771– 777.

Campbell R. N., Sim S.T. and Lecoq H. 1995. Virus transmission by host-specific strains of *Olpidium bornovanus* and *Olpidium brassicae*. Europ. J. Pl. Pathol. 101: 273–282.

Candole B.L. and Rothrock C.S. 1997. Characterization of the suppressiveness of hairy vetch-amended soils to *Thielaviopsis basicola*. Phytopathology 87: 197–202.

Caplan A., Herrera-Estrella L., Inze D., Van Haute E., Van Montagu M., Schell J. and Zambryski P. 1983. Introduction of genetic material into plant cells. Science 222: 815–821.

Cappaert M.R., Powelson M.L., Christensen N. W., Stevenson W.R. and Rouse D.I. 1994. Assessment of irrigation as a method of managing potato early dying. Phytopathology 84: 792–800.

Castejon-Munoz M. and Bollen G.J. 1993. Induction of heat resistance in *Fusarium oxysporum* and *Verticillium dahliae* caused by exposure to sublethal heat treatments. Neth. J. Plant Pathol. 99: 77–84.

Čatska V. and Vančura V. 1980. Volatile and gaseous metabolites released by germinating seeds of lentil and maize cultivars with different susceptibilities to fusariosis and smut. Folia Microbiol. 25: 177 –181.

Čatska V., Vančura V., Hudska G. and Prikryl Z. 1982. Rhizosphere micro-organisms in relation to the apple replant problem. Plant Soil 69: 187–197.

Cavalcante V.A. and Döbereiner J. 1988. A new acid-tolerant nitrogen-fixing bacterium associated with sugarcane. Plant Soil 108: 23–31.

Cavelier N., Lucas P. and Boulch G. 1985. Evolution du complexe parasitaire constitué par *Rhizoctonia cerealis* Van der Hoeven et *Pseudocercosporella herpotrichoides* (Fron) Deighton, champignons parasites de la base des tiges de céréales. Agronomie 5: 693–700.

Cayrol J.-C. and Quilès C. 1985. Les nématodes libres: auxiliaires utiles en agriculture. P.H.M.-Rev. Hortic. 260: 27–29.

Cerny G. 1976. Method for the distinction of Gramnegative from Grampositive Bacteria. Eur. J. Appl. Microbiol. 3: 223–225.

Chakly M. and Berthelin J. 1982. Rôle d'une ectomycorhize 'Pisolithus tinctorius-Pinus caribea' et d'une bactérie rhizosphérique sur la mobilisation du phosphore de phosphates minéraux et organiques insolubles. In S. Gianinazzi et al. (eds.). Les Mycorhizes, partie intégrante de la plante: biologie et perspectives d'utilisation. INRA 'les Colloques' 13: 215–220.

Chao W.L., Li R.K. and Chang W.T. 1988. Effect of root agglutinin on microbial activities in the rhizosphere. Appl. Environ. Microbiol. 54: 1838–1841.

Chen W., Hoitink H.A.J. and Madden L.V. 1988. Microbial activity and biomass in container media for predicting suppressiveness to damping-off caused by *Pythium ultimum*. Phytopathology 78: 1447–1450.

Chet I. 1987. *Trichoderma.* Application, mode of action, and potential as a biocontrol agent of soilborne plant pathogenic fungi. In I. Chet (ed.). Innovative Approaches to Plant Disease Control, pp. 137–160. John Wiley and Sons, New York, NY.

Chinn S.H.F. and Tinline R.D. 1964. Inherent germinability and survival of spores of *Cochliobolus sativus.* Phytopathology 54: 349–352.

Chou C.-H. and Muller C.H. 1972. Allelopathic mechanisms of *Arctostaphylos glandulosa* var. *zacaensis.* Am. Midl. Nat. 88: 324–347.

Clarholm M. 1985. Interactions of bacteria, protozoa and plants leading to mineralization of soil nitrogen. Soil Biol. Biochem. 17: 181–187.

Clerjeau M. 1976. Exigences thermiques de croissance et d'agressivité de divers isolats de *Pyrenochaeta lycopersici* Schneider et Gerlach. Ann. Phytopathol. 8: 9–15.

Colbach N. and Saur L. 1998. Influence of crop management on eyespot development and infection cycles of winter wheat. Europ. J. Pl. Pathol. 104: 37–48.

Cole J.S. and Zvenyika Z. 1988. Integrated control of *Rhizoctonia solani* and *Fusarium solani* in tobacco transplants with *Trichoderma harzianum* and triadimenol. Plant Pathol. 37: 271–277.

Coley-Smith J.R. 1987. Alternative methods of controlling white rot disease of *Allium.* In I. Chet (ed.). Innovative Approaches to Plant Disease Control, pp. 161–177. John Wiley and Sons, New York, NY.

Coley-Smith J.R. 1990. White rot disease of *Allium*: problems of soil-borne diseases in microcosm. Plant Pathol. 39: 214–222.

Conway K.E. 1986. Use of fluid-drilling gels to deliver biological control agents to soil. Plant Dis. 70: 835–839.

Cook R.J. 1981. The influence of rotation crops on take-all decline phenomenon. Phytopathology 71: 189–192.

Cook R.J. and Baker K.F. 1983. The Nature and Practice of Biological Control of Plant Pathogens. Am. Phytopathol. Soc., St. Paul, MN.

Cook R.J. and Veseth R.J. 1991. Wheat Health Management. Am. Phytopathol. Soc., St. Paul, MN.

Cook R.J. and Weller D.M. 1987. Management of take-all in consecutive crops of wheat or barley. In I. Chet (ed.). Innovative Approaches to Plant Disease Control, pp. 41–76. John Wiley and Sons, New York, NY.

Cooke R.C. and Rayner A.D.M. 1984. Ecology of Saprotrophic Fungi. Longman, New York, NY.

Coosemans J. 1995. Control of algae in hydroponic systems. In A. Vanachter (ed.). Fourth International Symposium on soil and substrate infestation and disinfestation, Leuven, Belgique. Acta Hortic. 382: 263 –268.

Corbett D.C.M. and Webb R.M. 1970. Plant and soil nematode population changes in wheat grown continuously in ploughed and in unploughed soil. Ann. Appl. Biol. 65: 327–335.

Corman A., Couteaudier Y., Zegerman M. and Alabouvette C. 1986. Réceptivité des sols aux fusarioses vasculaires: méthode statistique d'analyse des résultats. Agronomie 6: 751–757.

Couteaux M.M. 1967. Une technique d'observation des Thécamoebiens du sol pour l'estimation de leur densité absolue. Rev. Ecol. Biol. Sol 4: 593–596.

Couteaux M.M. and Bottner P. 1994. Biological interactions between fauna and the microbial community in soils. In K. Ritz, J. Dighton and K.E. Giller (eds.). Beyond the Biomass, pp. 159–172. John Wiley & Sons, New York, NY.

Crozat Y., Cleyet-Marel J.-C., Giraud J.-J. and Obaton M. 1982. Survival rates of *Rhizobium japonicum* populations introduced into different soils. Soil Biol. Bochem 14: 401–405.

Csinos A. and Hendrix J.W. 1978. Parasitic and nonparasitic pathogenesis of tomato plants by *Pythium myriotylum*. Can. J. Bot. 56: 2334–2339.

da Luz W.C., Bergstrom G.C. and Stockwell C.A. 1998. Seed-applied bioprotectants for control of seedborne *Pyrenophora tritici-repentis* and agronomic enhancement of wheat. Can. J. Plant Pathol. 20: 384–386.

Danso S.K.A., Keya S.O. and Alexander M. 1975. Protozoa and the decline of *Rhizobium* populations added to soil. Can. J. Microbiol. 21: 884–895.

Dashman T. and Stotzky G. 1986. Microbial utilization of amino acids and a peptide bound on homoionic montmorillonite and kaolinite. Soil Biol. Biochem. 18: 5–14.

Davet P. 1973. Distribution et évolution du complexe parasitaire des racines de tomate dans une région du Liban où prédomine le *Pyrenochaeta lycopersici* Gerlach et Schneider. Ann. Phytopathol. 5: 53–63.

Davet P. 1976a. Etude de quelques interactions entre les champignons associés à la maladie des racines liégeuses de la tomate. I. Phase non parasitaire. Ann. Phytopathol. 8: 171–182.

Davet P. 1976b. Etude de quelques interactions entre les champignons associés à la maladie des racines liégeuses de la tomate. II. Phase parasitaire. Ann. Phytopathol. 8: 183–190.

Davet P. 1976c. Etude d'interactions entre le *Fusarium oxysporum* et le *Pyrenochaeta lycopersici* sur les racines de la tomate. Ann. Phytopathol. 8: 191–202.

Davet P. 1979. Technique pour l'analyse des populations de *Trichoderma* et de *Gliocladium virens* dans le sol. Ann. Phytopathol. 11: 529–533.

Davet P. 1991. La solarisation: une technique non polluante de lutte contre les champignons du sol. Bul. Sem. 116: 2–3.

Davet P. 1995. Biodégradation accélérée des produits phytosanitaires. P.H.M.-Rev. Hortic. 362: 64 –67.

Davet P. and Camporota P. 1986. Etude comparative de quelques méthodes d'estimation de l'aptitude à la compétition saprophytique dans le sol des *Trichoderma*. Agronomie 6: 575–581.

Davet P. and Ravisé A. 1965. Les maladies de racines du riz en Afrique occidentale. Congrès de la Protection des cultures tropicales. C.R. publiés par la Chambre de Commerce et d'Industrie de Marseille, 23–27 mars 1965, pp. 809–812.

Davet P. and Serieys H. 1987. Relation entre la teneur en sucres réducteurs des tissus du tournesol (*Helianthus annuus* L.) et leur invasion par *Macrophomina phaseolina* (Tassi) Goid. J. Phytopathol. 118: 212–219.

Davet P., Herbach M., Rabat M. and Piquemal G. 1986. Effet de quelques facteurs intrinsèques ou extrinsèques d'affaiblissement des tournesols sur leur sensibilité au dessèchement précoce. Agronomie 6: 803–810.

De Mot R. and Vanderleyden J. 1991. Purification of a root-adhesive outer membrane protein of root-colonizing *Pseudomonas fluorescens*. FEMS Microbiol. Lett. 81: 323–328.

de Souza M.P., Huang C.P.A., Chee N. and Terry N. 1999. Rhizosphere bacteria enhance the accumulation of selenium and mercury in wetland plants. Planta 209: 259–263.

Deacon J.W. and Donaldson S.P. 1993. Molecular recognition in the homing responses of zoosporic fungi, with special reference to *Pythium* and *Phytophthora*. Mycol. Res. 97: 1153–1171.

Deacon J.W. and Fox F. 1988. Delivery of microbial inoculants into the root zone of transplant crops. Proc. British Crop Protection Conference (Pests and Diseases), Brighton 2: 645–653.

Deacon J.W. and Lewis S.J. 1982. Natural senescence of the root cortex of spring wheat in relation to susceptibility to common root rot (*Cochliobolus sativus*) and growth of a free-living nitrogen-fixing bacterium. Plant Soil 66: 13–20.

Deacon J.W. and Mitchell R.T. 1985. Toxicity of oat roots, oat root extracts, and saponins to zoospores of *Pythium* spp. and other fungi. Trans. Br. mycol. Soc. 84: 479–487.

Deacon J.W. and Scott D.B. 1985. *Rhizoctonia solani* associated with crater disease (stunting) of wheat in South Africa. Trans. Br. mycol. Soc. 85: 319–327.

Del Gallo M. and Haegi A. 1990. Characterization and quantification of exocellular polysaccharides in *Azospirillum brasilense* and *Azospirillum lipoferum*. Symbiosis 9: 155–161.

Dénarié J. and Joly P.B. 1994. La fixation de l'azote. II. Les enjeux de la recherche. Biofutur 133: 30 –34.

Deransart C., Chaumat E., Cleyet-Marcel J.C., Mousain D. and Labarère J. 1990. Purification assay of phosphatases secreted by *Hebeloma cylindrosporum* and preparation of polyclonal antibodies. Symbiosis 9: 185–194.

Devine K.J. and Jones P.W. 2000. Response of *Globodera rostochiensis* to exogenously applied hatching factors in soil. Ann. Appl. Biol. 137: 21–29.

Didelot D., Muchembled C., Lambert C.L. and Mai Dao. 1994. Fatigue des sols en culture betteravière. Importance des nécroses radicellaires et identification des agents responsables. Quatrième Conf. Internat. Maladies des Plantes, Bordeaux, 6-8 décembre 1994, A.N.P.P., Paris, pp. 359–366.

Dijst G. 1990. Effect of volatile and unstable exudates from underground potato plant parts on sclerotium formation by *Rhizoctonia solani* AG-3 before and after haulm destruction. Neth. J. Plant Pathol. 96: 155–170.

Dillard H.R. and Grogan R.G. 1985. Relationship between sclerotial spatial pattern and density of *Sclerotinia minor* and the incidence of lettuce drop. Phytopathology 75: 90–94.

Dobbs C.G. and Hinson W.H. 1953. A widespread fungistasis in soils. Nature, Lond., 172: 197– 199.

Dommergues Y. 1962. Contribution à l'étude de la dynamique microbienne des sols en zone semi-aride et en zone tropicale sèche. Ann. agron. 13: 265–324 and 391–468.

Dommergues Y. and Mangenot F. 1970. Ecologie microbienne du sol. Masson, Paris.

Dommergues Y., Jacq V. and Beck G. 1969. Influence de l'engorgement sur la sulfato-réduction rhizosphérique dans un sol salin. C.R. Acad. Sci., Ser. D: Sci. nat. FRA, Paris 268: 605–608.

Domsch K.H. 1985. Interactions with microflora. In I.M. Smith (ed.). Fongicides et protection des plantes: 100 ans de progrès, pp. 143–148. BCPC Publ., Croydon.

Domsch K.H., Jagnow G. and Anderson T.-H. 1983. An ecological concept for the assessment of side-effects of agrochemicals on soil microorganisms. Residue Rev. 86: 65–105.

Doupnik B. and Boosalis M.G. 1980. Ecofallow, a reduced tillage system, and plant diseases. Plant Dis. 64:31–35.

Ducel G. 1987. La décontamination cutanée. Antibiosept 4: 1–8.

Duchaufour P. 1980. Ecologie de l'humification et pédogénèse des sols forestiers. In P. Pesson (ed.). Actualités d'écologie forestière: sol, flore, faune, pp. 177–201. Gauthier-Villars, Paris.

Duchesne L.C., Ellis B.E. and Peterson R.L. 1989. Disease suppression by the ectomycorrhizal fungus *Paxillus involutus*: contribution of oxalic acid. Can. J. Bot. 67: 2726–2730.

Duffy B.K. and Défago G. 1997. Zinc improves biocontrol of *Fusarium* crown and root rot of tomato by *Pseudomonas fluorescens* and represses the production of pathogen metabolites inhibitory to bacterial antibiotic biosynthesis. Phytopathology 87: 1250–1257.

Duijff B.J., Pouhair D., Olivain C., Alabouvette C. and Lemanceau P. 1998. Implication of systemic induced resistance in the suppression of fusarium wilt of tomato by *Pseudomonas fluorescens* WCS417r and by nonpathogenic *Fusarium oxysporum* Fo 47. Eur. J. Plant Pathol. 104: 903–910.

Duponnois R. and Garbaye J. 1991. Effect of dual inoculation of Douglas fir with the ectomycorrhizal fungus *Laccaria laccata* and mycorrhization helper bacteria (MHB) in two bare-root forest nurseries. Plant Soil 138: 169–176.

Durand A. 1983. Les potentialités de la culture à l'état solide en vue de la production de microorganismes filamenteux. In B. Dubos and J.M. Olivier (eds.). Les antagonismes microbiens. Modes d'action et application à la lutte biologique contre les maladies des plantes. INRA, 'les Colloques' 18: 263–277.

Durand G. 1966. Thèse de Doctorat ès Sciences, Université de Toulouse.

Dutta B.K. and Isaac I. 1979. Effects of inorganic amendments (N, P and K) to soil on the rhizosphere microflora of *Antirrhinum* plants infected with *Verticillium dahliae* Kleb. Plant Soil 52: 561–569.

Eady R.R., Robson R.L. and Smith B.E. 1988. Alternative and conventional nitrogenases. In J.A. Cole and S.J. Ferguson (eds.). The Nitrogen and Sulphur Cycles, pp. 363–382. Cambridge University Press, Cambridge, England.

Ebben M.H. and Last F.T. 1975. Incidence of root rots—their prediction and relation to yield losses: a glasshouse study. In G.W. Bruehl (ed.). Biology and Control of Soil-borne Plant Pathogens, pp. 6–10. Am. Phytopathol. Soc., St. Paul, MN.

Edel V., Steinberg C., Gautheron N. and Alabouvette C. 1997. Populations of nonpathogenic *Fusarium oxysporum* associated with roots of four plant species compared to soilborne populations. Phytopathology 87: 693–697.

Egener T., Hurek T. and Reinhold-Hurek B. 1998. Use of green fluorescent protein to detect expression of *nif* genes of *Azoarcus* sp. BH72, a grass-associated diazotroph, on rice roots. Mol. Plant Microbe Interact. 11: 71–75.

El Aziz R., Angle J.S. and Chaney R.L. 1991. Metal tolerance of *Rhizobium meliloti* isolated from heavy-metal contaminated soils. Soil Biol. Biochem. 23: 795–798.

Elarosi H. 1958. Fungal associations. III. The role of pectic enzymes on the synergistic relation between *Rhizoctonia solani* Kühn and *Fusarium solani* Snyder and Hansen, in the rotting of potato tubers. Ann. Bot. N.S. 22: 399–416.

Elmer W.H., Johnson D.A. and Mink G.I. 1996. Epidemiology and management of the diseases causal to Asparagus decline. Plant Dis. 80: 117–125.

Entwistle A.R., Merriman P.R., Munasinghe H.L. and Mitchell P. 1982. Diallyl-disulphide to reduce the numbers of sclerotia of *Sclerotium cepivorum* in soil. Soil Biol. Biochem. 14: 229–232.

Fawe A., Abou-Zaid M., Menzies J.G. and Bélanger R.R. 1998. Silicon-mediated accumulation of flavonoid phytoalexins in cucumber. Phytopathology 88: 396–401.

Fehrmann H. 1988. *Pseudocercosporella herpotrichoides* (Fron) Deighton. In I.M. Smith, J. Dunez, R.A. Lelliott, D.H. Phillips and S.A. Archer (eds.). European Handbook of Plant Diseases, pp. 409–412. Blackwell Sci. Publ., Oxford, UK.

Ferriss R.S. 1982. Relationship of infection and damping-off of soyabean to inoculum density of *Pythium ultimum*. Phytopathology 72: 1397–1403.

Forsberg J.L. 1959. Relationship of the bulb mite *Rhizoglyphus echinopus* to bacterial scab of gladiolus. Phytopathology 49: 538.

Fravel D.R. and Roberts D.P. 1991. *In situ* evidence for the role of glucose oxidase in the biocontrol of Verticillium wilt by *Talaromyces flavus*. Biocontrol. Sci. Technol. 1: 91–99.

Fredrickson J.K. and Elliott L.F. 1985. Effects on winter wheat seedling growth by toxin-producing rhizobacteria. Plant Soil 83: 399–409.

Fries N. 1973. Effects of volatile organic compounds on the growth and development of fungi. Trans. Br. mycol. Soc. 60: 1–21.

Fries N., Serck-Hanssen K., Dimberg L.H. and Theander O. 1987. Abietic acid, an activator of basidiospore germination in ectomycorrhizal species of the genus *Suillus* (Boletaceae). Experiment. Mycol. 11: 360–363.

Furuya H., Takahashi T. and Matsumoto T. 1999. Suppression of *Fusarium solani* f. sp. *phaseoli* on bean by aluminum in acid soils. Phytopathology 89: 47–52.

Gadd G.M. and Griffiths A.J. 1978. Microorganisms and heavy metal toxicity. Microb. Ecol. 4: 303–317.

Gamliel A. and Stapleton J.J. 1993. Characterization of antifungal volatile compounds evolved from solarized soil amended with cabbage residues. Phytopathology 83: 899–905.

Garbaye J. 1991. Utilisation des mycorhizes en sylviculture. In D.G. Strullu (ed.). Les mycorhizes des arbres et plantes cultivées, pp. 197–248. 'Technique et Documentation', Lavoisier, Paris.

Garbaye J. 1994. Helper bacteria: a new dimension to the mycorrhizal symbiosis. New Phytol. 128: 197–210.

Garbaye J. and Bowen G.D. 1989. Stimulation of ectomycorrhizal infection of *Pinus radiata* by some microorganisms associated with the mantle of ectomycorrhizas. New Phytol. 112: 383–388.

Garber R.H., Jorgenson E.C., Smith S. and Hyer A.H. 1979. Interaction of population levels of *Fusarium oxysporum* f. sp. *vasinfectum* and *Meloidogyne incognita* on cotton. J. Nematol. 11: 133–137.

Gardner J.M., Chandler J.L. and Feldman A.W. 1984. Growth promotion and inhibition by antibiotic-producing fluorescent pseudomonads on citrus roots. Plant Soil 77: 103–113.

Gardner J.M., Chandler J.L. and Feldman A.W. 1985. Growth responses and vascular plugging of citrus inoculated with rhizobacteria and xylem-resident bacteria. Plant Soil 86: 333–345.

Garrett S.D. 1960. Biology of Root-infecting Fungi. Cambridge University Press, Cambridge, UK.

Garrett S.D. 1966. Cellulose-decomposing ability of some cereal foot-rot fungi in relation to their saprophytic survival. Trans. Br. mycol. Soc. 49: 57–68.

Garrett S.D. 1970. Pathogenic Root-infecting Fungi. Cambridge University Press, Cambridge, UK.

Gebhard F. and Smalla K. 1998. Transformation of *Acinetobacter* sp. strain BD413 by transgenic sugar beet DNA. Appl. Environ. Microbiol. 64: 1550–1554.

Gerdemann J.W. and Nicolson T.H. 1963. Spores of mycorrhizal *Endogone* species extracted from soil by wet sieving and decanting. Trans. Br. mycol. Soc. 46: 235–244.

Gerlagh M. 1968. Introduction of *Ophiobolus graminis* into new polders and its decline. Neth. J. Plant Pathol. 74, suppl. 2: 1–97.

Gianinazzi-Pearson V. 1992. Recent research into the cellular, molecular and genetical bases of compatible host-fungus interactions in (vesicular) arbuscular endomycorrhiza: approaches and advances. In: Interactions plantes—microorganismes (Congrès de Dakar, 17–22 fév 1992), pp. 253–263. Fondation Internationale pour la science, Stockholm.

Gilbert G.S., Handelsman J. and Parke J.L. 1994. Root camouflage and disease control. Phytopathology 84: 222–225.

Gillespie-Sasse L.M.J., Almassi F., Ghisalberti E.L. and Sivasithamparam K. 1991. Use of a clean seedling assay to test plant growth promotion by exudates from a sterile red fungus. Soil Biol. Biochem. 23: 95–97.

Gilligan C.A. 1979. Modeling rhizosphere infection. Phytopathology 69: 782–784.

Gilligan C.A. 1987. Epidemiology of soil-borne plant pathogens. In: M.S. Wolfe and C.E. Caten (eds.). Populations of Plant Pathogens: Their Dynamics and Genetics, pp. 119–133. Blackwell Sci. Publ., Oxford, UK.

Gilligan C.A. 1990. Mathematical models of infection. In: J.M. Lynch (ed.). The Rhizosphere, pp. 207–232. John Wiley and Sons, Chichester, UK.

Glass D.J. 1993. Commercialization of soil microbial technologies. In: F.B. Metting, Jr. (ed.). Soil Microbial Ecology, pp. 595–618. Marcel Dekker Inc., New York, NY.

Golden J.K. and Van Gundy S.D. 1975. A disease complex of okra and tomato involving the nematode *Meloidogyne incognita*, and the soil-inhabiting fungus, *Rhizoctonia solani*. Phytopathology 65: 265–273.

Goos R.J. and Fairlie T.E. 1988. Effect of ammonium thiosulfate and liquid fertilizer droplet size on urea hydrolysis. Soil Sci. Soc. Am. J. 52: 522–524.

Goudriaan J. 1992. Où va le gaz carbonique? Le rôle de la végétation. La Recherche 23: 597–604.

Gough C., Bonfante P. and Dénarié J. 2000. Can the study of endomycorrhizae open new avenues of research in symbiotic nitrogen fixation? In F.O. Pedrosa, M. Hungria, M.G. Yates and W.E. Newton (eds.). Nitrogen Fixation: from Molecules to Crop Productivity, pp. 653–658. Kluwer Acad. Publ., Dordrecht, The Netherlands.

Gregory P.A. 1948. The multiple infection transformation. Ann. Appl. Biol. 35: 412–417.

Grente J. 1965. Les formes hypovirulentes d'*Endothia parasitica* et les espoirs de lutte contre le chancre du châtaignier. C.R. Acad. Agric. 51: 1033–1037.

Griffin D.M. 1972. Ecology of Soil Fungi. Chapman and Hall, London.

Grime J.P., Mackey J.M.L., Hillier S.I I. and Read D.J. 1987. Floristic diversity in a model system using experimental microcosms. Nature 328: 420–422.

Grimes H.D. and Mount M.S. 1984. Influence of *Pseudomonas putida* on nodulation of *Phaseolus vulgaris*. Soil Biol. Biochem. 16: 27–30.

Guckert A., Choné T. and Jacquin F. 1975. Microflore et stabilité structurale des sols. Rev. Ecol. Biol. Sol. 12: 211–223.

Gupta V.V.S.R. and Germida J.J. 1988. Distribution of microbial biomass and its activity in different soil aggregate size classes as affected by cultivation. Soil Biol. Biochem. 20: 777–786.

Gur A. and Cohen Y. 1989. The peach replant problem. Some causal agents. Soil Biol. Biochem. 21: 829–834.

Guyon P., Petit A., Tempé J. and Dessaux Y. 1993. Transformed plants producing opines specifically promote growth of opine-degrading *Agrobacteria*. Mol. Plant-Microbe Interact. 6: 92–98.

Haahtela K., Laakso T., Nurmiaho-Lassila E.L., Rönkkö R. and Korhonen T.K. 1988. Interactions between N$_2$-fixing enteric bacteria and grasses. Symbiosis 6: 139–150.

Hadas R. and Okon Y. 1987. Effect of *Azospirillum brasilense* inoculation on root morphology and respiration in tomato seedlings. Biol. Fertil. Soils 5: 241–247.

Hagnère C. and Harf C. 1992. Bactéries résistantes au mercure hébergées par des amibes libres de l'environnement hydrique. C.R. 3ème Colloque Soc. Fr. Microbiol, Lyon, 21–24 Avril 1992, p. 90.

Harman G.E. 2000. Myths and dogmas of biocontrol. Changes in perceptions derived from research on *Trichoderma harzianum* T-22. Plant Dis. 84: 377–393.

Harman G.E. and Taylor A.G. 1988. Improved seedling performance by integration of biological control agents at favorable pH levels with solid matrix priming. Phytopathology 78: 520–525.

Harman G.E., Nedrow B. and Nash G. 1978. Stimulation of fungal spore germination by volatiles from aged seeds. Can. J. Bot. 56: 2124–2127.

Harman G.E., Mattick L.R., Nash G. and Nedrow B.L. 1980. Stimulation of fungal spore germination and inhibition of sporulation in fungal vegetative thalli by fatty acids and their volatile peroxidation products. Can. J. Bot. 58: 1541–1547.

Harrison L.A., Letendre L., Kovacevich P., Pierson E. and Weller D. 1993. Purification of an antibiotic effective against *Gaeumannomyces graminis* var. *tritici* produced by a biocontrol agent, *Pseudomonas aureofaciens*. Soil Biol. Biochem. 25: 215–221.

Hartman R.E., Keen N.T. and Long M. 1972. Carbon dioxide fixation by *Verticillium albo-atrum*. J. Gen. Microbiol. 73: 29–34.

Harvey R.G. 1990. Biodegradation of butylate, EPTC, and extenders in previously treated soils. Weed Sci. 38: 237–242.

Hashiba T. 1987. An improved system for biological control of damping-off by using plasmids in fungi. In I. Chet (ed.). Innovative Approaches to Plant Disease Control, pp. 337–351. John Wiley and Sons, New York, NY.

Hattori T. and Hattori R. 1976. The physical environment in soil microbiology: an attempt to extend principles of microbiology to soil microorganisms. CRC Crit. Rev. Microbiol. 4: 423–461.

Hauter R. and Mengel K. 1988. Measurement of pH at the root surface of red clover (*Trifolium pratense*) grown in soils differing in proton buffer capacity. Biol. Fertil. Soils 5: 295–298.

Hawes M.C., Brigham L.A., Wen F., Woo H.H. and Zhu Y. 1998. Function of root border cells in plant health: pioneers in the rhizosphere. Annu. Rev. Phytopathol. 36: 311–327.

Hayman D.S. 1975. Phosphorus cycling by soil microorganisms and plant roots. In N. Walker (ed.). Soil Microbiology, pp. 67–91. Butterworths, London, UK.

Hedger J.N. and Hudson H.J. 1974. Nutritional studies of *Thermomyces lanuginosus* from wheat straw compost. Trans. Br. mycol. Soc. 62: 129–143.

Hedlund K., Boddy L. and Preston C.M. 1991. Mycelial responses of the soil fungus, *Mortierella isabellina*, to grazing by *Onychiurus armatus* (Collembola). Soil Biol. Biochem. 23: 361–366.

Heijnen C.E. and Van Veen J.A. 1991. A determination of protective microhabitats for bacteria introduced into soil. FEMS Microbiol. Ecol. 85: 73–80.

Heijnen C.E., Van Elsas J.D., Kuikman P.J. and Van Veen J.A. 1988. Dynamics of *Rhizobium leguminosarum* biovar *trifolii* introduced into soil: the effect of bentonite clay on predation by protozoa. Soil Biol. Biochem. 20: 483–488.

Hénin S. 1944. Influence des phénomènes microbiens sur la formation d'une structure stable. C.R. Acad. Agric. F. 30: 373–375.

Henry C.M. and Deacon J.W. 1981. Natural (non-pathogenic) death of the cortex of wheat and barley seminal roots, as evidenced by nuclear staining with acridine orange. Plant Soil 60: 255–274.

Hepper C.M. and Smith G.A. 1976. Observations on the germination of *Endogone* spores. Trans. Br. mycol. Soc. 66: 189–194.

Hepper C.M., Azcon-Aguilar C., Rosendahl S. and Sen R. 1988. Competition between three species of *Glomus* used as spatially separated introduced and indigenous mycorrhizal inocula for leek (*Allium porrum* L.). New Phytol. 110: 207–215.

Hering T.F., Cook R.J. and Tang W.H. 1987. Infection of wheat embryos by *Pythium* species during seed germination and the influence of seed age and soil matric potential. Phytopathology 77: 1104–1108.

Heulin T., Guckert A. and Balandreau J. 1987. Stimulation of root exudation of rice seedlings by *Azospirillum* strains: carbon budget under gnotobiotic conditions. Biol. Fertil. Soils 4: 9–14.

Hill T.C.J., McPherson E.F., Harris J.A. and Birch P. 1993. Microbial biomass estimated by phospholipid phosphate in soils with diverse microbial communities. Soil Biol. Biochem. 25: 1779–1786.

Hirsch P.R., Jones M.J., Mc-Grath S.P. and Giller K.E. 1993. Heavy metals from past applications of sewage sludge decrease the genetic diversity of *Rhizobium leguminosarum* biovar *trifolii* populations. Soil Biol. Biochem. 25: 1485–1490.

Hoitink H.A.J. and Fahy P.C. 1986. Basis for the control of soilborne plant pathogens with composts. Annu. Rev. Phytopathol. 24: 93–114.

Horio T., Kawabata Y., Takayama T., Tahara S., Kawabata J., Fukushi Y., Nishimura H. and Mizutani J. 1992. A potent attractant of zoospores of *Aphanomyces cochlioides* isolated from its host, *Spinacia oleracea*. Experientia 48: 410–414.

Hornby D. 1979. Take-all decline: a theorist's paradise. In B. Schippers and W. Gams (eds.). Soil-borne Plant Pathogens, pp. 133–156. Academic Press, London, UK.

Hornby D. 1990. Root diseases. In J.M. Lynch (ed.). The Rhizosphere, pp. 233–258. John Wiley and Sons, Chichester, UK.

Howell C.R. and Stipanovic R.D. 1983. Gliovirin, a new antibiotic from *Gliocladium virens*, and its role in the biological control of *Pythium ultimum*. Can. J. Microbiol. 29: 321–324.

Howell C.R. and Stipanovic R.D. 1984. Phytotoxicity to crop plants and herbicidal effects on weeds of viridiol produced by *Gliocladium virens*. Phytopathology 74: 1346–1349.

Howell C.R., Hanson L.E., Stipanovic R.D. and Puckhaber L.S. 2000. Induction of terpenoid synthesis in cotton roots and control of *Rhizoctonia solani* by seed treatment with *Trichoderma virens*. Phytopathology 90: 248–252.

Hsu S.C. and Lockwood J.L. 1973. Chlamydospore formation in Fusarium in sterile salt solutions. Phytopathology 63: 597–602.

Huang H.C. and Kozub G.C. 1991. Monocropping to sunflower and decline of Sclerotinia wilt. Bot. Bull. Acad. Sin. 32: 163–170.

Huber D.M. and Watson R.D. 1970. Effect of organic amendment on soil-borne plant pathogens. Phytopathology 60: 22–26.

Huber D.M. and McKay-Buis T.S. 1993. A multiple component analysis of the take-all disease of cereals. Plant Dis. 77: 437–447.

Huisman O.C. 1982. Interrelations of root growth dynamics to epidemiology of root-invading fungi. Annu. Rev. Phytopathol. 20: 303–327.

Hussain A. and Vancura V. 1970. Formation of biologically active substances by rhizosphere bacteria and their effect on plant growth. Folia Microbiol. 15: 468–478.

Hütsch B.W., Webster C.P. and Powlson D.S. 1993. Long-term effects of nitrogen fertilization on methane oxidation in soil of the Broadbalk wheat experiment. Soil Biol. Biochem. 25: 1307–1315.

Huysman F. and Verstraete W. 1993. Water-facilitated transport of bacteria in unsaturated soil columns: influence of cell surface hydrophobicity and soil properties. Soil Biol. Biochem. 25: 83–90.

Ikediugwu F.E.O. and Webster J. 1970. Hyphal interference in a range of coprophilous fungi. Trans. Br. mycol. Soc. 54: 205–210.

Inbar J. and Chet I. 1992. Biomimics of fungal cell-cell recognition by use of lectin-coated nylon fibres. J. Bacteriol. 174: 1055–1059.

Ioannou N., Schneider R.W. and Grogan R.G. 1977. Effect of flooding on the soil gas composition and the production of microsclerotia by *Verticillium dahliae* in the field. Phytopathology 67: 651–656.

Isopi R., Fabbri P., Del Gallo M. and Puppi G. 1995. Dual inoculation of *Sorghum bicolor* (L.) Moench ssp. *bicolor* with vesicular arbuscular mycorrhizas and *Acetobacter diazotrophicus*. Symbiosis 18: 43–55.

Jabaji-Hare S.H. and Kendrick W.B. 1987. Response of an endomycorrhizal fungus in *Allium porrum* L. to different concentrations of the systemic fungicides, Metalaxyl (Ridomil) and Fosetyl-Al (Aliette). Soil Biol. Biochem. 19: 95–99.

Jackson M.B. 1985. Ethylene and responses of plants to soil waterlogging and submergence. Annu. Rev. Plant Physiol. 36: 145–174.

Jacobs M.J., Bugbee W.M. and Gabrielson D.A. 1985. Enumeration, location, and characterization of endophytic bacteria within sugar beet roots. Can. J. Bot. 63: 1262–1265.

Jatala P. 1986. Biological control of plant-parasitic nematodes. Annu. Rev. Phytopathol. 24: 453–489.

Jenkinson D.S. and Powlson D.S. 1976. The effects of biocidal treatments on metabolism in soil. V. A method for measuring soil biomass. Soil Biol. Biochem. 8: 209–213.

Jensen H.J. 1967. Do saprobic nematodes have a significant role in epidemiology of plant diseases? Plant Dis. Rep. 51: 98–102.

Johnson L.F., Chambers A.Y. and Reed H.E. 1967. Reduction of root-knot of tomatoes with crop residue amendments in field experiments. Plant Dis. Rep. 51: 219–222.

Johnson N.C., Copeland P.J., Crookston R.K. and Pfleger F.L. 1992. Mycorrhizae: possible explanation for yield decline with continuous corn and soybean. Agron. J. 84: 387–390.

Johnson S.R. and Berger R.D. 1972. Nematode and soil fungi control in celery seedbeds on muck soil. Plant Dis. Rep. 56: 661–664.

Jones D.A., Ryder M.H., Clare B.G., Farrand S.K. and Kerr A. 1988: Construction of a Tra⁻ deletion mutant of pAgK84 to safeguard the biological control of crown gall. Mol. Gen. Genet. 212: 207–214.

Jones J.P., Engelhard A.W. and Woltz S.S. 1989. Management of Fusarium wilt of vegetables and ornamentals by macro- and microelement nutrition. In A.W. Engelhard (ed.). Soilborne Plant Pathogens. Management of Diseases with Macro- and Microelements, pp. 18–32. Am. Phytopathol. Soc., St. Paul, MN.

Jones R.W. and Hancock J.G. 1990. Soilborne fungi for biological control of weeds. A.C.S. Symp. ser. 439: 276–286.

Jost J.L., Drake J.F., Fredrickson A.G. and Tsuchiya H.M. 1973. Interactions of *Tetrahymena pyriformis, Escherichia coli, Azotobacter vinelandii,* and glucose in a minimal medium. J. Bacteriol. 113: 834–840.

Jouan B. and Lemaire J.M. 1974. Modifications des biocénoses du sol. I. Etude préliminaire de l'influence de l'incorporation de substrats nutritifs au sol et ses conséquences pour l'évolution d'agents phytopathogènes d'origine tellurique. Ann. Phytopathol. 6: 297–308.

Jouper-Jaan A., Goodman A.E. and Kjelleberg S. 1992. Bacteria starved for prolonged periods develop increased protection against lethal temperatures. FEMS Microbiol. Ecol. 101: 229–236.

Kao C.W. and Ko W.H. 1986. The role of calcium and microorganisms in suppression of cucumber damping-off caused by *Pythium splendens* in a Hawaiian soil. Phytopathology 76: 221–225.

Kasenberg T.R. and Traquair J.A. 1989. Lettuce siderophores and biocontrol of fusarium rot in greenhouse tomatoes. Can. J. Plant Pathol. 11: 192.

Katan J. 1981. Solar heating (solarization) of soil for control of soilborne pests. Annu. Rev. Phytopathol. 19: 211–236.

Keel C., Schnider U., Maurhofer M., Voisard C., Laville J., Burger U., Wirthner P., Haas D. and Défago G. 1992. Suppression of root diseases by *Pseudomonas fluorescens* CHAO: importance of the bacterial secondary metabolite 2,4-diacetylphloroglucinol. Mol. Plant-Microb. Interact. 5: 4–13.

Keerthisinghe D.G. and Freney J.R. 1994. Inhibition of urease activity in flooded soils: effect of thiophosphorictriamides and phosphorictriamides. Soil Biol. Biochem. 26: 1527–1533.

Kerry B.R. and Crump D.H. 1998. The dynamics of the decline of the cereal cyst nematode, *Heterodera avenae,* in four soils under intensive cereal production. Fundam. Appl. Nematol. 21: 617–625.

Klironomos J.N., Widden P. and Deslandes I. 1992. Feeding preferences of the collenbolan *Folsomia candida* in relation to microfungal successions on decaying litter. Soil Biol. Biochem. 24: 685–692.

Kloepper J.W. and Schroth M.N. 1981. Relationship of *in vitro* antibiosis of plant growth-promoting rhizobacteria to plant growth and the displacement of root microflora. Phytopathology 71: 1020–1024.

Kloepper J.W., Schippers B. and Bakker P.A.H.M. 1992. Proposed elimination of the term *endorhizosphere.* Phytopathology 82: 726–727.

Kloepper J.W., Schroth M.N. and Miller T.D. 1980. Effects of rhizosphere colonization by plant growth-promoting rhizobacteria on potato plant development and yield. Phytopathology 70: 1078–1082.

Kluepfel D.A. and Tonkyn D.W. 1992. The ecology of genetically altered bacteria in the rhizosphere. In E.C. Tjamos, G.C. Papavizas and R.J. Cook (eds.). Biological Control of Plant Diseases. Progress and Challenges for the Future, pp. 407–413. Plenum Press, New York, NY.

Knowlton S., Berry A. and Torrey J.G. 1980. Evidence that associated soil bacteria may influence root hair infection of actinorhizal plants by *Frankia.* Can. J. Microbiol. 26: 971–977.

Ko W.H. and Chow F.K. 1977. Characteristics of bacteriostasis in natural soils. J. Gen. Microbiol. 102: 295–298.

Ko W.H. and Lockwood J.L. 1970. Mechanism of lysis of fungal mycelia in soil. Phytopathology 60: 148–154.

Ko W.H., Hora F.K. and Herlicska E. 1974. Isolation and identification of a volatile fungistatic factor from alkaline soil. Phytopathology 64: 1398–1400.

Kobayashi N. and Ko W.H. 1985. Nature of suppression of *Rhizoctonia solani* in Hawaiian soils. Trans. Br. mycol. Soc. 84: 691–694.

Komada H. 1975. Development of a selective medium for quantitative isolation of *Fusarium oxysporum* from natural soil. Rev. Plant Protect. Res. 8: 114–124.

Koske R.E. 1982. Evidence for a volatile attractant from plant roots affecting germ tubes of a VA mycorrhizal fungus. Trans. Br. mycol. Soc. 79: 305–310.

Kothari S.K., Marschner H. and Römheld V. 1991. Contribution of the VA mycorrhizal hyphae in acquisition of phosphorus and zinc by maize grown in a calcareous soil. Plant Soil 131: 177–185.

Krupa S. and Nylund J. 1972. Studies on ectomycorrhizae of pine. III. Growth inhibition of two root pathogenic fungi by organic volatile constituents of ectomycorrhizae root systems of *Pinus sylvestris* L. Eur. J. For. Pathol. 2: 88–94.

Kucey R.M.N., Janzen H.H. and Leggett M.E. 1989. Microbially mediated increases in plant-available phosphorus. Adv. Agron. 42: 199–228.

Kuikman P.J. and Van Veen J.A. 1989. The impact of protozoa on the availability of bacterial nitrogen to plants. Biol. Fertil. Soils 8: 13–18.

Lacombe J.P. and Garcin C. 1988. Résultats récents obtenus avec l'aldicarbe contre les nématodes sur céréales à paille. C.R. Deuxième Conf. Internat. Mal. Plantes, Bordeaux, 8–10 nov. 1988. A.N.P.P., Paris, pp. 437–444.

Lähdesmäki P. and Piispanen R. 1992. Soil enzymology: role of protective colloid systems in the preservation of exoenzyme activities in soil. Soil Biol. Biochem. 24: 1173–1177.

Landi L., Barraclough D., Badalucco L., Gelsomino A. and Nannipieri P. 1999. L-methionine-sulphoximine affects N mineralization-immobilization in soil. Soil Biol. Biochem. 31: 253–259.

Lartey R.T., Curl E.A., Peterson C.M. and Harper J.D. 1989. Mycophagous grazing and food preference of *Proisotoma minuta* (Collembola: Isotomidae) and *Onychiurus encarpatus* (Collembola: Onychiuridae). Environ. Entomol. 18: 334–337.

Latour X., Philippot L., Corberand T. and Lemanceau P. 1999. The establishment of an introduced community of fluorescent pseudomonads in the soil and in the rhizosphere is affected by the soil type. FEMS Microbiol. Ecol. 30: 163–170.

Latour X., Corberand T., Laguerre G., Allard F. and Lemanceau P. 1996. The composition of fluorescent pseudomonad populations associated with roots is influenced by plant and soil type. Appl. Environ. Microbiol. 62: 2449–2456.

Lattudy F. 1990. La précipitation des métaux lourds. Biofutur 93: 36–37.

Lechappé J., Rouxel F. and Sanson M.T. 1988. Le complexe parasitaire du pied du haricot. I. Mise en évidence des principaux champignons responsables de la maladie: *Fusarium solani* f. sp. *phaseoli, Thielaviopsis basicola*. Agronomie 8: 451–457.

Ledingham R.J. and Chinn S.H.F. 1955. A flotation method for obtaining spores of *Helminthosporium sativum* from soil. Can. J. Bot. 33: 298–303.

Lemaire J.-M. and Coppenet M. 1968. Influence de la succession céréalière sur les fluctuations de la gravité du piétin échaudage (*Ophiobulus graminis* Sacc.). Ann. Epiphyt. 19: 589–599.

Lemanceau P. and Alabouvette C. 1993. Suppression of Fusarium wilts by fluorescent pseudomonads: mechanisms and applications. Biocontrol Sci. Technol. 3: 219–234.

Lemanceau P., Alabouvette C. and Couteaudier Y. 1988. Recherches sur la résistance des sols aux maladies. XIV. Modification du niveau de réceptivité d'un sol résistant et d'un sol sensible aux fusarioses vasculaires en réponse à des apports de fer ou de glucose. Agronomie 8: 155–162.

emanceau P., Bakker P.A.H.M., De Kogel W.J., Alabouvette C. and Schippers B. 1993. Antagonistic effect of nonpathogenic *Fusarium oxysporum* strain Fo 47 and pseudobactin 358 upon pathogenic *Fusarium oxysporum* f. sp. *dianthi.* Appl. Environ. Microbiol. 59: 74–82.

eong J. 1986. Siderophores: their biochemistry and possible role in the biocontrol of plant pathogens. Annu. Rev. Phytopathol. 24: 187–209.

erouge P., Roche P., Promé J.C., Faucher C., Vasse J., Maillet F., Camut S., De Billy F., Barker D.G., Dénarié J. and Truchet G. 1990. *Rhizobium meliloti* nodulation genes specify the production of an alfalfa-specific sulfated lipo-oligosaccharide signal. In P.M. Gresshoff, L.E. Roth, G. Stacey and W.E. Newton (eds.). Nitrogen Fixation: Achievements and Objectives, pp. 177–186. Chapman and Hall, New York, NY.

evrat P., Pussard M. and Alabouvette C. 1992. Enhanced bacterial metabolism of a *Pseudomonas* strain in response to the addition of culture filtrate of a bacteriophagous Amoeba. Eur. J. Protistol. 28: 79–84.

ewis B.G. 1970. Effects of water potential on the infection of potato tubers by *Streptomyces scabies* in soil. Ann. Appl. Biol. 66: 83–88.

ewis J.A. and Papavizas G.C. 1971. Effect of sulfur-containing volatile compounds and vapors from cabbage decomposition on *Aphanomyces euteiches.* Phytopathology 61: 208–214.

ewis J.A., Lumsden R.D., Papavizas G.C. and Kantzes J.G. 1983. Integrated control of snap bean diseases caused by *Pythium* spp. and *Rhizoctonia solani.* Plant Dis. 67: 1241–1244.

iebman J.A. and Epstein L. 1992. Activity of fungistatic compounds from soil. Phytopathology 82: 147–153.

ifshitz R., Guilmette H. and Kozlowski M. 1988. Tn 5-mediated cloning of a genetic region from *Pseudomonas putida* involved in the stimulation of plant root elongation. Appl. Environ. Microbiol. 54: 3169–3172.

ifshitz R., Kloepper J.W., Kozlowski M., Simonson C., Carlson J., Tipping E.M. and Zaleska I. 1987. Growth promotion of canola (rapeseed) seedlings by a strain of *Pseudomonas putida* under gnotobiotic conditions. Can. J. Microbiol. 33: 390–395.

inderman R.G. and Gilbert R.G. 1973. Influence of volatile compounds from alfalfa hay on microbial activity in soil in relation to growth of *Sclerotium rolfsii.* Phytopathology 63: 359–362.

ingappa B.T. and Lockwood J.L. 1962. Fungitoxicity of lignin monomers, model subtances, and decomposition products. Phytopathology 52: 295–299.

ockwood J.L. 1964. Soil fungistasis. Annu. Rev Phytopathol. 2: 341–362.

ockwood J.L. 1977. Fungistasis in soils. Biol. Rev. 52: 1–43.

ockwood J.L. 1981. Exploitation competition. In D.T. Wicklow and G.C. Carroll (eds.). The Fungal Community: Its Organization and Role in the Ecosystem, pp. 319–349. Marcel Dekker, New York, NY.

öffler H.J.M., Van Dongen M. and Schippers B. 1986. Effect of NH_3 on chlamydospore formation of *Fusarium oxysporum* f. sp. *dianthi* in an NH_3-flow system. J. Phytopathol. 117: 43–48.

oper J.E. 1988. Role of fluorescent siderophore production in biocontrol of *Pythium ultimum* by a *Pseudomonas fluorescens* strain. Phytopathology 78: 166–172.

opez Aguillon R. and Garbaye J. 1989. Some aspects of a double symbiosis with ectomycorrhizal and VAM fungi. Agric. Ecosystems Environ. 29: 263–266.

ouvet J. and Bulit J. 1964. Recherches sur l'écologie des champignons parasites dans le sol. I. Action du gaz carbonique sur la croissance et l'activité parasitaire de *Sclerotinia minor* et de *Fusarium oxysporum* f. sp. *melonis.* Ann. Epiphyt. 15: 21–44.

Lumsden R.D., Millner P.N. and Lewis J.A. 1986. Suppression of lettuce drop caused by *Sclerotinia minor* with composted sewage sludge. Plant Dis. 70: 197–201.

Lung-Escarmant B. and Taris B. 1988. Etude du comportement d'une gamme d'hôte feuillus et résineux vis-à-vis d'*Armillaria obscura*: approche méthodologique e perspectives. C.R. Deuxième Conf. Internat. Mal. Plantes, Bordeaux, 8–10 nov 1988. A.N.P.P., Paris, pp. 1185–1191.

Lutchmeah R.S. and Cooke R.C. 1984. Aspects of antagonism by the mycoparasit *Pythium oligandrum*. Trans. Br. mycol. Soc. 83: 696–700.

Lynch J.M. 1981. Promotion and inhibition of soil aggregate stabilization by micro organisms. J. Gen. Microbiol. 126: 371–375.

Maertens C. and Bosc M. 1981. Etude de l'évolution de l'enracinement du tournesc (variété Stadium). Informations Techniques CETIOM, 73: 3–11.

Mai W.F. and Abawi G.S. 1987. Interactions among root-knot nematodes and *Fusariur* wilt fungi on host plants. Annu. Rev. Phytopathol. 25: 317–338.

Mangenot F. and Diem H.G. 1979. Fundamentals of biological control. In S.V. Krup and Y.R. Dommergues (eds.). Ecology of Root Pathogens, pp. 207–265. Elsevie Sci. Publ. Company, Amsterdam.

Mao W., Carroll R.B. and Whittington D.P. 1998. Association of *Phoma terrestris Pythium irregulare*, and *Fusarium acuminatum* in causing red root rot of corn. Plan Dis. 82: 337–342.

Marschner H., Römheld V. and Ossenberg-Neuhaus H. 1982. Rapid method fo measuring changes in pH and reducing processes along roots of intact plants. Z Pflanzenphysiol. 105: 407–416.

Marshall K.C. 1968. Interaction between colloidal montmorillonite and cells of *Rhizobiun* species with different ionogenic surfaces. Biochim. Biophys. Acta 156: 179–186

Martens D.A. and Bremner J.M. 1984. Effectiveness of phosphoroamides for retarda tion of urea hydrolysis in soils. Soil Sci. Soc. Am. J. 48: 302–305.

Martin C. and Thicoïpé J.P. 1994. La solarisation. Une désinfection des sols alternativ ou complémentaire des fumigants chimiques? Phytoma 464: 34–36.

Martin C., Davet P., Dubois M. and Lagier J. 1995. Prospects of integrated control o lettuce drop caused by *Sclerotinia* spp. In J.M. Whipps and T. Gerlagh (eds.) Biological Control of Sclerotium-forming Pathogens. Bull. OILB SROP 18: 57–59

Martin C., Véga D., Bastide J. and Davet P. 1990. Enhanced degradation of iprodione in soil after repeated treatments for controlling *Sclerotinia minor*. Plant Soil 127 140–142.

Martino E., Turnau K., Girlanda M., Bonfante P. and Perotto S. 2000. Ericoid mycorrhiza fungi from heavy metal polluted soils: their identification and growth in the presence of zinc ions. Mycol. Res. 104: 338–344.

Marty D. and Bianchi M. 1992. Présence simultanée et paradoxale de bactéries aérobie strictes (nitrifiantes) et anaérobies strictes (méthanogènes) dans des particules C.R. 3ème Congrès S.F.M., 21–24 avril 1992, Lyon, p. 71.

Marx D.H. and Davey C.B. 1969. The influence of ectotrophic mycorrhizal fungi on the resistance of pine roots to pathogenic infections. III. Resistance of aseptically formed mycorrhizae to infection by *Phytophthora cinnamomi*. Phytopathology 59 549–558.

Maurhofer M., Keel C., Schnider U., Voisard C., Haas D. and Défago G. 1992. Influence of enhanced antibiotic production in *Pseudomonas fluorescens* strain CHAO on its disease suppressive ability. Phytopathology 82: 190–195.

Mavingui P., Laguerre G., Bergé O. and Heulin T. 1992. Genetic and phenotypic diversity of *Bacillus polymyxa* in soil and in the wheat rhizosphere. Appl. Environ Microbiol. 58: 1894–1903.

Mazzola M. 1998. Elucidation of the microbial complex having a causal role in the development of apple replant disease in Washington. Phytopathology 88: 930–938.

Mazzola M. and Gu Y.-H. 2000. Impact of wheat cultivation on microbial communities from replant soils and apple growth in greenhouse trials. Phytopathology 90: 114–119.

Mazzola M., Stahlman P.W. and Leach J.E. 1995. Application method affects the distribution and efficacy of rhizobacteria suppressive of downy brome (*Bromus tectorum*). Soil Biol. Biochem 27: 1271–1278.

Megee R.D., Drake J.F., Fredrickson A.G. and Tsuchiya H.M. 1972. Studies in intermicrobial symbiosis, *Saccharomyces cerevisiae* and *Lactobacillus casei*. Can. J. Microbiol. 18: 1733–1742.

Messiaen C.M., Mampouya P.C. and Belliard-Alonzo L. 1976. Effet des composés azotés solubles sur la croissance mycélienne de *Sclerotium rolfsii* Sacc. dans le sol. Ann. Phytopathol. 8: 17–23.

Messiaen C.M., Blancard D., Rouxel F. and Lafon R. 1991. Les maladies des plantes maraîchères. INRA, Paris.

Metting B. 1987. Dynamics of wet and dry aggregate stability from a three-year microalgal soil conditioning experiment in the field. Soil Sci. 143: 139–143.

Meyer J.R. and Linderman R.G. 1986. Response of subterranean clover to dual inoculation with vesicular-arbuscular mycorrhizal fungi and a plant growth-promoting bacterium, *Pseudomonas putida*. Soil Biol. Biochem. 18: 185–190.

Meyer J.R., Shew H.D. and Harrison U.J. 1994. Inhibition of germination and growth of *Thielaviopsis basicola* by aluminum. Phytopathology 84: 598–602.

Mihail J.D. and Alcorn S.M. 1987. *Macrophomina phaseolina*: spatial patterns in a cultivated soil and sampling strategies. Phytopathology 77: 1126–1131.

Mihuta-Grimm L. and Rowe R.C. 1986. *Trichoderma* spp. as biocontrol agents of *Rhizoctonia* damping-off of radish in organic soil and comparison of four delivery systems. Phytopathology 76: 306–312.

Millar W.N. and Casida L.E. 1970. Evidence for muramic acid in soil. Can. J. Microbiol. 16: 299–304.

Miller R.M. and Jastrow J.D. 1990. Hierarchy of root and mycorrhizal fungal interactions with soil aggregation. Soil Biol. Biochem. 22: 579–584.

Mondal S. N. and Hyakumachi M. 1998. Carbon loss and germinability, viability, and virulence of chlamydospores of *Fusarium solani* f. sp. *phaseoli* after exposure to soil at different pH levels, temperatures, and matric potentials. Phytopathology 88: 148–155.

Montegano B. and Vergniaud P. 1987. Essai de lutte contre la pourriture blanche de l'ail (*Sclerotium cepivorum* Berk.). Rapport interne SRIV-INRA, Centre de recherches agronomiques d'Avignon.

Montenecourt B.S. and Eveleigh D.E. 1979. Production and characterization of high yielding cellulase mutants of *Trichoderma reesei*. Proc. TAPPI Annual Meeting, 12–14 March 1979, New York, pp. 101–108.

Montuelle B. 1966. Synthèse bactérienne de substances de croissance intervenant dans le métabolisme des plantes. Ann. Inst. Pasteur Paris 111 (suppl. 3): 136–146.

Morandi D., Bailey J.A. and Gianinazzi-Pearson V. 1984. Isoflavonoid accumulation in soybean roots infected with vesicular-arbuscular mycorrhizal fungi. Physiol. Plant Pathol. 24: 357–364.

Morel R. 1989. Les sols cultivés. Lavoisier, Paris.

Morris B.M. and Gow N.A.R. 1993. Mechanism of electrotaxis of zoospores of phytopathogenic fungi. Phytopathology 83: 877–882.

Morris, P.F. and Ward E.W.B. 1992. Chemoattraction of zoospores of the soybean pathogen, *Phytophthora sojae*, by isoflavones. Physiol. Mol. Plant Pathol. 40: 17–22.

Mousain D., Plassard C., Argillier C., Sardin T., Leprince F., El Karkouri K., Arvieu J.C. and Cleyet-Marel J.-C. 1994. Stratégie d'amélioration de la qualité des plants forestiers et des reboisements méditerranéens par utilisation de la mycorhization contrôlée en pépinière. Acta bot. Gallica 141: 571–580.

Mowe G., King B. and Senn S.J. 1983. Tropic responses of fungi to wood volatiles. J. Gen. Microbiol. 129: 779–784.

Muchembled C. and Richard-Molard M. 1991. Lutte génétique en culture de betteraves. C.R. 3ème Conf. Internat. Mal. Plantes, Bordeaux, 3–5 déc. 1991. A.N.P.P., Paris, pp. 745–751.

Nagtzaam M.P.M., Termorshuizen A.J. and Bollen G.J. 1997. The relationship between soil inoculum density and plant infection as a basis for a quantitative bioassay of *Verticillium dahliae*. Eur. J. Plant Pathol. 103: 597–605.

Neal J.L., Larson R.I. and Atkinson T.G. 1973. Changes in rhizosphere populations of selected physiological groups of bacteria related to substitution of specific pairs of chromosomes in spring wheat. Plant Soil 39: 209–212.

Nelson E.B. 1987. Rapid germination of sporangia of *Pythium* species in response to volatiles from germinating seeds. Phytopathology 77: 1108–1112.

Nelson E.B., Harman G.E. and Nash G.T. 1988. Enhancement of *Trichoderma*-induced biological control of *Pythium* seed rot and pre-emergence damping-off of peas. Soil Biol. Biochem. 20: 145–150.

Newsham K.K. 1999. *Phialophora graminicola*, a dark septate fungus, is a beneficial associate of the grass *Vulpia ciliata* ssp. *ambigua*. New Phytol. 144: 517–524.

Niemira B.A., Safir G.R. and Hawes M.C. 1996. Arbuscular mycorrhizal colonization and border cell production: a possible correlation. Phytopathology 86: 563–565.

Nishijima F., Evans W.R. and Vesper S.J. 1988. Enhanced nodulation of soybean by *Bradyrhizobium* in the presence of *Pseudomonas fluorescens*. Plant Soil 111: 149–150.

Noguera G.R. 1982. Alterations in the production of rishitin in roots and tyloses in stems and the interaction of *Meloidogyne-Fusarium* in tomato plants. Agron. Trop. Maracay 32: 303–308.

Norstadt F.A. and McCalla T.M. 1969. Microbial populations in stubble-mulched soil. Soil Sci. 107: 188–193.

Norton J.M. and Harman G.E. 1984. Responses of soil microorganisms to volatile exudates from germinating pea seeds. Can. J. Bot. 63: 1040–1045.

Obrigawitch T., Roeth F.W., Martin A.R. and Wilson Jr. R.G. 1982. Addition of R-33865 to EPTC for extended herbicide activity. Weed Sci. 30: 417–422.

Odum E.P. and Odum H.T. 1953. Fundamentals of Ecology. Saunders, Philadelphia, PA, 2nd ed. (1959), revised ed. (1987).

Ohr H.D., Sims J.J., Grech N.M., Becker J.O. and McGiffen M.E. 1996. Methyl iodide, an ozone-safe alternative to methyl bromide as a soil fumigant. Plant Dis. 80: 731–735.

Old K.M. and Chakraborty S. 1986. Mycophagous soil Amoebae: their biology and significance in the ecology of soil-borne plant pathogens. In: Progress in Protistology. vol. I, pp. 163-194. Biopress Ltd., Bristol, UK.

Olivain C. and Alabouvette C. 1997. Colonization of tomato root by a non-pathogenic strain of *Fusarium oxysporum*. New Phytol. 137: 481–494.

Olsen P.E., Rice W.A. and Collins M.M. 1995. Biological contaminants in North American legume inoculants. Soil Biol. Biochem. 27: 699–701.

Orchard V.A. and Cook F.J. 1983. Relationship between soil respiration and soil moisture. Soil Biol. Biochem. 15: 447–453.

Owen Evans G., Sheals J.G. and Macfarlane D. 1961. The Terrestrial Acari of the British Isles. An Introduction to Their Morphology, Biology and Classification, vol. I: Introduction and Biology. British Museum of Natural History, London.

Owens L.D., Guggenheim S. and Hilton J.L. 1968. Rhizobium-synthesized phytotoxin: an inhibition of β-cystathionase in *Salmonella typhimurium*. Biochim. Biophys. Acta 158: 219–225.

Ozawa T. and Yamagushi M. 1986. Fractionation and estimation of particle-attached and unattached *Bradyrhizobium japonicum* strains in soils. Appl. Environ. Microbiol. 52: 911–914.

Pace N.R. 1997. A molecular view of microbial diversity and the biosphere. Science 276: 734–740.

Panagopoulos C.G., Psallidas P.G. and Alivizatos A.S. 1979. Evidence of a breakdown in the effectiveness of biological control of crown-gall. In B. Schippers and W. Gams (eds.). Soil-borne Plant Pathogens, pp. 569–578. Academic Press, London, UK.

Panchaud-Mattei E. 1990. Propriétés nématicides de quelques plantes. P.H.M.-Revue Horticole 309: 29–31.

Papavizas G.C. 1992. Biological control of selected soilborne plant pathogens with *Gliocladium* and *Trichoderma*. In E.C. Tjamos, G.C. Papavizas and R.J. Cook (eds.). Biological Control of Plant Diseases, pp. 223–230. Plenum Press, New York, NY.

Papendick R.I. and Cook R.J. 1974. Plant water stress and development of *Fusarium* foot rot in wheat subjected to different cultural practices. Phytopathology 64: 358–363.

Pares R.D. and Gunn L.V. 1989. The role of non-vectored soil transmission as a primary source of infection by pepper mild mottle and cucumber mosaic viruses in glasshouse-grown Capsicum in Australia. J. Phytopathology 126: 353–360.

Patrick Z.A., Toussoun T.A. and Koch L.W. 1964. Effect of crop-residue decomposition products on plant roots. Annu. Rev. Phytopathol. 2: 267–292.

Patrick Z.A., Toussoun T.A. and Snyder W.C. 1963. Toxic substances in arable soils associated with decomposing plant residues. Phytopathology 53: 152–161.

Paul E.A. and Clark F.E. 1989. Soil Microbiology and Biochemistry. Academic Press, San Diego, CA.

Paulitz T.C. 1991. Effect of *Pseudomonas putida* on the stimulation of *Pythium ultimum* by seed volatiles of pea and soybean. Phytopathology 81: 1282–1287.

Paustian K. and Schnurer J. 1987. Fungal growth response to carbon and nitrogen limitations: application of a model to laboratory and field data. Soil Biol. Biochem. 19: 621–629.

Pedersen C.T., Safir G.R., Siqueira J.O. and Parent S. 1991. Effect of phenolic compounds on asparagus mycorrhiza. Soil Biol. Biochem. 23: 491–494.

Penalva M.A., Moya A., Dopazo J. and Ramon D. 1990. Sequence of isopenicillin N synthetase genes suggests horizontal transfer genes from prokaryotes to eukaryotes. Proc. R. Soc. London 241: 164–168.

Penn D.J. and Lynch J.M. 1982. The effect of bacterial fermentation of couch grass rhizomes and *Fusarium culmorum* on the growth of barley seedlings. Plant Pathol. 31: 39–43.

Peters G.A. and Meeks J.C. 1989. The *Azolla-Anabaena* symbiosis: basic biology. Annu. Rev. Plant Physiol. Plant Mol. Biol. 40: 193–210.

Phillips D.A. 1992. Flavonoids: plant signals to soil microbes. In H.A. Stafford and R.K. Ibrahim (eds.). Phenolic Metabolism in Plants, pp. 201–231. Plenum Press, New York, NY.

Pieczarka D.J. and Abawi G.S. 1978. Effect of interaction between *Fusarium, Pythium,* and *Rhizoctonia* on severity of bean root rot. Phytopathology 68: 403–408.

Pierson E.A. and Weller D.M. 1994. Use of mixtures of fluorescent pseudomonads to suppress take-all and improve the growth of wheat. Phytopathology 84: 940–947.

Pierson, L.S. III. 2000. Expanding the club: engineering plants to talk to bacteria. Trends Plant Sci. 5: 89–91.

Pierson L.S. III and Pierson E.A. 1996. Phenazine antibiotic production in *Pseudomonas aureofaciens*: role in rhizosphere ecology and pathogen suppression. FEMS Microbiol. Lett. 136: 101–108.

Pierson L.S. III, Wood D.W., Pierson E.A. and Chancey S.T. 1998. N-acyl-homoserine lactone-mediated gene regulation in biological control by fluorescent pseudomonads: current knowledge and future work. Eur. J. Plant Pathol. 104: 1–9.

Pilet P.E., Versel J.M. and Mayor G. 1983. Growth distribution and surface pH patterns along maize roots. Planta 158: 398–402.

Plesofsky-Vig N. and Brambl R. 1985. Topical review: the heat shock response of fungi. Experiment. Mycol. 9: 187–194.

Pline M., Diez J.A. and Dusenbery D.B. 1988. Extremely sensitive thermotaxis of the nematode *Meloidogyne incognita*. J. Nematol. 20: 605–608.

Prévost D., Angers D.A. and Nadeau P. 1991. Determination of ATP in soils by high performance liquid chromatography. Soil Biol. Biochem. 23: 1143–1146.

Pugh K.B. and Waid J.S. 1969. The influence of hydroxamates on ammonia loss from various soils treated with urea. Soil Biol. Biochem. 1: 207–217.

Pullman G.S., De Vay J.E. and Garber R.H. 1981. Soil solarization and thermal death: a logarithmic relationship between time and temperature for four soilborne plant pathogens. Phytopathology 71: 959–964.

Punja Z.K. and Jenkins S.F. 1984. Influence of temperature, moisture, modified gaseous atmosphere, and depth in soil on eruptive sclerotial germination of *Sclerotium rolfsii*. Phytopathology 74: 749–754.

Punja Z.K., Smith V.L., Campbell C.L. and Jenkins S.F. 1985. Sampling and extraction procedures to estimate numbers, spatial pattern and temporal distribution of sclerotia of *Sclerotium rolfsii* in soil. Plant Dis. 69: 469–474.

Pussard M., Alabouvette C. and Levrat P. 1994. Protozoan interactions with the soil microflora and possibilities for biocontrol of plant pathogens. In J.F. Darbyshire (ed.). Soil Protozoa, pp. 123–146. CAB International, Wallingford, UK.

Pussard M., Alabouvette C. and Pons R. 1979. Etude préliminaire d'une amibe mycophage *Thecamoeba granifera* s. sp. *minor* (Thecamoebidae, Amoebida). Protistologica 15: 139–149.

Raaijmakers J.M., Bonsall R.F. and Weller D.M. 1999. Effect of population density of *Pseudomonas fluorescens* on production of 2,4-diacetylphloroglucinol in the rhizosphere of wheat. Phytopathology 89: 470–475.

Rao J.R., Fenton M. and Jarvis B.D.W. 1994. Symbiotic plasmid transfer in *Rhizobium leguminosarum* biovar *trifolii* and competition between the inoculant strain ICMP 2163 and transconjugant soil bacteria. Soil Biol. Biochem. 26: 339–351.

Reid C.P.P. 1974. Assimilation, distribution, and root exudation of ^{14}C by ponderosa pine seedlings under induced water stress. Plant Physiol. 54: 44–49.

Reinhold-Hurek B. and Hurek T. 1998. Life in grasses: diazotrophic endophytes. Trends Microbiol. 6: 139–144.

Reis E.M., Cook R.J. and Mcneal B.L. 1983. Elevated pH and associated reduced trace-nutrient availability as factors contributing to take-all upon soil liming. Phytopathology 73: 411–413.

Renner E.D. and Becker G.E. 1970. Production of nitric oxide and nitrous oxide during denitrification by *Corynebacterium nephridii*. J. Bacteriol. 101: 821–826.

Reuveni R., Krikun J. and Shani U. 1983. The role of *Monosporascus eutypoides* in a collapse of melon plants in an arid area of Israel. Phytopathology 73: 1223–1226.

Revellin C., Leterme P. and Catroux G. 1993. Effect of some fungicide seed treatments on the survival of *Bradyrhizobium japonicum* and on the nodulation and yield of soybean (*Glycine max* (L.) Merr.). Biol. Fertil. Soils 16: 211–214.

Reynolds K.M., Benson D.M. and Bruck R.I. 1985. Epidemiology of *Phytophthora* root rot of Fraser fir: rhizosphere width and inoculum efficiency. Phytopathology 75: 1010–1014.

Ribbe M., Gadkari D. and Meyer O. 1997. N_2 fixation by *Streptomyces thermoautotrophicus* involves a molybdenum-dinitrogenase and a manganese-superoxide oxido-reductase that couple N_2 reduction to the oxidation of superoxide produced from O_2 by a molybdenum-CO dehydrogenase. J. Biol. Chem. 272: 26627–26633.

Ricci P. 1972. Moyens d'étude de l'inoculum du *Phytophthora nicotianae* f. sp. *parasitica* (Dastur) Waterh., parasite de l'oeillet, dans le sol. Ann. Phytopathol. 4: 257–276.

Ricci P. 1974. Mesure de la densité d'inoculum d'un agent pathogène dans le sol à l'aide d'une technique d'isolement par "tout ou rien". Ann. Phytopathol. 6: 441–453.

Ricci P., Bonnet P. and Blein J.P. 1994. Induction d'une réponse hypersensible et de résistance acquise: l'exemple des élicitines. Rapport interne INRA, 10 pp.

Rice E.L. 1968. Inhibition of nodulation of inoculated legumes by pioneer plant species from abandoned fields. Bull. Torrey bot. Club 95: 346–358.

Rice W.A., Olsen P.E. and Leggett M.E. 1995. Co-culture of *Rhizobium meliloti* and a phosphorus-solubilizing fungus (*Penicillium bilaii*) in sterile peat. Soil Biol. Biochem. 27: 703–705.

Richard-Molard M. 1983. La fatigue des terres à betteraves. In La fatigue des sols—Diagnostic de la fertilité dans les systèmes culturaux. INRA, "les Colloques", 17: 23–28.

Rittenhouse C.M. and Griffin G.J. 1985. Pattern of *Thielaviopsis basicola* in tobacco field soil. Can. J. Plant Pathol. 7: 377–381.

Rivière J. 1960. Etude de la rhizosphère du blé. Ann. Agron. 11: 397–440.

Rivière J. and Chaussat R. 1966. Destruction de la coumarine dans la rhizosphère. Ann. Inst. Pasteur 111: 155–167.

Rizzo D.M., Blanchette R.A. and Palmer M.A. 1992. Biosorption of metal ions by *Armillaria* rhizomorphs. Can. J. Bot. 70: 1515–1520.

Robert M. and Schmit J. 1982. Rôle d'un exopolysaccharide (le xanthane) dans les associations organo-minérales. C.R. Acad. Sci., Paris, 294, série II: 1031–1036.

Robinson R.K. 1972. The production by roots of *Calluna vulgaris* of a factor inhibitory to growth of some mycorrhizal fungi. J. Ecol. 60: 219–224.

Roger P.A. and Reynaud P.A. 1976. Dynamique de la population algale au cours d'un cycle de culture dans une rizière sahélienne. Rev. Ecol. Biol. Sol 13: 545–560.

Rolfe B.G. and Gresshoff P.M. 1988. Genetic analysis of Legume nodule initiation. Annu. Rev. Plant Physiol. Plant Mol. Biol. 39: 297–319.

Romig W.R. and Sasser M. 1972. Herbicide predisposition of snapbeans to *Rhizoctonia solani*. Phytopathology 62: 785–786.

Romine M. and Baker R. 1973. Soil fungistasis: evidence for an inhibiting factor Phytopathology 63: 756–759.

Rousseau J.V.D., Reid C.P.P. and English R.J. 1992. Relationship between biomass o the mycorrhizal fungus *Pisolithus tinctorius* and phosphorus uptake in loblolly pine seedlings. Soil Biol. Biochem. 24: 183–184.

Rouxel F. 1991. Natural suppressiveness of soils to plant diseases. In A.B.R. Beemster G.J. Bollen, M. Gerlagh, M.A. Ruissen, B. Schippers and A. Tempel (eds.). Biotic Interactions and Soil-borne Diseases, pp. 287–296. Elsevier, Amsterdam.

Rouxel F. and Bouhot D. 1971. Recherches sur l'écologie des champignons parasites dans le sol. IV. Nouvelles mises au point concernant l'analyse sélective e quantitative des *Fusarium oxysporum* et *Fusarium solani* dans le sol. Ann Phytopathol. 3: 171–188.

Rovira A.D., Newman E.I., Bowen H.J. and Campbell R. 1974. Quantitative assessmen of the rhizoplane microflora by direct microscopy. Soil Biol. Biochem. 6: 211–216

Rowland C.Y., Kurtböke D.I., Shankar M. and Sivasithamparam K. 1994. Nutritiona and biological activities of a sterile red fungus which promotes plant growth and suppresses take-all. Mycol. Res. 98: 1453–1457.

Runia W.T., Van Os E.A. and Bollen G.J. 1988. Disinfection of drainwater from soilless cultures by heat treatment. Neth. J. Agric. Sci. 36: 231–238.

Sacks L.E., King Jr. A.D. and Schade. 1986. A note on pH gradient plates for fungal growth studies. J. Appl. Bacteriol. 61: 235–238.

Safir G.R. 1994. Involvement of cropping systems, plant produced compounds and inoculum production in the functioning of VAM fungi. In F.L. Pfleger and R.G Linderman (eds.). Mycorrhizae and Plant Health, pp. 239–259. Am. Phytopathol Soc., St. Paul, MN.

Salamanca C.P., Herrera M.A. and Barea J.M. 1992. Mycorrhizal inoculation o micropropagated woody legumes used in revegetation programmes for desertified Mediterranean ecosystems. Agronomie 12: 869–872.

Sarig S., Blum A. and Okon Y. 1988. Improvement of the water status and yield of field-grown sorghum (*Sorghum bicolor*) by inoculation with *Azospirillum brasilense*. J Agric. Sci., Camb., 110: 271–277.

Sarig S., Okon Y. and Blum A. 1990. Promotion of leaf area development and yield ir *Sorghum bicolor* inoculated with *Azospirillum brasilense*. Symbiosis 9: 235–245.

Savka M.A. and Farrand S.K. 1997. Modification of rhizobacterial populations by engineering bacterium utilization of a novel plant-produced resource. Nature Biotechnol. 15: 363–368.

Schenck S. and Stotzky G. 1975. Effect on microorganisms of volatile compounds released from germinating seeds. Can. J. Microbiol. 21: 1622–1634.

Schiller C.T., Ellis M.A., Tenne F. and Sinclair J. 1977. Effect of *Bacillus subtilis* on soybean seed decay, germination, and stand inhibition. Plant Dis. Rep. 61: 213–217.

Schippers B., Bakker A.W. and Bakker P.A.H.M. 1987. Interactions of deleterious and beneficial rhizosphere microorganisms and the effect of cropping practices. Annu. Rev. Phytopathol. 25: 339–358.

Schippers B., Meijer J.W. and Liem J.I. 1982. Effect of ammonia and other soil volatiles on germination and growth of soil fungi. Trans. Br. mycol. Soc. 79: 253–259.

Schisler D.A. and Linderman R.G. 1989. Selective influence of volatiles purged from coniferous forest and nursery soils on microbes of a nursery soil. Soil Biol. Biochem. 21: 389–396.

Schmit J., Prior P., Quiquampoix H. and Robert M. 1990. Studies on survival and localization of *Pseudomonas solanacearum* in clays extracted from vertisols. In Z. Klement (ed.). Plant Pathogenic Bacteria, pp. 1001–1009. Akadémiai Kiado, Budapest.

Scholte K. 2000. Effect of potato used as a trap crop on potato cyst nematodes and other soil pathogens and on the growth of a subsequent main potato crop. Ann. Appl. Biol. 136: 229–238.

Schroth M.N. and Hendrix F.F. 1962. Influence of non-susceptible plants on the survival of *Fusarium solani* f. *phaseoli* in soil. Phytopathology 52: 906–909.

Schroth M.N., Weinhold A.R. and Hayman D.S. 1966. The effect of temperature on quantitative differences in exudates from germinating seeds of bean, pea, and cotton. Can. J. Bot. 44: 1429–1432.

Schumann G.L. 1991. Plant Diseases: Their Biology and Social Impact. Am. Phytopathol. Soc., St. Paul, MN.

Sequerra J., Capellano A., Faure-Raynard M. and Moiroud A. 1994. Root hair infection process and myconodule formation on *Alnus incana* by *Penicillium nodositatum*. Can. J. Bot. 72: 955–962.

Serra-Wittling C., Houot S. and Alabouvette C. 1996. Increased soil suppressiveness to *Fusarium* wilt of flax after addition of municipal solid waste compost. Soil Biol. Biochem. 28: 1207–1214.

Shew H.D. and Beute M.K. 1979. Evidence for the involvement of soilborne mites in *Pythium* pod rot of peanut. Phytopathology 69: 204–207.

Siegel M.R., Latch G.C.M. and Johnson M.C. 1987. Fungal endophytes of grasses. Annu. Rev. Phytopathol. 25: 293–315.

Simon A. and Sivasithamparam K. 1989. Pathogen suppression: a case study in biological suppression of *Gaeumannomyces graminis* var. *tritici* in soil. Soil Biol. Biochem. 21: 331–337.

Sivasithamparam K. 1998. Root cortex. The final frontier for the biocontrol of root-rot with fungal antagonists: a case study on a sterile red fungus. Annu. Rev. Phytopathol. 36: 439–452.

Skujins J.J. and McLaren A.D. 1969. Persistence of enzymatic activities in stored and geologically preserved soils. Enzymologia 34: 213–225.

Smiley R.W. and Cook R.J. 1972. Use and abuse of the soil pH measurement. Phytopathology 62: 193–194.

Smiley R.W. and Cook R.J. 1973. Relationship between take-all of wheat and rhizosphere pH in soils fertilized with ammonium vs. nitrate-nitrogen. Phytopathology 63: 882–890.

Smit G., Kijne J.W. and Lugtenberg B.J.J. 1987. Involvement of both cellulose fibrils and a Ca^{2+}-dependent adhesin in the attachment of *Rhizobium leguminosarum* to pea root hair tips. J. Bacteriol. 169: 4294–4301.

Smith S.E., Walker N.A. and Tester M.A. 1986. The apparent width of the rhizosphere of *Trifolium subterraneum* for vesicular-arbuscular mycorrhizal infection: effects of time and other factors. New Phytol. 104: 547–558.

Sneh B., Ichielevich-Auster M. and Plaut Z. 1989. Mechanism of seedling protection induced by a hypovirulent isolate of *Rhizoctonia solani*. Can. J. Bot. 67: 2135–2141.

Sparling G.P. and Searle P.L. 1993. Dimethyl sulphoxide reduction as a sensitive indicator of microbial activity in soil: the relationship with microbial biomass and mineralization of nitrogen and sulphur. Soil Biol. Biochem. 25: 251–256.

Stacey G., Sanjuan J., Luka S., Dockendorff T. and Carlson R.W. 1995. Signal exchange in the *Bradyrhizobium*-soybean symbiosis. Soil Biol. Biochem. 27: 473–483.

Stanghellini M.E. and Miller R.M. 1997. Biosurfactants. Their identity and potential efficacy in the biological control of zoosporic plant pathogens. Plant Dis. 81: 4–12.

Staphorst J.L. and Strijdom B.W. 1976. Effects on Rhizobia of fungicides applied to legume seed. Phytophylactica 8: 47–54.

Steinberg C. 1987. Dynamique d'une population bactérienne introduite dans le sol: régulation par les protozoaires et modélisation mathématique de la relation de prédation *Bradyrhizobium japonicum*-amibes indigènes. Thèse, Université de Lyon I, Lyon, France.

Steinberg C., Moulin F., Gaillard P., Gautheron N., Stawiecki K., Bremeersch P. and Alabouvette C. 1994. Disinfection of drain water in greenhouses using a wet condensation heater. Agronomie 14: 627–635.

Steinberg R.A. 1951. Occurrence of *Bacillus cereus* in Maryland soils with frenched tobacco. Plant Physiol. 26: 807–811.

Stotzky G. 1974. Activity, ecology, and population dynamics of microorganisms in soil. In A. Laskin and H. Lechevalier (eds.). Microbial Ecology, pp. 231–247. CRC Press, Cleveland, OH.

Stotzky G. 1980. Surface interactions between clay minerals and microbes, viruses and soluble organics, and the probable importance of these interactions to the ecology of microbes in soil. In R.C.W. Berkeley, J.M. Lynch, J. Melling, P.R. Rutter and B. Vincent (eds.). Microbial Adhesion to Surfaces, pp. 231–247. Ellis Horwood Ltd., Chichester, UK.

Stout J.D. and Heal O.W. 1967. Protozoa. In A. Burges and F. Raw (eds.). Soil Biology, pp. 149–195. Academic Press, New York, NY.

Stover R.H. 1962. Fusarial wilt (Panama disease) of bananas and other *Musa* species. Paper no. 4, CMI, Kew, Surrey, UK.

Strullu D.G. 1991. Les mycorhizes des arbres et plantes cultivées. "Technique et Documentation", Lavoisier, Paris.

Suslow T.V. and Schroth M.N. 1982. Role of deleterious rhizobacteria as minor pathogens in reducing crop growth. Phytopathology 72: 111–115.

Sylvia D.M. 1999. Fundamentals and applications of arbuscular mycorrhizae: a "biofertilizer" perspective. In J.O. Siqueira, F.M.S. Moreira, A.S. Lopes, L.R.G. Guilherme, V. Faquin, A.E. Furtini Neto and J.G. Carvalho (eds.). Inter-relação fertilidade, biologia do solo e nutrição de plantas, pp. 705–722. Sociedade Brasileira de Ciência do Solo, Lavras, Brazil.

Tabak H.H. and Cooke W.B. 1968. Growth and metabolism of fungi in an atmosphere of nitrogen. Mycologia 60: 115–140.

Taylor C.E. 1980. Nematodes. In K.F. Harris and K. Maramorosch (eds.). Vectors of Plant Pathogens, pp. 375–416. Academic Press, New York, NY.

Taylor G.S. and Parkinson D. 1961. The growth of saprophytic fungi on root surfaces. Plant Soil 15: 261–267.

Taylor J.M. 1995. Molecular phylogenetic classification of fungi. Arch. Med. Res. 26: 307–314.

Teplitski M., Robinson J.B. and Bauer W.D. 2000. Plants secrete substances that mimic bacterial N-acyl homoserine lactone signal activities and affect population density-dependent behaviors in associated bacteria. Mol. Plant-Microbe Interact. 13: 637–648.

Thies J.E., Singleton P.W. and Bohlool B.B. 1991. Influence of the size of indigenous rhizobial populations on establishment and symbiotic performance of introduced rhizobia on field-grown legumes. Appl. Environ. Microbiol. 57: 19–28.

Thomashow L.S., Weller D.M., Bonsall R.F. and Pierson L.S. 1990. Production of the antibiotic phenazine-l-carboxylic acid by fluorescent *Pseudomonas* species in the rhizosphere of wheat. Appl. Environ. Microbiol. 56: 908–912.

Tien T.M., Gaskins M.H. and Hubbell D.H. 1979. Plant growth substances produced by *Azospirillum brasilense* and their effect on the growth of pearl millet (*Pennisetum americanum* L.). Appl. Environ. Microbiol. 37: 1016–1024.

Tisdall J.M. 1991. Fungal hyphae and structural stability of soil. Aust. J. Soil Res. 29; 729–743.

Tivoli B., Corbière R. and Lemarchand E. 1990. Relations entre le pH des sols et leur niveau de réceptivité à *Fusarium solani* var. *coeruleum* et *Fusarium roseum* var. *sambucinum*, agents de la pourriture sèche des tubercules de pomme de terre. Agronomie 10: 63–68.

Tjepkema J.D., Schwintzer C.R. and Benson D.R. 1986. Physiology of actinorhizal nodules. Annu. Rev. Plant Physiol. 37: 209–232.

Tomimatsu G.S. and Griffin G.J. 1982. Inoculum potential of *Cylindrocladium crotalariae*: infection rates and microsclerotial density-root infection relationships on peanut. Phytopathology 72: 511–517.

Toussoun T.A. 1970. Nutrition and pathogenesis of *Fusarium solani* f. sp. *phaseoli*. In T.A. Toussoun, R.V. Bega and P.E. Nelson (eds.). Root Diseases and Soil-borne Pathogens, pp. 95–98. Univ. California Press, Berkeley, CA.

Toussoun T.A. and Patrick Z.A. 1963. Effect of phytotoxic substances from decomposing plant residues on root rot of bean. Phytopathology 53: 265–270.

Trappe J.M. 1977. Selection of fungi for ectomycorrhizal inoculation in nurseries. Annu. Rev. Phytopathol. 15: 203–222.

Trevors J.T. 1984. Dehydrogenase activity in soil: a comparison between the INT and TTC assay. Soil Biol. Biochem. 16: 673–674.

Trigalet A. 1994. Advances in biological control of bacterial wilt caused by *Pseudomonas solanacearum*. In M. Lemattre, S. Fraigoun, K. Rudolph and J.G. Swings (eds.). Plant Pathogenic Bacteria. INRA, "les Colloques", 66: 885–890.

Turlier M.F., Eparvier A. and Alabouvette C. 1995. Early dynamic interactions between *Fusarium oxysporum* f. sp. *lini* and the roots of *Linum usitatissimum* as revealed by transgenic Gus-marked hyphae. Can. J. Bot. 72: 1605–1612.

Utkhede R.S. and Li T.S.C. 1989. Chemical and biological treatments for control of apple replant disease in British Columbia. Can. J. Plant Pathol. 11: 143–147.

Vaartaja O. 1977. Responses of *Pythium ultimum* and other fungi to a soil extract containing an inhibitor with low molecular weight. Phytopathology 67: 67–71.

Van Bruggen A.H.C., Grogan R.G., Bogdanoff C.P. and Waters C.M. 1988. Corky root of lettuce in California caused by a gram-negative bacterium. Phytopathology 78: 1139–1145.

Van Gestel M., Merckx R. and Vlassak K. 1993. Microbial biomass responses to soil drying and rewetting: the fate of fast- and slow-growing microorganisms in soils from different climates. Soil Biol. Biochem. 25: 109–123.

Van Gundy S.D., Bird A.F. and Wallace H.R. 1967. Aging and starvation in larvae of *Meloidogyne javanica* and *Tylenchulus semipenetrans*. Phytopathology 57: 559–571.

van Loon L.C., Bakker P.A.H.M. and Pieterse C.M.J. 1998. Systemic resistance induced by rhizosphere bacteria. Annu. Rev. Phytopathol. 36: 453–483.

Vaughan D., Sparling G.P. and Ord B.G. 1983. Amelioration of the phytotoxicity of phenolic acids by some soil microbes. Soil Biol. Biochem. 15: 613–614.

Vierheilig H. and Ocampo J.A. 1990. Role of root extract and volatile substances of non-host plants on vesicular-arbuscular mycorrhizal spore germination. Symbiosis 9: 199–202.

Vigouroux A. 1997. A first approach to assessing soil oxygenation in the field using the Tensionic ceramic tensiometer. Agronomie 17: 389–394.

Voisard C., Keel C., Haas D. and Défago G. 1989. Cyanide production by *Pseudomonas fluorescens* helps suppress black root rot of tobacco under gnotobiotic conditions. EMBO J. 8: 351–358.

Voland R.P. and Epstein A.H. 1994. Development of suppressiveness to diseases caused by *Rhizoctonia solani* in soils amended with composted and noncomposted manure. Plant Dis. 78: 461–466.

Wainwright M. 1988. Metabolic diversity of fungi in relation to growth and mineral cycling in soil—A review. Trans. Br. mycol. Soc. 90: 159–170.

Walton B.T. and Anderson T.A. 1990. Microbial degradation of trichloroethylene in the rhizosphere: potential application to biological remediation of waste sites. Appl. Environ. Microbiol. 56: 1012–1016.

Warembourg F.R. and Billès G. 1979. Estimating carbon transfers in the plant rhizosphere. In J.L. Harley and R.S. Russel (eds.). The Soil-Root Interface, pp. 183–196. Academic Press, London, UK.

Warembourg F.R., Esterlich D.H. and Lafont F. 1990. Carbon partitioning in the rhizosphere of an annual and a perennial species of bromegrass. Symbiosis 9: 29–36.

Watanabe I. and Furusaka C. 1980. Microbial ecology of flooded rice soils. In M. Alexander (ed.). Advances in Microbial Ecology, vol. 4, pp. 125–168. Plenum Press, New York, NY.

Weaver D.B., Rodríguez-Kábana R. and Carden E.L. 1998. Velvetbean and bahiagrass as rotation crops for management of *Meloidogyne* spp. and *Heterodera glycines* in soybean. J. Nematol. (suppl.) 30 (4S): 563–568.

Wei G., Kloepper J.W. and Tuzun S. 1996. Induced systemic resistance to cucumber diseases and increased plant growth by plant growth-promoting rhizobacteria under field conditions. Phytopathology 86: 221–224.

Weste G. 1984. Damage and loss caused by *Phytophthora* species in forest crops. In R.K.S. Wood and G.J. Ellis (eds.). Plant Diseases. Infection, Damage and Loss, pp. 273–284. Blackwell Sci. Publ., Oxford, UK.

Whipps J.M. 1990. Carbon economy. In J.M. Lynch (ed.). The Rhizosphere, pp. 59–97. John Wiley and Sons, Chichester, UK.

Whittaker R.H. 1969. New concepts of kingdoms of organisms. Science 163: 150–160.

Wicklow D.T. 1992. Interference competition. In G.C. Carroll and D.T. Wicklow (eds.). The Fungal Community: Its Organization and Role in the Ecosystem, pp. 265–274. Marcel Dekker, New York, NY.

Widmer T.L. and Abawi G.S. 2000. Mechanism of suppression of *Meloidogyne hapla* and its damage by a green manure of Sudan grass. Plant Dis. 84: 562–568.

Wilhelm S. 1959. Parasitism and pathogenesis of root-disease fungi. In C.S. Holton, G.W. Fischer, R.W. Fulton, H. Hart and S.E.A. McCallan (eds.). Plant Pathology, Problems and Progress, 1908–1958, pp. 356–366. Univ. Wisconsin Press, Madison, WI.

Wilkinson H.T., Alldredge J.R. and Cook R.J. 1985. Estimated distances for infection of wheat roots by *Gaeumannomyces graminis* var. *tritici* in soils suppressive and conductive to take-all. Phytopathology 75: 557–559.

Wilkinson T.G., Topiwala H.H. and Hamer G. 1974. Interactions in a mixed bacterial population growing on methane in continuous culture. Biotechnol. Bioeng. 16: 41–47.

Williams P.M. 1979. Vegetable crop protection in the People's Republic of China. Annu. Rev. Phytopathol. 17: 311–324.

Vinogradski S. 1890. Recherches sur les organismes de la nitrification. Ann. Inst. Pasteur 4: 213-231 and 257–275.

Woese C.R., Kandler O. and Wheelis M.L. 1990. Towards a natural system of organisms: proposal for the domains Archaea, Bacteria, and Eucarya. Proc. Natl. Acad. Sci. USA 87: 4576–4579.

Woltz S.S. 1978. Nonparasitic plant pathogens. Annu. Rev. Phytopathol. 16: 403–430.

Workneh F. and Van Bruggen A.H.C. 1994. Suppression of corky root of tomatoes in soils from organic farms associated with soil microbial activity and nitrogen status of soil and tomato tissue. Phytopathology 84: 688–694.

Workneh F., Van Bruggen A.H.C., Drinkwater L.E. and Shennan C. 1993. Variables associated with corky root and *Phytophthora* root rot of tomatoes in organic and conventional farms. Phytopathology 83: 581–589.

Wu J., Joergensen R.G., Pommerening B., Chaussod R. and Brookes P.C. 1990. Measurement of soil microbial biomass C by fumigation-extraction. An automated procedure. Soil Biol. Biochem. 22: 1167–1169.

Yanni Y., Rizk R., de Bruijn F., Squartini A. and Dazzo F. 1998. Natural beneficial endophytic association between *Rhizobium leguminosarum* bv. *trifolii* and rice roots and its relevance to sustainable agriculture. 16th World Congress of Soil Science, Montpellier, 20–26 Aug. 1998, Summaries, vol. 1, p. 195.

Yeoh H.T., Bungay H.R. and Krieg N.R. 1968. A microbial interaction involving a combined mutualism and inhibition. Can. J. Microbiol. 14: 491–492.

Young C.C. and Chou T.C. 1985. Autointoxication in residues of *Asparagus officinalis* L. Plant Soil 85: 385–393.

Young I.M. 1998. Biophysical interactions at the root-soil interface: a review. J. Agricult. Sci. 130: 1–7.

Zantua M.I. and Bremner J.M. 1977. Stability of urease in soils. Soil Biol. Biochem. 9: 135–140.

Zelles L., Bai Q.Y., Beck T. and Beese F. 1992. Signature fatty acids in phospholipids and lipo-polysaccharides as indicators of microbial biomass and community structure in agricultural soils. Soil Biol. Biochem. 24: 317–323.

Zhang L., Mitra A., French R.C. and Langenberg W.G. 1994. Fungal zoospore-mediated delivery of a foreign gene to wheat roots. Phytopathology 84: 684–687.

Zimmerman W.J. 1993. Microalgal biotechnology and applications in agriculture. In F.B. Metting (ed.). Soil Microbial Ecology, pp. 457–479. Marcel Dekker Inc., New York, NY.

Index